#71
Westwood Branch
1246 Glendon Ave.
Los Angeles, CA 90024

# THE
# BACKYARD
# ASTRONOMER'S
# GUIDE

FIREFLY BOOKS

# THE
# BACKYARD
# ASTRONOMER'S
# GUIDE

REVISED EDITION

## TERENCE DICKINSON & ALAN DYER

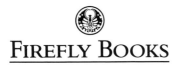

# FIREFLY BOOKS

Published by Firefly Books Ltd. 2002

Second Edition, First Printing 2002

### Publisher Cataloging-in-Publication Data (U.S.)

Dickinson, Terence.
    The backyard astronomer's guide /
Terence Dickinson and Alan Dyer. – 2nd ed.
[336] p. : col. ill. , photos. ; cm.
Includes index.

Summary: An illustrated guide to equipment and techniques for amateur astronomers. Includes charts, how to use your telescope, observation tips, how to use your camera to shoot the night sky and updated information on astronomy software and computerized equipment.
ISBN 1-55209-507-X
1. Astronomy.   2. Astronomy – Observers' manuals.    I. Dyer, Alan.    II. Title.
520 21   CIP   QB44.3.D53   2002

Published in the United States in 2002 by
Firefly Books (U.S.) Inc.
P.O. Box 1338, Ellicott Station
Buffalo, New York 14205

Produced by
Bookmakers Press Inc.
12 Pine Street
Kingston, Ontario K7K 1W1
(613) 549-4347
tcread@sympatico.ca

Design by
Robbie Cooke

Printed and bound in Canada by
Friesens
Altona, Manitoba

Printed on acid-free paper

### National Library of Canada Cataloguing in Publication Data

Dickinson, Terence
    The backyard astronomer's guide /
Terence Dickinson, Alan Dyer. – 2nd ed.

Includes index.
ISBN 1-55209-507-X

1. Astronomy – Amateurs' manuals.
I. Dyer, Alan, 1953 –     II. Title.

QB64.D513 2002        522        C2002-902225-8

Published in Canada in 2002 by
Firefly Books Ltd.
3680 Victoria Park Avenue
Willowdale, Ontario M2H 3K1

Front Cover: CCD image of Andromeda Galaxy (M31) by Robert Gendler

Other books by Terence Dickinson:
    NightWatch
    The Universe and Beyond
    Exploring the Night Sky
    Exploring the Sky by Day
    From the Big Bang to Planet X
    Other Worlds
    Summer Stargazing
    Splendors of the Universe
        (with Jack Newton)

*The Publisher acknowledges the financial support of the Government of Canada through the Book Publishing Industry Development Program for its publishing activities.*

To Susan, who inspires me always.
                —T.D.

For all the friends I've met under the stars.
                —A.D.

# FOREWORD

## Astronomy: Getting Started Right

Do you suppose there's an "astronomy gene" coded into human nature?

I don't know whether such a thing exists —probably doesn't—but you do have to wonder. Everybody feels a tug from the stars under the right circumstances. Even the most committed urban dwellers, totally at home in the world of concrete, are stunned when they confront an inky-black sky strewn with stars. There's something in that sight to make any person grow very quiet.

The everyday world we live in is a much noisier place, audibly, visually and informationally. Too many of us live our entire lives in that world without getting more than brief glimpses of what lies on the other side of sunset. The grand spectacle of night parallels our human world, dwarfing it, in fact. But for many people, the only personal experience they have of astronomy is what they see on vacation far from civilization— the north woods, perhaps, or the desert or far out at sea.

This unfamiliarity with the sky sometimes has a funny side. Tales are told of people who see the Milky Way clearly for the first time and worriedly demand to know what's happened—there's all that smoke in the sky! (Confession time: Even experienced skygazers can fall for this. While observing on a dark, moonless night, I once mistook the rising Milky Way for oncoming clouds and prepared to pack up the telescope.) Others are astonished to learn that you can see the Moon almost every day of the month and that stars can be found in the daytime.

The night sky's remoteness from human affairs creates almost a cultural barrier for anyone caught by an interest in astronomy. Not surprisingly, the science of astronomy—astrophysics, really—is forbiddingly complex and requires years of professional training. But even astronomy as practiced by amateur astronomers lies very far from most people's everyday knowledge. For them, even finding Square One, let alone moving off it, is a big challenge.

Planetariums and science museums offer informative, even thrilling programs. But the shows and displays are always mediated by professionals, and the voyage away from Earth is scripted tightly. Even the best of them is no substitute for exploring the heavens on your own.

This is by far the best book I know of for helping anyone become a knowledgeable skygazer. Terry and Alan have spent most of their lives as backyard astronomers. And their jobs regularly bring them into contact with beginners who stand in astronomy's doorway, wondering how to come in. The guidebook they have written has such newcomers firmly in mind, and if you, too, feel the tug of the stars, there's no better place to begin than here.

— Robert Burnham
former Editor in Chief,
*Astronomy* magazine

# CONTENTS

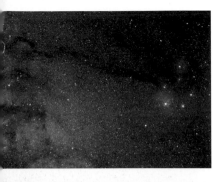

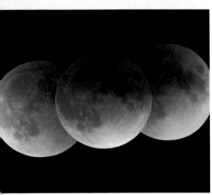
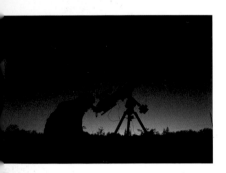
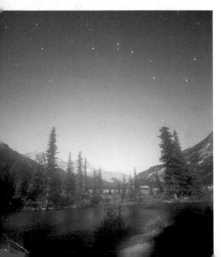
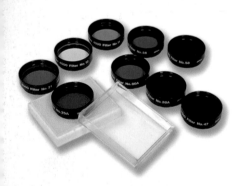

## PART 3

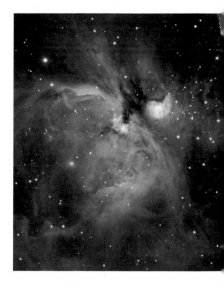

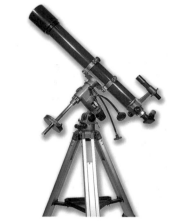

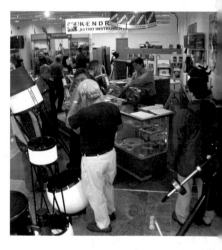

Facing page, top to
bottom: Mike Mayerchak,
Steve Barnes, Terence
Dickinson, Alan Dyer (both).
This page, top to bottom:
Robert Gendler, Alan Dyer,
Terence Dickinson.

# INTRODUCTION

Since the first edition of *The Backyard Astronomer's Guide* appeared in 1991, amateur astronomy has become ever more popular and its equipment ever more varied. Recent technology has ushered in completely new ways of conducting the hobby. One of the most visible of these innovations, but by no means the only one, is affordable computerized telescopes that can locate any of thousands of celestial targets.

In many respects, this book is a sequel to coauthor Dickinson's *NightWatch: A Guide to Viewing the Universe*, which emphasizes preliminary material for the absolute beginner. In *The Backyard Astronomer's Guide*, we provide more in-depth commentary, guidance and reference material for the enthusiast.

In preparing this Second Edition, we decided that a general tune-up would not be enough. We ended up taking the old book entirely apart for a complete overhaul. We rewrote every page, revising and expanding every topic to accommodate new equipment and techniques. The biggest visual change is the use of color throughout, as well as double the number of illustrations. For more real-world focus, we have used photos of equipment that were taken in the field and in our own studios rather than relying on stock shots from manufacturers. We've really used this equipment. (Whenever prices are mentioned, they are average dealer prices in U.S. dollars.)

We also realize that astronomy as a hobby has grown so much, there's far more information we could have included in the book. Sometimes, it's just as difficult deciding what to leave out as what to put in. But this time, our task was made easier because the book now has its own website (www.backyardastronomy.com), where supplementary material, updates and links to other informative sites can be found. The website is also the exclusive home for resources and lists of manufacturers that we had placed in the Appendix in the original edition. Please pay a visit to the site. We hope you'll bookmark it as a favorite.

Terence Dickinson
NightWatch Observatory

Alan Dyer
Calgary Science Centre

website:
www.backyardastronomy.com

Big Dipper photo
by Ken Hewitt-White.

9

# Amateur Astronomy Comes

There is something deeply compelling about the starry night sky. Those fragile flickering points of light in the blackness beckon to the inquisitive mind. So it was in antiquity, and so it remains today.

But only in the past two decades or so have large numbers of people chosen to delve into stargazing—recreational astronomy—as a leisure activity. Today, more than half a million people in North America call themselves amateur astronomers.

The magic moment when you know you're hooked usually comes with your eye at a telescope eyepiece. It often takes just one exposure to Saturn's stunningly alien, yet serenely beautiful ring system or a steady view of an ancient lunar crater frozen in time on the edge of a rumpled, airless plain.

......................................

Spectacular celestial images, such as this amateur photo of the Orion Nebula taken from a Connecticut driveway, inspire thousands of people to seek out objects like this for them- selves using backyard telescopes. Image by Robert Gendler.

Age

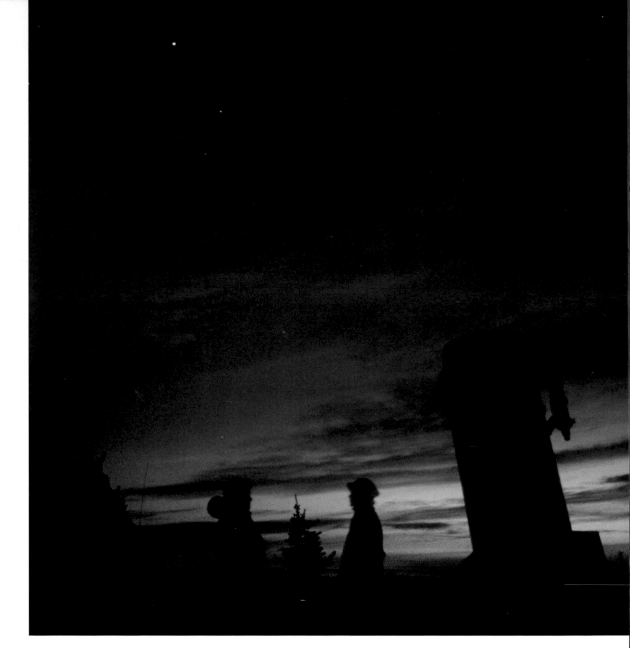

# Naturalists of the Night

American 19th-century poet and essayist Ralph Waldo Emerson once wrote: "The man on the street does not know a star in the sky." Of course, he was right then and now. Well, almost. In recent years, a growing number of people want to become acquainted with the stars. Sales of astronomy books, telescopes and astronomy software have reached all-time highs. More people than ever before are enrolling in the astronomy courses offered by colleges, universities and planetariums. Summer weekend gatherings of astronomy enthusiasts for telescope viewing and informative talks (known among the participants as "star parties") now attract thousands of fans. There is no mistaking the signals: Astronomy has come of age as a mainstream interest and recreational activity.

Not coincidentally, the growth of interest in astronomy has paralleled the rise in our awareness of the environment. The realization that we live on a planet with finite resources and dwindling access to wilderness areas has generated a sharp increase in activities which involve observing and appreciating nature: birding, nature walks, hiking, scenic drives, camping and nature photography. Recreational astronomy is in this category too. Amateur astronomers are naturalists of the night, captivated by the mystique of the vast universe that is accessible only under a dark sky.

In recent decades, the darkness that astronomy enthusiasts seek has been beaten back by the ever-growing domes of artificial light over cities and towns and by the increased use of security lighting everywhere. In many places, the luster of the Milky Way arching across a star-studded sky has been obliterated forever. Yet amateur astronomy flourishes as never before. Why? Perhaps it is an example of that well-known human tendency to ignore the historic or acclaimed tourist sights in one's own neighborhood while attempting to see everything when traveling to distant lands. Most people now perceive a starry sky as foreign and enchanting rather than something that can be seen from any sidewalk, as it was when our grandparents were young.

That is certainly part of the answer, but consider how amateur astronomy has changed in two generations. The typical 1960s amateur astronomer was usually male and a loner, with a strong interest in physics, mathematics and optics. In high school, he spent his weekends grinding a 6-inch f/8 Newtonian telescope mirror from a kit sold by Edmund Scientific, in accordance with the instructions in *Scientific American* telescope-making books. The four-foot-long telescope was mounted on what was affectionately called a plumber's nightmare—an equatorial mount made of pipe fittings. In some cases, it was necessary to keep the telescope out of sight to be brought out only under cover of darkness to avoid derisive commentary from the neighbors.

Practical reference material was almost nonexistent in the 1960s. Most of what there was came from England, and virtually all of it was written by one man, Patrick Moore. Amateur astronomy was like a secret religion—so secret, it was almost unknown.

Thankfully, that is all history. Current astronomy hobbyists represent a complete cross section of society, encompassing men and women of all ages, occupations and levels of education. Amateur astronomy has finally come into its own as a legitimate recreational activity, not the pastime of perceived lab-coated rocket scientists and oddballs. Indeed, it has emerged as a leisure activity with a certain prestige. Unlike some hobbies, it is not possible to buy your way into astronomy. Astronomical knowledge and experience take time to accumulate. But be forewarned: Once you gain that knowledge and experience, astronomy can be addictive.

## AMATEUR ASTRONOMY TODAY

Amateur astronomy has become incredibly diversified. No individual can master the field entirely. It is simply too large; it has too many activities and choices. In general, though, amateur astronomers divide fairly easily into three groups: the observers, the techno-enthusiasts and the armchair astronomers. The last category refers to people who pursue the hobby mainly vicariously—through books, magazines, lectures, discussion groups or conversations with other aficionados. Armchair astronomers are often self-taught experts on nonobservational aspects of the subject, such as cosmology or astronomical history.

The techno-enthusiast category includes telescope makers and those fascinated by the technical side of the hobby, especially the application of computers to astronomical imaging and telescope use and the application of technological innovations related to amateur-astronomy equipment. It can also involve crafting optics, though

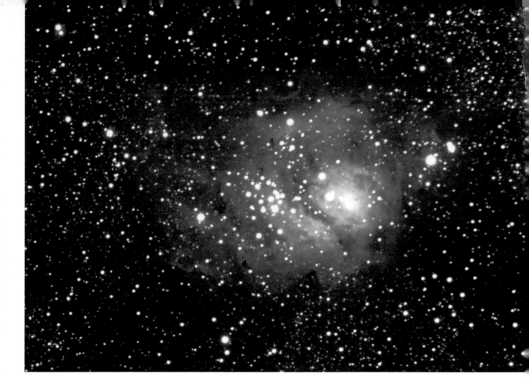

**▲ Star-Formation Nebula**
At a distance of 5,000 light-years, the Lagoon Nebula is faintly visible to the naked eye. Four-inch refractor photo by Terence Dickinson.

**▼ Classic Comet**
Some backyard astronomers were first attracted to the night sky by Comet Hale-Bopp's appearance in 1997. Photo by Dan Falk.

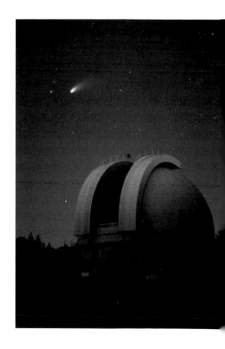

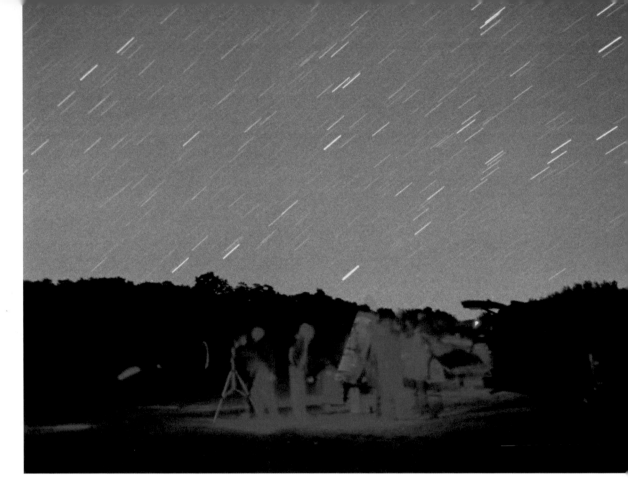

**Red-Light District ▶**
Amateur astronomy has few regulations and little formality, but show up at a star party like this with white lights blaring, and you'll be in for a rude greeting. To preserve night vision, red lights are the rule. (Stars are streaked because of the Earth's rotation during the exposure.) Photo by Terence Dickinson.

**Red-Light District ▶**
Amateur astronomy has few regulations and little formality, but show up at a star party like this with white lights blaring, and you'll be in for a rude greeting. To preserve night vision, red lights are the rule. (Stars are streaked because of the Earth's rotation during the exposure.) Photo by Terence Dickinson.

**Looking Up ▼**
A special event such as a comet can awaken interest, but the sky presents something new and wonderful to see every night. Photo by Ken Hewitt-White.

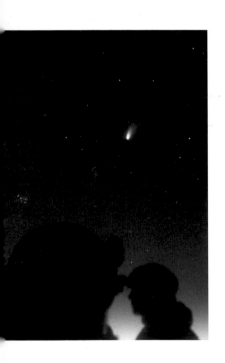

this type of telescope making is less prevalent than it was a few decades ago. With the vast array of commercial equipment available today, "rolling your own" is not the common activity it once was.

This book is written primarily for the third kind of amateur astronomer, the observer, one whose dominant interest in astronomy is to explore the visible universe with eye and telescope. Observing, we believe, is what it is all about. The exhilaration of exploring the sky, of seeing for yourself the remote planets, galaxies, clusters and nebulas—real objects of enormous dimensions at immense distances—is the essence of backyard astronomy.

## GETTING IN DEEPER

Amateur astronomy can range from an occasional pleasant diversion to a full-time obsession. Some amateur astronomers spend more time and energy on the hobby than do all but the most dedicated research astronomers at mountaintop observatories. Such "professional amateurs" are the exception, but they are indeed the true amateur astronomers—that is, they

have selected an area which professional astronomers, either by choice or through lack of human resources, have neglected. They are, in the purest sense, amateurs: unpaid researchers.

In the past, such impassioned individuals were often independently wealthy and able to devote much time and effort to a single-minded pursuit. This is seldom the case anymore. For instance, Australian Robert Evans is a pastor of three churches, has a family with four daughters and is by no means a man of wealth or leisure. Yet he has spent almost every clear night since 1980 searching for supernovas in galaxies up to 100 million light-years away. He discovered 18 within a decade—more than were found during the same period by a team of university researchers using equipment designed exclusively for that purpose.

Similarly, most bright comets in recent years have been found by committed amateur astronomers. However, the persistent supernova or comet hunter represents just a tiny fraction of those who call themselves amateur astronomers. The vast majority—at least 99 percent—are more accurately described as recreational backyard astronomers. Although this term has not gained wide usage, it more precisely describes what most amateur astronomers do. They are out enjoying themselves under the stars, engaging in a personal exploration of the universe that has no scientific purpose beyond self-edification. It's challenging and fun.

Backyard astronomy was neatly summed up a few years ago in *Astro Notes*, the newsletter of the Ottawa Centre of The Royal Astronomical Society of Canada: "The objective is to explore strange new phenomena, to seek out new celestial objects and new nebulosities, to boldly look where no human has looked before... and mainly to have fun."

Tom Williams, a chemist by profession and an astronomy hobbyist from Houston, Texas, has taken an interest in the distinction between the vast majority of casual stargazers and the handful of scientific amateurs. Williams points out some parallels with ornithology: "There are 15 million bird watchers in North America, but they call themselves birders, not amateur ornithologists. The real amateur ornithologists are the few thousand people involved in migration analyses, and so on." Similarly, he notes, "Of the 500,000 astronomy hobbyists,

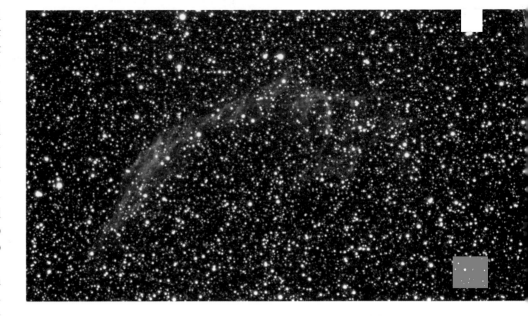

the same small percentage are scientific amateur astronomers who contribute in some way to research. The rest are recreational astronomers. The majority are in these activities for pure enjoyment, nothing more." The somewhat confusing aspect is that both groups—the scientific amateurs and the recreational amateurs—call themselves the same thing: amateur astronomers.

That is not to say there is no place for systematic and potentially scientifically valuable observing. Quite the contrary. But it is not every backyard astronomer's duty. Some choose to take a more rigorous approach to the hobby; most do not. Our book is dedicated to the latter group.

## REACHING FOR THE STARS

Some of the activities of astronomy buffs totally baffle those not afflicted with the bug. Take the arrival of Comet West, for instance, one of the brightest comets visible from midnorthern latitudes in the past century. Comet West was at its best in early March 1976, but the weather over much of North America was terrible. Astronomy addicts were having severe withdrawal symptoms as they stared at the clouds each night, knowing that the comet was out there, just beyond reach. In Vancouver, several young enthusiasts decided that they had had enough. "The comet was peaking in brightness. We had to do something," recalls Ken Hewitt-White, then a producer

▲ **Shock Wave of Star-Stuff**
Backyard telescopes can show us objects such as the Veil Nebula, blown into space by a supernova explosion. Objects like these seed space with the elements that make planets, plants . . . and people. Photo by Terence Dickinson.

◀ **Aware of the Night**
You know you are an amateur astronomer when, upon stepping outside at night, you automatically look up just to see what the sky contains. In this case, the distinctive constellation Orion greets the observer in a moonlit evening sky. Photo by Terence Dickinson.

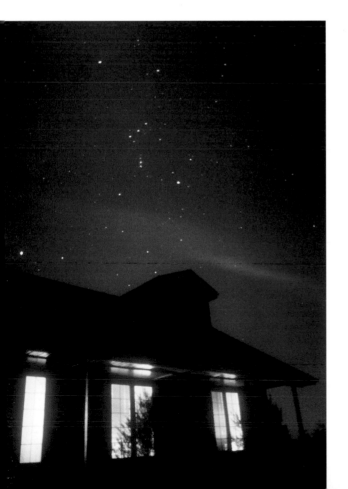

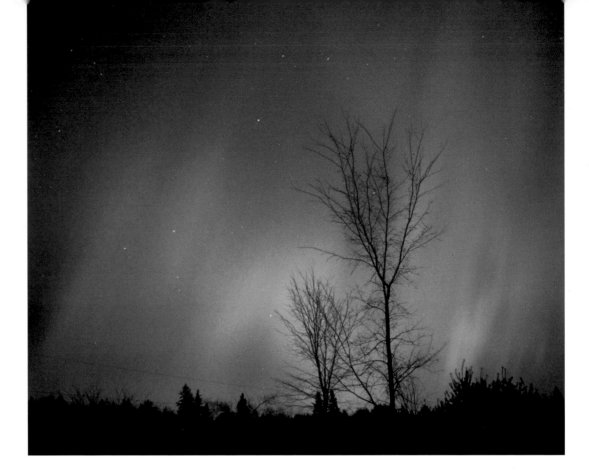

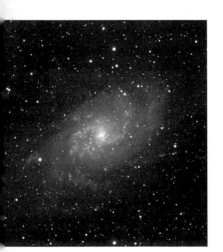

at Vancouver's MacMillan Planetarium and the mastermind of the Great Comet Chase.

They rented a van and began driving inland over the mountains, which the forecast predicted would be clear of cloud cover by 4:30 a.m., the time when the comet was to be in view. The outlook for Vancouver was continued rain. "There were five of us with our telescopes, cameras and binoculars all packed in the van," says Hewitt-White. "A sixth member of our group had to get up early for work and reluctantly stayed behind.

"It was a nightmare from the start—a blinding snowstorm. 'It's got to clear up,' we told each other. We drove 200 miles, and it was still snowing. After a few close calls on the treacherous mountain road, we finally turned back. Then, as we crossed the high point in the Coast Mountains, the sky miraculously began to clear. It was exactly 4:30. We pulled over and immediately got stuck. But we had not gone far enough—a mountain peak blocked the view.

"Five comet-crazed guys in running shoes started scrambling up the snowdrifts on the nearest cliff to gain altitude. By the time we reached a point where the comet should have been in view, twilight was too bright for us to see it. Half-frozen and drip-

ping wet with snow, we pushed the van out and headed back to Vancouver. Within minutes, we drove out of the storm area and saw cloudless blue sky over the city. When we got home, we heard the worst: The guy who stayed behind had seen the comet from a park bench one block from his home."

The eclipse chasers, another subgroup of recreational astronomers, spend countless evenings planning every detail of an eclipse expedition—a trip, sometimes to remote regions of the globe, for the express purpose of standing in the Moon's shadow to watch a total eclipse of the Sun. Given the vagaries of the weather and the inevitable glitches in foreign countries, probably half of these pilgrimages are partial or complete failures. Ventures have been foiled by dust storms blowing away tents, lost luggage, broken-down rental cars and balky camera equipment.

Regardless of the outcome, though, as soon as they get home, the eclipse stalkers whip out maps and start planning the next year's expedition. For anyone who has not seen a total solar eclipse, the behavior may seem odd. But for veteran eclipse chaser Robert May of Scarborough, Ontario, it is "the greatest of all natural spectacles, a truly awesome phenomenon. I want to see every one I can while I am still physically able to do so." May says that for him, eclipse chasing has added a new dimension and a real purpose to foreign travel.

## ARE YOU READY?

As we said previously, astronomy is not an instant-gratification hobby. It takes time and effort to appreciate what you are looking at and to coax the best performance out of your telescope or binoculars. Moreover, backyard astronomers come to know how enjoyable it is to hear the "oohs" and "aahs" from people who are looking through a telescope for the first time. The ultimate thrill, though, is to be uttering the oohs and aahs yourself. With this in mind, we offer the backyard astronomer's Aah Factor, a 1-to-10 scale of celestial exclamation.

Factor 1 on the scale is a detectable smile, a mild ripple of satisfaction or contentment. Factor 10 is speechless rapture, an overwhelming rush of awe and astonishment. Here are a few examples to aid in developing your own Aah Factor list.
• One: Any routine celestial view through binoculars or a telescope; a faint meteor; a well-turned phrase in a good astronomy book.
• Two: Finding the planet Mercury; sunspots; the Moon's surface through a telescope; discovering how clear things look through binoculars mounted on a tripod; cloud belts on Jupiter.
• Three: Saturn or the Orion Nebula through a telescope, even if you have seen them umpteen times before; the starry dome on a clear, dark night in the country; Jupiter's Red Spot; a colored double star.
• Four: A beautiful sunset or sunrise; seeing a bright Earth satellite for the first time; a partial eclipse of the Moon; a close conjunction of two planets or of the Moon and Venus; Earthshine in binoculars; finding the Andromeda Galaxy for the first time.
• Five: Identifying Jupiter's moons through binoculars for the first time; a moderately bright comet in binoculars; telescopic detail on Mars; a meteor shower.
• Six: Recognizing your first constellation; a bright meteor; a good telescopic view of a galaxy or a globular cluster; the shadow of one of Jupiter's moons slowly crawling across the planet's face; your initial look at your first successful astrophoto.
• Seven: A first view of the Moon through a telescope; a first view of the Milky Way with binoculars; a total eclipse of the Moon; a bolide or a fireball meteor.
• Eight: A rare all-sky multicolored auroral display; the moment you begin to realize how immense the universe is.
• Nine: A bright comet with a naked-eye tail; your *first* view of Saturn's rings through a telescope; a meteor storm.
• Ten: A perfect view of a total eclipse of the Sun; discovering a comet or a nova.

It is nice to log a two or a three on the scale each night. Soon, you will be climbing the scale of celestial aahs. It is captivating and addictive. It can even get out of hand. For instance, one rabid enthusiast we know became physically ill while attending a concert with his wife and friends because he had noticed a spectacular aurora brewing when they were parking the car. He felt tortured by not seeing it but did not want to spoil the evening for the others. Such is the power of the night sky. How far you are taken by its spell depends on you.

Of course, there is always the frustration of being clouded out after preparing for an eclipse or other major celestial event for weeks—or even years. This is an activity with frustration minefields along with the rapture. It's not for everybody. But with the help of this book, you will soon know whether it's for you.

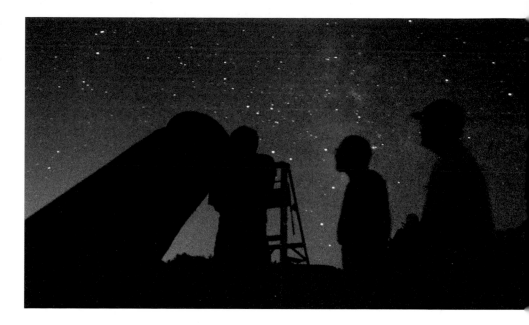

▲ **Seeing by Starlight**
Any dark sky can hold sights that rate a 6, even an 8, on our scale.

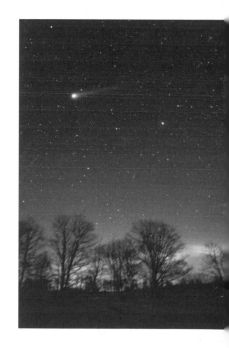

▲ **Everyone's Sky**
Few will get their name on a comet, like Comet Hyakutake seen in 1996, but the joy of stargazing still brings "personal best" discoveries to share with friends and family.

## PART 1

# Binoculars for the Beginner a

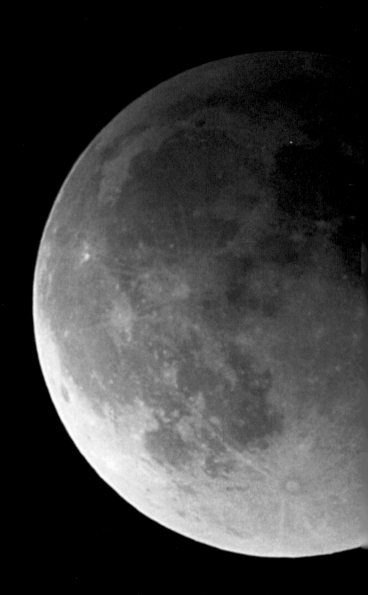

Veteran backyard astronomers always have binoculars within easy reach. Why? Binoculars are midway between unaided eyes and telescopes in power, field of view and convenience. Of all the equipment an amateur astronomer uses, binoculars are the most versatile and the most essential. Yet the capabilities of good binoculars are often underrated by backyard astronomers, especially beginners. This is a pity, because binoculars are so much easier to use than a small telescope.

Admittedly, binoculars are not nearly as exotic as a telescope, but they can be found in almost any home. Even so, many people ignore them when they think of celestial observing. They purchase a telescope without ever turning their binoculars to the night sky, thinking that only a telescope can truly reveal the universe.

Binoculars are ideal for viewing a total lunar eclipse, seen here in a triple exposure showing the event over a period of one hour. Photos by Steve Barnes.

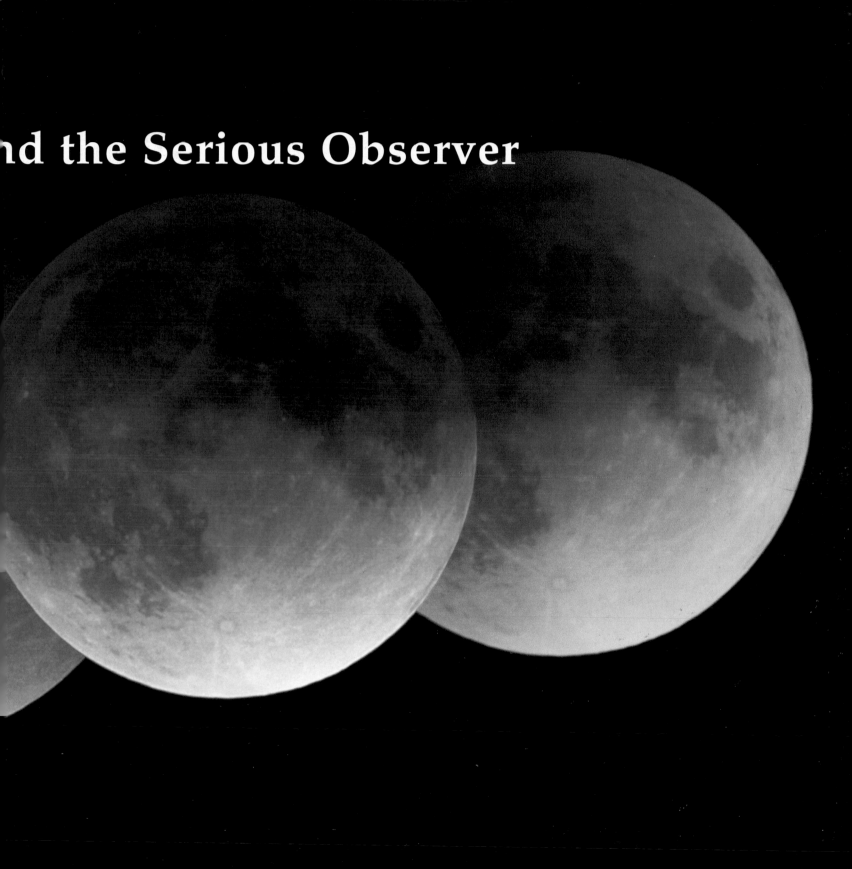

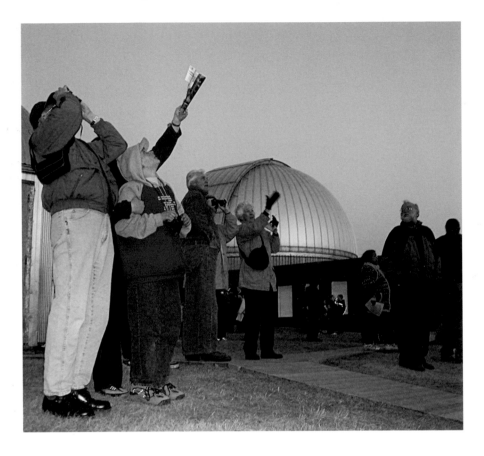

# Consider the Humble Binocular

Here are some of the celestial objects binoculars will reveal:

• In a dark, moonless sky, ordinary bird-watcher binoculars can pick up more than 100,000 stars, compared with the 4,000 or so visible to the unaided eye. The hazy band of the Milky Way breaks up into countless thousands of stars—one of the great treats in amateur astronomy.

• Star colors, ranging from blue to yellow to rusty orange, are more evident with binoculars than without.

• Any night that Jupiter is visible, two to four of its large moons can be seen close beside the brilliant planet.

• The planets Uranus and Neptune are easy targets with binoculars when you know where to look.

• The Andromeda Galaxy, a huge city of stars larger than our entire Milky Way Galaxy, is plainly visible as an oval smudge near overhead in autumn and early winter for northern-hemisphere observers.

• Star clusters of exquisite beauty, such as the Pleiades and Hyades, are seen in their entirety in binoculars, whereas most telescopes (because of their smaller fields of view) can show only portions of them.

• On the Moon, at least 100 craters and mountain ranges are visible, as well as subtle shadings on the flat plains that 17th-century astronomers thought were seas.

The utility of binoculars goes far beyond this list. Planets hidden in twilight glow are most often first detected by sweeping with binoculars. Earthshine on the Moon (the faint illumination of the Moon's nightside) is greatly enhanced by binoculars. There is no better instrument for watching a lunar eclipse, for monitoring a planet's motion through a constellation over weeks or months or for observing a bright comet. Nearly every celestial object visible to the eyes alone will be improved by binoculars.

Moreover, binoculars can reveal a multitude of objects completely invisible to the naked eye: nebulas (star-forming regions); wispy remnants of ancient supernovas; and star clusters ranging from bright stellar splashes to dim patches of starlight. Most challenging are the galaxies, great islands of stars like our Milky Way Galaxy that dot the void of deep space. With practice, you can detect several dozen galaxies up to 30 million light-years from Earth. They are not easy to find, but just seeing them with binoculars is astonishing. For 30 million years, the galaxy's light has been on its way to Earth, ending its journey by entering the eyes of a curious observer. Not bad for binoculars.

Easier quarry for beginners are star clusters in the Milky Way. These range from naked-eye collections of stars like the Pleiades to such glittering jewels as the Double Cluster in Perseus or M7 in Scorpius. Hundreds of celestial sights await observers with binoculars, enough to keep a backyard astronomer busy for years. Far from being a substitute for a small telescope, binoculars are indispensable partners in the exploration of the universe.

One further advantage: Using two eyes for celestial viewing allows you to see more. Your body is more comfortable, and the

# Celestial Showcase

A telescope is not necessary for examining many celestial objects, and sometimes binoculars or even the unaided eyes can provide a better view. This inventory shows the versatility of humble equipment—or no equipment at all.

| Naked Eye | Binoculars | Telescope |
|---|---|---|
| constellations* | star clouds of Milky Way* | hundreds of double and multiple stars* |
| meteors* | Earthshine on the Moon* | hundreds of variable stars* |
| auroras* | planetary motion* | hundreds of galaxies* |
| Earth satellites* | bright comets* | hundreds of star clusters* |
| solar and lunar halos* | lunar eclipses* | dozens of nebulas* |
| a few double stars | details of constellations* | planetary detail* |
| five planets | moons of Jupiter | planetary satellites* |
| planetary motion | dozens of lunar craters | planetary phases* |
| bright comets | dozens of variable stars | thousands of lunar features* |
| a few star clusters | dozens of double stars | sunspots and solar detail* |
| three galaxies | dozens of star clusters | solar eclipses |
| a few nebulas | several galaxies | comets |
| a few variable stars | several nebulas | lunar eclipses |
| solar eclipses | seven planets | planetary motion |
| lunar eclipses | solar eclipses | lunar occultations |
| largest sunspots | sunspots | asteroids |
| Milky Way | bright asteroids | |

*Asterisked items in each column indicate the viewing targets most easily seen.

brain is at ease receiving messages from both eyes. When observed with two eyes, objects at the threshold of vision register as real, whereas one-eyed detection produces fleeting and uncertain cerebral messages. How much more can be seen? Most experts estimate 40 percent above single-eye viewing.

## SELECTING BINOCULARS

Binoculars are, in essence, miniature telescopes—a pair of prismatic spotting scopes reduced in size and linked together in parallel for viewing with two eyes. The prism system has a threefold purpose: reducing the length of the optical system by folding the light path; reducing the overall weight; and finally, producing a right-side-up image for convenient terrestrial viewing. Binoculars come in a bewildering array of sizes, magnifications, models and prices. Virtually useless toy binoculars with plastic lenses can be had for a few dollars; at the other end of the scale, the colossal Fujinon

6-inch refractor binoculars (25x150) cost as much as a new car. In between, there is something for everybody.

There are two basic types of prism binoculars: porro prism and roof prism. Roof prism binoculars have straight tubes and are generally smaller and more expensive than porro prism models of otherwise equivalent optical size (two 8x40s, for instance). Although roof prism binoculars are available with main lenses up to 63mm in diameter, the primary advantage of the design—compactness—is defeated in sizes over 50mm. Porro prism binoculars, with the familiar humped, N-shaped light-path design, are available in all sizes. For astronomy, we've found good optical performance in roof prism binoculars only in the premium models with features such as phase-corrected coatings. Porro prism models that provide top-notch images are much less costly.

Binoculars have two numbers engraved on the body, such as 7x50, usually near the eyepiece end. The first number is the mag-

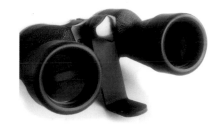

▲ **Tripod Adapter**
Sky observers consider the inexpensive L-shaped binocular tripod adapter as an essential accessory. A threaded hole at the foot of the L accepts the tripod head screw. The steadied view afforded by the tripod significantly increases the detail visible. When purchasing a new binocular, be sure it has a threaded hole to accept the adapter.

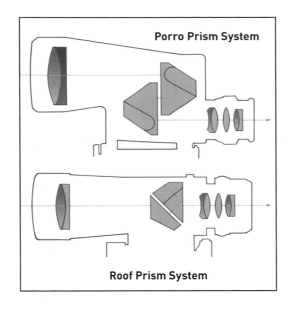

**Porro Prism System**

**Roof Prism System**

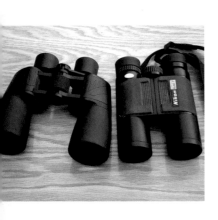

**Two Types of Binoculars** ▲
Roof prism binoculars are more compact than porro prism in sizes under 42mm. They are generally more costly too. On the other hand, top-of-the-line roof prism glasses are superb in every way. For general astronomy binoculars at a reasonable price, the authors recommend porro prism models in the 7x50 and 10x50 sizes.

nification (the "x" means magnification, or power); the second number is the diameter of the front lenses in millimeters. Thus 7x50 means 7 magnification and 50mm-diameter objectives (main lenses). There are dozens of combinations, from tiny 6x16 binoculars to the 25x150 monsters. There are, predictably, advocates of every combination of size and magnification for different purposes. The optimum magnification for astronomical binoculars is a subject of ongoing debate. We'll try to guide you through it.

Acceptable-quality binoculars can be purchased for about $100. First-class glasses are in the $200-to-$500 range. For connoisseurs of fine optics, the sky's the limit. Price is a major guide in this competitive market. There is usually a good reason why one binocular is three times the price of another, even though they may look the same on the outside. We have tested superb binoculars with famous labels, but we've also discovered equally good glasses marketed with labels we'd never heard of before. One thing we didn't find were any zoom binoculars that met our standards of acceptable optical quality for astronomy.

Weight is also an important consideration. Binoculars in the popular 7x50 and 10x50 sizes can range from 26 to 50 ounces. Every ounce counts in astronomy, because binoculars are held tilted above horizontal, a more tiring position than horizontal or lower, as in most terrestrial viewing. We recommend forgoing ruggedness for light weight. Astronomers tend to use their equip-

ment in low-impact environments, so "armor" cladding or "military specs" only add extra baggage. A good binocular weight is 22 to 32 ounces. Most people can hold this weight long enough for a satisfying observation before returning to an arms-down position. In this weight category are the 8x40, 7x42 and 8x42 glasses traditionally used by birders. We regard these sizes as the minimum for astronomy, so if you already own a pair, they could serve perfect double duty.

## EXIT PUPIL

In many observing references, you will read that for maximum efficiency under typical low-light astronomical conditions, the light cone exiting the binocular (or telescope) eyepiece—called the exit pupil—should be the same size as that of the dilated pupil of the eye. The theory is that all the light from the instrument should enter the pupil rather than some light falling uselessly on the surrounding iris (the colored part of the eye). Most people under the age of 30 have pupil diameters of seven to eight millimeters in dark conditions. After 30, everyone generally loses one millimeter every 20 years or so throughout life, as the eye muscles become less flexible. So applying the 7mm rule to all observers ignores age variations. Furthermore, the outer edges of the lens of everybody's eyes have some inherent optical aberrations.

For these reasons and as a result of tests we have done on binocular performance using our own eyes with many different binoculars, our conclusion is that the glasses which actually reveal the most detail on all kinds of celestial objects have exit pupils in the 2.5mm-to-4mm range. A binocular's exit pupil need not be measured directly; it can be quickly calculated by dividing the aperture in millimeters by the magnification. For example, 7x50 glasses have 7.1mm (50 ÷ 7) exit pupils.

An important caveat must be added here: The tests mentioned in the paragraph above were done with the binoculars tripod-mounted, to make sure it was an "apples and apples" comparison. But what about the 7mm conventional wisdom? For full details, we refer you to an important article on binoculars in astronomy that is reprinted each year in the annual *Observer's Handbook* of The Royal Astronomical Society of Canada. Written by retired physics professor Roy Bishop, who has had a long-standing interest in this question, the article argues for a new way of evaluating binocular performance.

The principal factors influencing what can be seen in the sky at night with binoculars are the amount of light entering the instrument and the amount it is magnified. No surprise there. But what is surprising is that the most meaningful way of gauging binocular performance on both stars and extended objects like the Moon and nebulas, says Dr. Bishop, is to multiply the aperture in millimeters by the magnification. He calls this the Visibility Factor (VF).

Thus the VF for 7x50s is 350 and for 10x50s is 500. For 11x80s, a common size of giant binocular, the VF is 880. And the VF for the new-generation 18x50 Canon image-stabilized binocular (see page 24) is 900! This is definitely unconventional wisdom. But is it true? Side-by-side tests (by TD) suggest it is. The 18x50s showed the same 10th-magnitude galaxies (NGC3077 and NGC2978) as the 11x80s, both handheld and tripod-mounted. On some celestial objects, 11x80s had the edge; but on others,

the 18x50s revealed more. Overall, it was a draw.

Another point raised by comparisons like this is emphasized by the huge difference in VF between 7x50s and 18x50s, both of which have the same aperture but a large difference in performance. So why not 50x50s? As Dr. Bishop explains, the VF applies to binoculars with exit pupils from 3mm to 7mm.

Now before anyone rushes out to purchase large and/or high-powered binoculars with a VF close to 1,000, a big however has to go in here: To retain the fundamental convenience of binoculars, they should be used handheld and carried on a strap around your neck.

Most people find that 10x is the limit for comfortably holding binoculars, because every quiver of the arms is also magnified by 10x. A steadier view can be had with 7x or 8x glasses. It is here that the 7mm exit pupils of 7x50s have their secret advantage. It's not so much the long-touted matching of the 7mm exit pupil to the eye, but the big exit beams from the 7x50s' eyepieces make it easier for the average observer to get fully illuminated exit pupils positioned where they need to be on the eyes. Even more im-

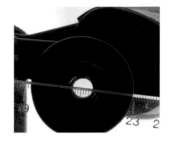

**▲ Exit Pupil**
The famous 7mm exit pupil is displayed on these 7x50 glasses for all to see. Although many astronomy guidebooks have stated otherwise for decades, the 7mm exit pupil is not necessarily ideal for astronomical binoculars.

# Focusing Tip

Most binoculars are brought to focus by turning a knurled knob on the central bar of the instrument. This focuses both eyepieces by the same amount. However, the precise focal point can be slightly different for each eye. To accommodate this, the right binocular eyepiece usually has a separate focusing capability with a scale and a zero point. To set this diopter adjustment, as it is called, focus with the left eye only (right eye closed), using the center focus wheel. Then focus with the right eye only (left eye closed) using *only* the right eyepiece. When the right eye is in sharp focus, note where the right-eyepiece scale is set. Try to remember this setting. Whenever someone else uses your binoculars, you will probably have to reset the scale. Once your diopter is set, just use the center focus wheel to focus for both eyes simultaneously. So-called perma-focus binoculars are just a gimmick and are especially unsuited to astronomy. We also avoid "quick-focus" rocker-focus models, which may be fine for other applications but are too coarse for astronomy.

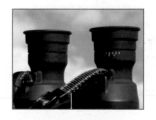

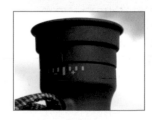

 *The diopter scale on the right eyepiece is of crucial importance for taking full advantage of your binoculars' optical capabilities. Carefully follow the focusing sequence described above to achieve quick, sure focus.*

portant, though, 7x50s have, by their nature, wide fields of view, typically seven degrees. This is the reason we also recommend this size: They are easy to look through and easy to aim, especially for beginners or in group-observing situations where the glasses are passed from person to person.

To summarize: Ideally, the objective (front) lenses of astronomy binoculars should be as large as possible for producing the brightest possible image. But hand-holding intervenes, limiting the weight that can be hoisted to the eyes. Fifty-millimeter binoculars tend to be the largest size that is convenient to hold for any length of time.

## FIELD OF VIEW

The diameter of the circle that can be seen through binoculars is called the field of view. This is often expressed as the number of feet that span the field when viewed from a distance of 1,000 yards (or meters at 1,000 meters). Thankfully, this awkward nomenclature seems to be fading in favor of the more convenient angular diameter in degrees. For conversion purposes, one degree is equivalent to 52.5 feet at 1,000 yards, or 17 meters at 1,000 meters. Most 7x50 binoculars have seven-degree fields; 10x50s typically have six-degree fields. While some models offer a wider field of view, it often comes at the expense of optical quality.

Binoculars designated as wide-angle or ultra-wide-angle can have 8-to-12-degree fields of view. Invariably, though, these models have severe optical distortions around the edge of the field. This is usually not objectionable in everyday terrestrial use, but in astronomy, when all the stars in

# Canon Image-Stabilized Binoculars

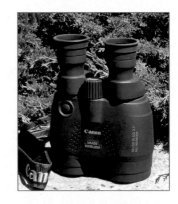

Introduced in 1996, these techno-breakthrough binoculars stand apart—way apart—from all other binoculars we have ever used. With our first look through these glasses, we immediately lost our socks in the distant trees somewhere, so explosively were they blown off by the views. Ever since, these glasses have been our personal workhorse astronomy binoculars. Because of their unique capabilities, we have singled them out for special treatment here.

Pressing a button atop the binoculars engages motion sensors coupled to tiny microprocessors that control the shape of accordionlike prisms in the light path. The prisms contain a high-refractive-index liquid that bends the light path to compensate for the quivers imparted by the user's arms. We'll dispense with further details. The fact is, the quivering and dancing images so familiar in handheld binocular views are compensated for instant by instant. The corrections smooth out the jerkiness in a unique, fluidlike manner that most observers find magical. The huge advantage resulting from the stabilization is that magnifications of up to 18x are possible with less image quavering than with standard, unstabilized 7x glasses.

Although Canon produces 10x30 and 12x36 image-stabilized models, it was the 15x45s, introduced in 1997, that caused the stir in astronomy circles. In 2000, the 15x45s were discontinued, replaced by 15x50 and 18x50 models. All are roof prism designs with superb optics and wide-angle eyepieces that are among the best we've seen in any binocular at any price. The 15x models have 4.5-degree fields; the 18x has a 3.7-degree field. Many astronomy users prefer the 15x glasses because they are easier to aim into a starry sky than are the narrower 18x ones. However, the higher power does reveal more detail. Both 50mm glasses have standard threaded holes for tripod mounting, which the earlier, smaller models do not. The 15x45s, at 36 ounces, are still available on the used-equipment market. The newer 50mm models, both 6 ounces heavier, are priced around $1,200.

 *The 15x50 and 18x50 image-stabilized models are identical in appearance. Depressing the button on top locks in the stabilizer. A second button push (or a five-minute default timer) disengages it. Two AA batteries power the stabilization feature.*

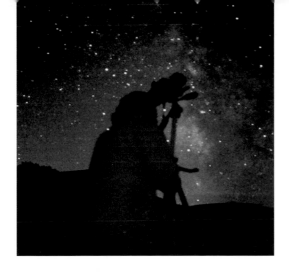

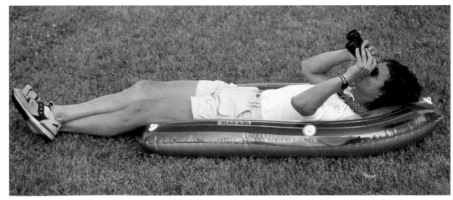

the outer field resemble comets or seagulls, there is no advantage. Frankly, we think the effect spoils the aesthetics of the view. When stars are sharp across the field, it's like looking through a porthole into the universe; but when stars turn into blobs at the edge of the field, it's like looking at the heavens through a glass ashtray. We'll sacrifice the wider field for a sharp one.

## EYEGLASSES AND BINOCULARS

Binocular users who must wear eyeglasses for correction of astigmatism or who prefer to keep their glasses on while observing will benefit from so-called high-eyepoint binoculars, designed to push the exit pupil 18mm to 26mm from the surface of the eyepiece lens, compared with 10mm to 15mm for normal binoculars. This allows room for the lens of the glasses to fit between the exit-pupil point and the eyepiece.

High-eyepoint binoculars come with oversized rubber eyecups that, when folded down, provide the extra distance for eyeglass users. When up, they act as a guide for positioning the eyes when the user is not wearing glasses. Achieving the high eyepoint requires larger and, consequently, more costly eyepieces. But as the population ages, accommodation for eyeglasses in binoculars has become more common.

## BINOCULAR TESTS

As a quick check for viewing comfort and optical quality, follow points one through six outlined here when shopping for binoculars. Point seven is a more rigorous test.

**1.** Weight. How heavy are the binocu-

lars? Eliminate all glasses built to withstand jungle warfare. You shouldn't need Arnold Schwarzenegger arms and shoulders to hold them up for a reasonable length of time.

**2.** Prisms. Hold the binoculars a few inches in front of your eyes, and look in the eyepieces. Aim the glasses at the sky or a window. The illuminated optical path should be completely round and evenly illuminated. Discard those glasses with a squarish rather than a circular appearance to the illuminated field.

**3.** Craftsmanship. Check all the moving parts. Moderate but even pressure should be required to adjust focus and interpupillary distance. Hold the binoculars with the eyepieces pointing up. With the index finger of each hand, push down on the eyepiece housings, alternating between left and right. Be careful not to touch the actual eye lenses. Look for excessive rocking of the two eyepieces and the bridge that connects them. If they are too sloppy, it will be hard to maintain focus on both sides simultaneously, especially when making contact with your eyebrows.

**4.** Optics Check. The central area of the field of view should be pin-sharp, with no evidence of fuzziness, false color or double imaging. Many glasses with perfectly acceptable image sharpness in the central region of the field quickly lose their definition toward the edge. We generally rate binoculars unacceptable if the image grows fuzzy less than 50 percent of the way from the center to the edge. This capability can be tested during the daytime by looking at sharp detail, such as the branches of a distant tree or the top of a building against the bright sky. Such testing will also reveal other potential problems. In high-contrast situa-

▲ **Observing Comforts**
Sailing the celestial seas is best done in a boat—on land. Try using a small inflatable child's boat (about $25) to scan the night sky in complete comfort. Head, legs and shoulders are supported more effectively than in the standard lawn chair. The head can be raised or lowered by varying the leg pressure. Above left: Observer using tripod-mounted binoculars.

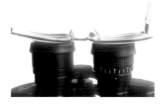

▲ **Eyeglass Mode**
Many binoculars have fold-down eyecups to accommodate people who wear their glasses while observing. Before buying, be sure to check that the eyecups provide full field-of-view access with eyeglasses on as well as comfortable eye position with eyeglasses off.

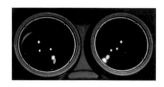

tions, a blue or green color fringe, called a chromatic aberration, may be obvious around the edge of objects, which is a sign of lower-quality optics.

5. Coatings Check. Light transmission is increased and flare and ghosting from internal reflections are reduced by coatings on the lenses. The best binoculars are multi-coated on all optical surfaces, including the prisms. It may be printed on the binoculars themselves or in the literature provided with them that the glasses are coated or multicoated. When held under bright light, coated lenses (meaning single-layer coatings) are generally pale blue, while multi-coated lenses usually give off a deep green or purple sheen.

Coating increases light transmission at each optical surface to about 97 percent, compared with 93 percent for no coating. Multicoating lets about 99 percent of the light through at each glass-to-air surface. The problem is determining whether all the optical surfaces are coated—often, they are not. To find out, shine a bright light into the binoculars from the objective (large lens) end. Looking down into the glasses, tilt them slowly back and forth and watch for the multiple reflections from the coated-lens surfaces. All should be roughly equally sub-

dued blue, green or purple, depending on the coatings used. Noticeably brighter white reflections are a sign of uncoated elements.

6. Collimation. If, after using the binoculars for a few minutes, you feel eyestrain and for some reason have to force the images to merge, the binoculars are probably out of collimation, which means that the two optical systems are not precisely parallel. This is the main thing to watch for when purchasing used binoculars. All it takes is accidentally dropping the binoculars from about eye height to the ground for the collimation to be knocked out. It then requires professional attention to repair.

7. Astronomical Testing. Optical perfection is a never-ending quest among amateur astronomers, and all but the very finest binoculars usually yield evidence of some optical imperfections when used to observe the stars. Viewing brilliant point sources on a black background is the most rigorous test of optics. In the center of the field, a bright star should show near-pointlike imagery, with small, irregular spikes emerging from the bright central point. The fewer spikes seen, the better; but the important thing is that they must be symmetrically arrayed around the point, with no obvious flaring in any direction. If you find there is flaring and

# Key Factors to Consider When Selecting Binoculars

• Larger main lenses mean brighter images, but for most people, a 50mm lens is a practical limit in handheld binoculars.

• Binoculars with 7mm exit pupils are easier to bring to correct position in front of the eye, an advantage for young people and beginners of any age.

• Higher magnification means better resolution, but it also means more stringent optical-quality standards to produce good images.

• Higher magnification results in amplified jiggling during handheld operation. This factor alone limits binocular magnification for handheld astronomical viewing to 10x.

• When we put all of this together, the most popular sizes are 7x50 and 10x50. For those who prefer somewhat smaller and lighter glasses, we recommend the 7x42 and 8x42 sizes.

• All other things being equal, aren't the 10x50s the obvious choice over the 7x50s? But all other things are never equal. Aiming and observing through binoculars at night is much easier for some people than for others. In our experience, 7x50s are easier to use. On the other hand, 10x50 glasses will yield fainter stars and more detail on the Moon and many other celestial objects than will 7x50s. More detail makes sense because of the higher power, but why are dimmer stars apparent if the aperture is the same 50mm? Part of the reason is that the smaller exit pupil helps avoid the edge-of-eye aberrations (producing sharper stars), but mainly, it is that the higher magnification in effect spreads out the sky background, darkening it in the process.

you normally wear glasses, put them on and see whether the asymmetry disappears. If it is still there, the binoculars are likely at fault and should be rejected. Now move the bright star toward the edge of the field. It will begin to grow wings, usually parallel to the edge of the field. This indicates astigmatism in the eyepieces, which is nearly always present to some degree because it is a very difficult defect to eliminate in short-focal-length systems such as those of binoculars. Compare carefully, because there are great differences among binoculars in the amount of astigmatism present.

## RECOMMENDATIONS

We have tested dozens of binoculars from several manufacturers. Among the ones we most frequently recommend are three nearly identical models of 10x50s, all in the $200-to-$250 range: the Vista marketed by Orion Telescopes, the Ultima by Celestron and the Adlerblick by Carton Optics of Japan. All are 27 ounces—an exceptionally light weight for 50mm binoculars—with very sharp 5.3-degree fields and good eye relief. These are excellent general-astronomy glasses.

In the $100-to-$150 category, the Bausch & Lomb 10x50 Legacy is a standard stock item in many camera stores. We rate these as acceptable glasses for astronomy, although the image quality falls off rapidly toward the field edge, as with all binoculars we looked at in this price range—with one exception: the Bausch & Lomb 7x50 Legacy. These binoculars have very good optics for the price (same price as the 10x50s) and a seven-degree field. They are an ideal beginner's binocular for astronomy. But although the Legacy 7x50s are sold in Canada and many other countries, they are not available in the United States.

Moving into the $250-to-$500 range, we found that many models of the best-known brand names were no more impressive than similarly priced models by less well-known manufacturers. For this reason, we urge anyone in the market for new binoculars to compare as many brands and models as possible. The limiting factor is usually finding a well-stocked dealer.

For the connoisseur of fine optics, the Nikon 7x50 IF SP Prostar (about $700) has the best optics we have seen in any binoculars at any price by any manufacturer. The images are astonishingly crisp and bright. But at 49 ounces, these are extraordinarily heavy brutes, too heavy to be held for any reasonable length of time. They score an A+ for optics but an F for weight and really must be tripod-mounted for proper viewing, which substantially reduces their attractiveness. In the same weight and price range and optically a close contender is the Fujinon FMT-SX 7x50, with a comfortable 20mm of eye relief.

Our personal favorites in the high-end class (in this case, the $1,000 range) are the 15x50 and 18x50 Canon image-stabilized glasses, already praised earlier. Conventional but superb glasses in this class are the top models of 7x42 or 8x42 roof prism binoculars from Leica, Zeiss, Nikon and Swarovski (8.5x42), all 25 to 29 ounces. Designed for birders who want the best, they are worth considering as optimum dual-purpose glasses.

## GIANT BINOCULARS

Sizes of giant binoculars used in backyard astronomy are 9x63, 10x70, 16x70, 11x80, 15x80, 20x80, 14x100 and 25x100. Binoculars in these sizes range from $300 to several thousand dollars. Because of the introduction in the 1990s and early in this century of a much larger selection of short-focal-length 70mm-to-100mm-aperture refractor telescopes, along with modern eyepieces that can match or approach the low powers and wide fields of giant binoculars, we are less enthusiastic about these big glasses than we were in earlier editions of this book. One of the continuing problems is overhead viewing, even in models with 45-degree eyepiece diagonals. A few of the more costly ($1,000 to $5,000) giant binoculars have angled eyepieces that make viewing close to the zenith easier on the neck, but most big binoculars don't.

Also available, usually from small cottage-industry suppliers, are counterweighted "parallelogram" holders for cantilevering the binoculars away from the tripod head so that the observer can get underneath them. Even so, you are forced to crane your neck way back to look up; neither of us has ever become a frequent user of these binocular stands.

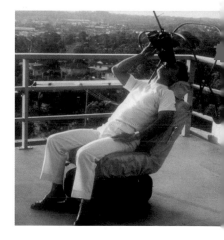

▲ **Giant-Binocular Chair**
This homebuilt observer's chair, made from discarded auto parts, has all-weather covering for year-round use. Note the counterweights to balance the 80mm giant binoculars.

# Telescopes for Recreational A

Until the 1970s, most backyard
astronomers made their telescopes from scratch or ordered
a few commercial parts, such as
a focuser and mirror cell, and
built a combination homemade/
commercial scope. Today, the
wide availability of commercial
telescopes at lower relative
prices has turned the old hobby
on its head, so now all but a
handful of amateur astronomers
purchase complete commercially
manufactured equipment.

With the increasing popularity of recreational astronomy,
telescope companies have
evolved from basement operations to publicly traded companies, with annual sales in
the tens of millions of dollars.
Competition is intense, with
leading manufacturers conducting lavish advertising campaigns
in the key astronomy magazines.

All too often, such advertisements are the only sources of
information many buyers use
to make a purchase decision. To
widen the information base, we

decided to include not only
general information on telescope designs but also our
opinions about many of the
models on the market, recommendations gleaned from our
use and testing of dozens of
telescopes. We have *not* relied
upon opinion surveys, Internet
"wisdom" or the secondhand
accounts of fellow amateurs. We
report on what we know from
firsthand experience. We hope
our advice will assist you in
choosing a telescope that's right
for you in what has become a
complex market.

A word about that market:
As is common in high-tech
fields, telescopes have become
a volatile commodity. By the
time you read this, some of the
telescopes discussed may have
been replaced with "this year's
model." Indeed, some of the
companies listed may no longer
exist or will have been absorbed
by new players. For updates,
please check our website:
www.backyardastronomy.com

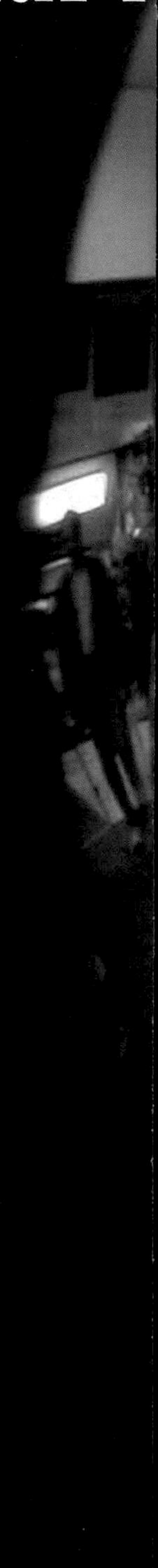

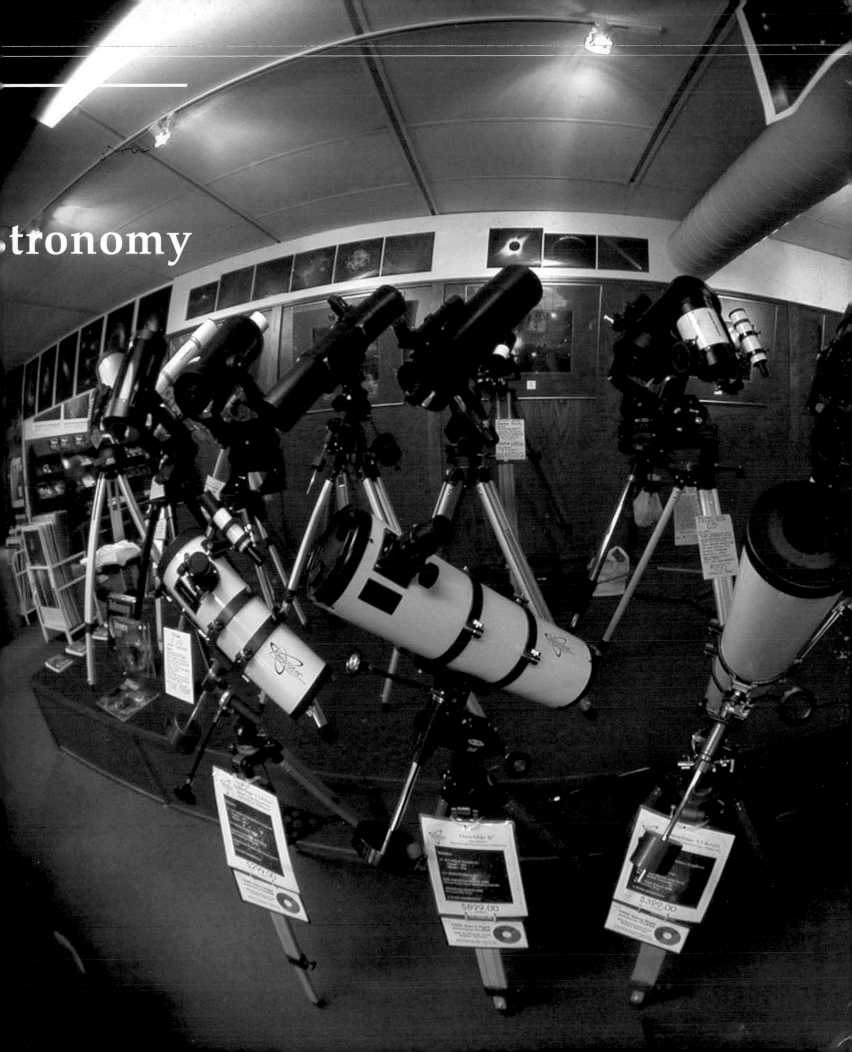

tronomy

# A Brief History of Telescopes

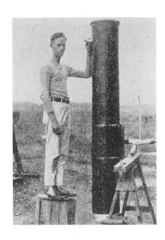

Life used to be simple. In the past, one type of telescope predominated, making the choice of which instrument to buy downright easy: You bought what everyone else was buying. Indeed, you had few other choices. Observing tastes also followed in lockstep fashion. Whatever type of observing the telescope in vogue excelled at was what most people concentrated on. It has long been our opinion that the history of amateur astronomy can be largely divided into periods during which a single type of telescope—and observing—defined the backyard astronomer's universe.

## BEFORE 1950: SMALL-REFRACTOR ERA

If you wanted a telescope before 1950, chances are, you made your own. A typical commercial telescope was a 2-to-3-inch brass-fitted refractor that looked good in a study beside an oak bookcase. Larger telescopes were available, but at a high price.

Commercial telescopes were expensive relative to the wages of the working person. They were made for the upper class, the genteel astronomers of wealth and leisure. The chief observing activities of such amateurs were the casual viewing of the Moon and planets, logging descriptions of the colors of double stars and scanning a handful of clusters and nebulas. A few advanced observers engaged in the more technical activity of measuring the positions of double stars and the brightnesses of variable stars—tasks well suited to a small refractor.

## 1950 to 1970: NEWTONIAN ERA

After World War II, small telescope companies, such as Cave Optical, Criterion, Optical Craftsman and Starliner, began to offer high-quality Newtonian reflectors at a more reasonable price. With relatively large apertures of 6 to 12 inches and costs well below that of sizable refractors, the Newtonian quickly became the most popular instrument of the day.

The Newtonians of the 1950s and 1960s were medium- or long-focal-length telescopes (f/7 to f/10). With the required equatorial mounts, they were big and awkward.

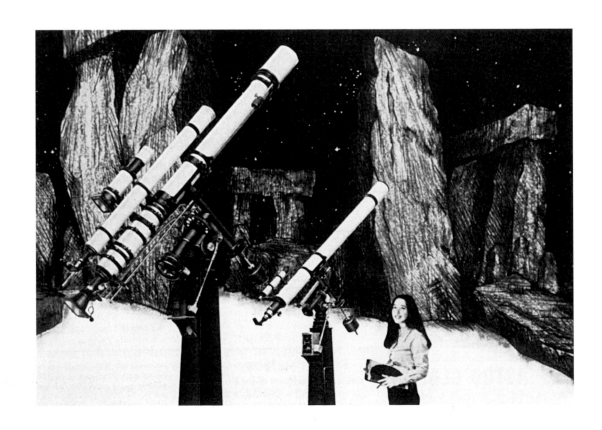

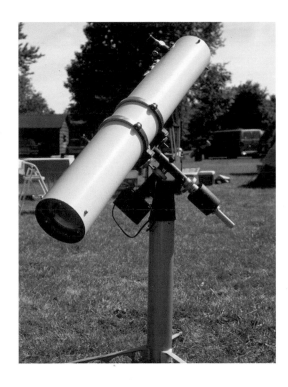

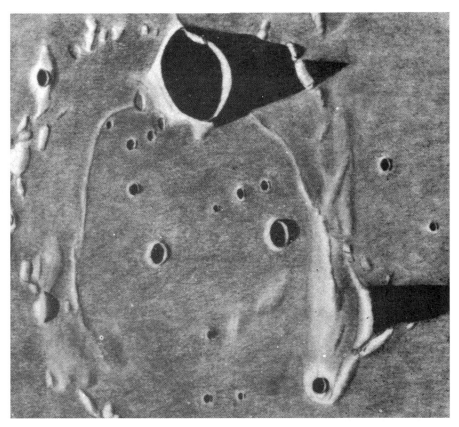

However, long-focus Newtonians were, and remain today, excellent for high-resolution planetary observing. As a result, we entered the golden era of planetary study by backyard astronomers. It was a time when the journal of the Association of Lunar and Planetary Observers was filled with wonderful drawings of the Moon, Venus, Mars, Jupiter and Saturn.

Deep-sky objects such as nebulas and galaxies remained a minor sideline. The books of the day bear this out. *Norton's Star Atlas*, the observer's bible of the 1950s and 1960s, lists only 75 deep-sky objects in the descriptive tables of the 1959 edition, yet thousands of deep-sky objects are within reach of a good 6-inch telescope. J.B. Sidgwick's classic *Observational Astronomy for Amateurs*, first published in 1957, devoted 270 of its 310 pages to solar system objects; the deep-sky realm of nebulas and galaxies was all but ignored, even though from the modestly light-polluted backyards of the day, telescopes could have revealed hundreds of them with no difficulty.

Then, in 1965, Mariner 4 became the first interplanetary probe to return close-up images of Mars, revealing a cratered surface unlike anything telescopic observers had imagined. The pictures from Mariner 4 and a string of follow-up Mars probes into the 1970s removed much of the romance of the

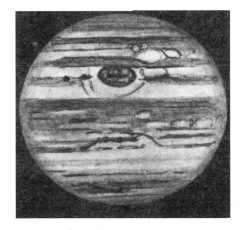

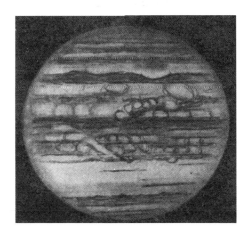

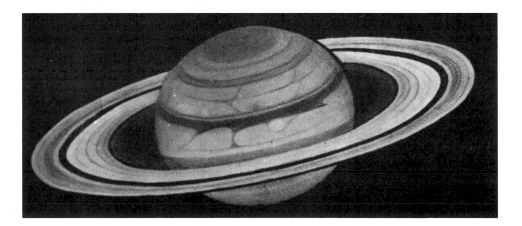

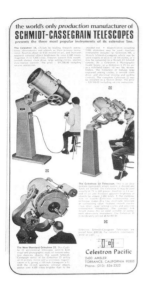

**Trendsetters** ▲
Celestron's early Schmidt-Cassegrains of the 1970s and Coulter Optical's first Dobsonian, seen below at its debut in 1980, changed the hobby forever.

enigmatic red planet. Meanwhile, robotic voyages to other planets and the Apollo Moon landings effectively removed the Moon and the planets as rich targets of opportunity for amateur astronomers, and consequently, planetary observing as a primary activity plummeted.

But, at the same time, the space program of the 1960s heightened public interest in astronomy and space. The hobby turned from a fringe pursuit into a mainstream pastime, setting the stage for the next era.

## 1970 to 1980:
## SCHMIDT-CASSEGRAIN BREAKTHROUGH

The many converts to astronomy during the Apollo era increased telescope sales to the extent that companies could introduce mass-production techniques for serious amateur telescopes. At that time, Californian Tom Johnson was a pioneering telescope maker. Lured by the theoretically near-perfect star images that a Schmidt-Cassegrain

could produce, Johnson built a 19-inch unit for himself. This prototype was the forerunner of all Celestrons. In 1964, Johnson renamed his electronics company Celestron Pacific and began making telescopes.

In 1970, Celestron introduced a compact 8-inch f/10 Schmidt-Cassegrain, the original orange-tube C8 (even the color was novel). The retail price for the basic telescope, without tripod, was $795. Considering that top-grade 8-inch Newtonians cost $600 at the time, the C8 was expensive. But because of its compact size and ease of use, many amateurs flocked to the new instrument, dubbed a catadioptric, or "cat."

The new mirror-lens telescope, along with its turnkey array of photo accessories, another innovation, was marketed by the first modern advertising in the field. In many ways, the hobby as we know it today began with this telescope.

Schmidt-Cassegrains are at their best when used for deep-sky observing and astrophotography. Their portability enabled amateur astronomers to transport telescopes of significant aperture to dark-sky sites, an uncommon practice prior to the rampant spread of light pollution in the 1970s. Despite the growth in light pollution, observing and photographing deep-sky objects began to soar in popularity. Refractors and cumbersome reflectors on heavy equatorial mounts were threatened with telescope extinction.

## 1980 to late 1980s:
## NEWTONIAN REBORN

The new interest in deep-sky observing propelled by the orange 8-inch Schmidt-Cassegrain created a hunger for even more aperture. As compact as the design is, a Schmidt-Cassegrain larger than 8 inches in aperture is still a hefty instrument. How could enthusiasts move to yet bigger telescopes and retain the portability so essential for observing at remote sites?

The solution: Return to the Newtonian, but sacrifice the automatic tracking of equatorial mounts by using simple, squat altazimuth mounts. Popularized by California amateur astronomer John Dobson, these telescopes are now universally called Dobsonians. Beginning in 1980, companies such as Coulter Optical offered lighter, thin-mir-

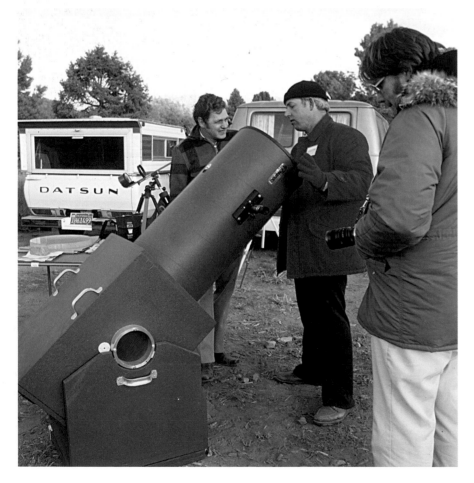

ror Dobsonians for as little as $500 for a 13-inch model. Aperture fever swept the land.

Once again, the instrumentation led observers into new territory. The big light buckets, as large Newtonians are irreverently called, were unsurpassed at revealing faint deep-sky targets. Objects that were barely perceptible smudges in an 8-inch Schmidt-Cassegrain became impressive spectacles in the 17-inch Dobsonian typical of the early 1980s. Armed with giant Dobsonians, observers pursued deep-sky targets previously thought inaccessible.

## Late 1980s to mid-1990s:
## REFRACTOR REBORN

Despite their reputation for sharp, contrasty optics, refractors as serious backyard-astronomy equipment were all but gone by the 1980s. Their inherent chromatic aberration (false color) plus high cost had removed them from contention. However, the Schmidt-Cassegrains and Dobsonian-mounted Newtonians that were dominating the market did not win over everyone. A substantial number of observers complained of the fuzzy views offered by the large but low-cost light buckets typical of the 1980s. Others were disappointed by a spate of poorly made Schmidt-Cassegrains that reached the market during the "Halley boom" in the mid-1980s. It was this situation, plus interesting new developments in glass technology, that led two telescope designers, working independently in the mid- to late 1980s, to revolutionize the refractor for recreational astronomy.

Al Nagler, a military optics expert and amateur astronomer, began developing ways to reduce the focal ratio without increasing chromatic aberration, the traditional flaw of refractors. His Tele Vue Renaissance and Genesis 4-inch models were instrumental in reestablishing the refractor as a serious tool for backyard astronomy.

Aerospace engineer and amateur astronomer Roland Christen also attacked chromatic aberration. In the mid-1980s, Christen's firm, Astro-Physics, marketed the first triplet-lens refractors priced for amateur astronomers. Christen's telescopes have a three-element objective lens. Each element is made from a different type of glass, and together, they produce virtually false-color-free, or apochromatic, images. His design breakthrough allowed reasonably priced 4-, 5-, 6- and even 7-inch apo refractors.

The race for bigger telescopes that had driven amateur astronomy since the 1950s gave way to the desire for better telescopes. Planetary observing, which requires first-class optics, saw a resurgence with the return of the refractor.

## The 1990s:
## QUALITY AND CHOICE

In the 1990s, the doors swung wide open to every type of observing interest and choice of equipment. The only common factor de-

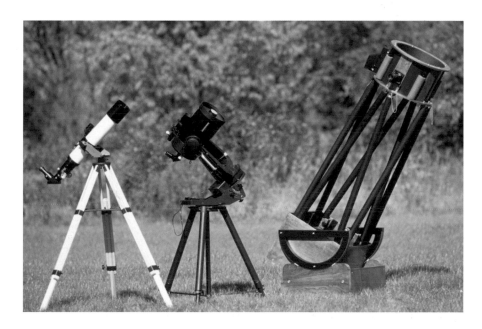

manded by backyard observers was quality. Regardless of the size and design of the instrument, telescope users looked for top-grade optics on solid mounts, and manufacturers were forced to deliver.

By the close of the 20th century, for the first time in the history of the hobby, the three main classes of telescope optics—refractor, reflector and catadioptric (the Schmidt-Cassegrain)—shared equal billing in popularity. No one telescope dominated. Similarly, for the first time, amateur astronomers pursued their eclectic observing interests, enjoying a wide range of celestial targets, from the nearby Moon to distant clusters of galaxies.

▲ **Telescopes of the 1990s** Three classes of telescopes were equally popular by the 1990s, each with its own merits: the new high-quality refractor, left, the compact Schmidt-Cassegrain, center, and the large-aperture Newtonian, right, now configured on a new generation of solid but portable Dobsonian mounts. From left to right: Tele Vue Genesis, Celestron Ultima 8 and Starsplitter 14.

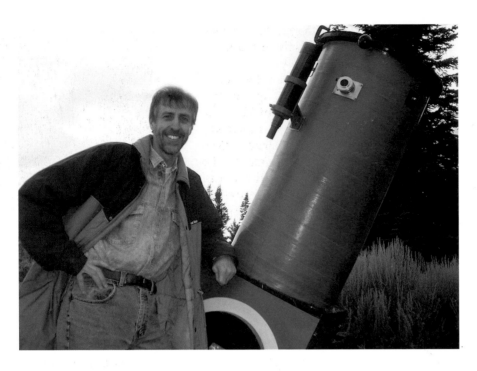

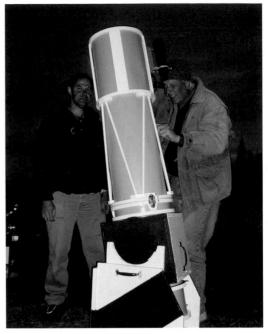

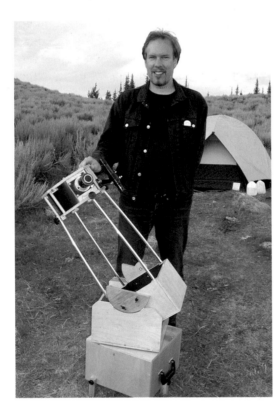

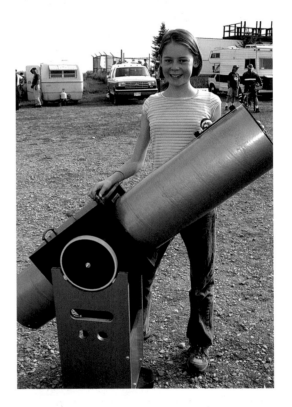

**Custom Telescopes**
Few amateurs today grind their own mirrors, but homemade telescopes, usually Dobsonians, remain popular. With store-bought telescopes now so affordable, people "roll their own" for the personal satisfaction, perhaps as a family project or for the desire to have a unique telescope with optics, fittings and finish a cut above mass-market scopes. Clockwise from top left: Ken Hewitt-White and a classic 17.5-inch; Les Disher (left) shows off his modular 12.5-inch; young Lauren Gibson and her well-crafted solid-tube 8-inch Dob; and Gary Seronik and his airline-portable 8 that collapses and nests into one neat box.

## The 2000s:
### COMPUTERS TAKE CONTROL

With optics perfected to such a high art and available in such a diverse array of telescope designs, what was next? Like most consumer products in the 1990s, telescopes were revolutionized by computers, a change that swept into the mass market of affordable telescopes as the 21st century opened.

While Celestron had introduced a computerized telescope as far back as 1987, with its now obsolete CompuStar models, the first high-tech scopes to make an impact were Meade Instruments' LX200 models, introduced in 1992. Meade's popular LX200 Schmidt-Cassegrains and the competitive models introduced by Celestron remained high-end telescopes throughout the 1990s, aimed at enthusiasts willing to spend several thousand dollars.

In 1999, led by company founder John Diebel, Meade broke the price barrier for computerized telescopes by introducing the ETX-90EC, with its well-designed Autostar computer. For $750, one could now own a telescope that aimed itself, solving the problem that discourages so many beginners: how to find celestial objects. In 2000, Meade shattered the price barrier again by offering "Go To" telescopes with Autostar technology for as little as $300. Promoted by widespread advertising (in *Popular Science* and *Discover* magazines, for example) and mass-marketed through chain-store outlets, the little ETX-60 and ETX-70 refractors and competitive telescopes from Celestron began to supplant the usual "Christmas trash telescope" as the entry-level instrument of choice, even for the most novice stargazers.

The appeal of the "Go To" scopes has put a lot more telescopes into the hands of people who might never have purchased a telescope or taken up the hobby. So, in that sense, these scopes are also expanding the hobby in new directions. With so many wonderful choices before us, this is an exciting time to be an amateur astronomer.

# Choosing a Telescope

Because no single telescope dominates the market anymore, the choices facing the prospective buyer can be daunting. Time and again, we are asked, What is the best telescope? or Which telescope would you buy? We provide recommendations further along, including models we feel are the best value in a beginner's scope. In many cases, these are telescopes *we* bought.

But the truth is, there is no single best telescope, just as there is no single best car or camera or computer. Furthermore, a telescope we would choose may not be the one best suited to your needs. Each telescope has its good and bad points, which we will outline. Many veterans who recognize the strengths of each type of telescope own more than one. That's the real answer to finding the perfect telescope.

Still, many just starting out in the hobby can't afford the luxury of owning multiple scopes. Faced with the task of picking one "best" telescope from a field of hundreds, buyers often fret with endless deliberation. Ed Ting, author and webmaster of the excellent ScopeReviews.com website, describes the syndrome nicely as "paralysis by analysis." He suggests that if you spend more than an hour a day reading telescope catalogs, you are afflicted. We agree. Pick one of our recommended telescopes in the price range that suits you, and buy it. You'll have fun with it, see a lot and discover where you might like to go next: perhaps a bigger scope for deep-sky exploring, a supersharp scope for planetary views or a scope designed for astrophotography—the avenues are many.

Our advice is to make your decision not on "refractor versus reflector" or "altazimuth versus equatorial" but on other factors far more important for ensuring enjoyable nights under the stars. We will outline these, then dive into the survey of models.

## THE MAGNIFICATION SCAM

First, we must point out the one telescope trait you can ignore: how much the telescope magnifies. Magnification is a meaningless specification. With the right eyepiece, any telescope can magnify hundreds of times. The question is, How does the image look at, say, 450x? Probably faint and blurry. Why? There are two reasons:

❖ Not Enough Light
The telescope simply is not collecting enough light to allow the image to be magnified to that extent. When an image is enlarged, it is spread out over a greater area and becomes too faint to be useful. In other words, the telescope has been pushed beyond its limits.

How much can a telescope magnify? The general magnification limit is 50 times the aperture in inches, or 2 times the aperture in millimeters. For example, the maximum usable power for a 60mm telescope is only 120x. Claims that such a telescope can magnify 400x are misleading, intended solely to lure the unsuspecting buyer.

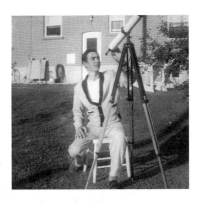

▲ **The Innocence of Youth**
Sporting the requisite Perry Como sweater, a 1950s teenager demonstrates a 60mm refractor, a classic junk telescope similar to those still sold today under the banner "250x Professional Model." Unfortunately, such misleading claims sell telescopes, usually to novices and well-intentioned parents and spouses. Aren't these instruments better than nothing? We think the money would be more wisely spent on good-quality 7x50 or 10x50 binoculars and star charts, complemented with time spent learning the sky.

▼ **Mirrors and Lenses**
All telescopes can be divided into two main classes: those which use mirrors to collect light (reflectors, as at left) and those which use lenses (refractors, as at right).

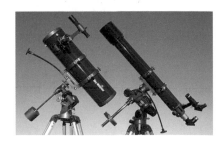

❖ Blurry Atmosphere

The Earth's atmosphere is always in motion, distorting the view through a telescope. Some nights are worse than others. At low power, the effect may not be noticeable. But at high power, atmospheric turbulence (poor seeing) can blur the image badly.

Most amateur astronomers find that a magnification of about 300x is the practical upper limit, even for a large-aperture telescope. People do not build or buy giant instruments to obtain highly magnified images but, rather, to get brighter, sharper images and to see fainter objects.

So Rule #1 when buying a telescope: Avoid any telescope sold on the basis of its magnification ("Powerful 475x Model") or any beginner telescope that promises powers over 300x. We've seen models advertised as providing 675x! All are sure signs that the telescope is just a junky toy in disguise.

## BEWARE OF APERTURE FEVER

The most important specification of a telescope is its aperture. When amateurs speak of a 90mm or an 8-inch telescope (both metric and imperial units are used in the hobby), they are referring to the diameter of the main lens or mirror.

With the magnification myth dispelled, you know you need as much aperture as you can afford. Or do you? If you are not careful, you may catch aperture fever. The first symptoms are longer and longer perusals of the ads in astronomy magazines, accompanied by imagining the views to be had with the "Colossal SuperScope."

However, be warned. Big telescopes do not always foster contentment among astronomers, for several reasons:

❖ Portability

One often-overlooked fact is that large telescopes do not fit into small cars. It is surprising how many amateur astronomers buy or make a huge instrument without considering how to transport it—or carry it. For a first telescope, in particular, any instrument you cannot easily carry out to the backyard in one or, at most, two pieces is unlikely to be used after the novelty wears off.

❖ The Shakes

Another problem with big telescopes is the shakes. A lightly mounted large-aperture telescope might be portable, but the images will dance about with every puff of wind and every touch of a hand. Such an instrument is simply not fun to use. Yet a mount sturdy enough to steady the telescope might also be too heavy and cumbersome. By having a smaller aperture, you can get a solid telescope that retains portability. Rule #2: A good mount is just as important as good optics.

❖ Price

Big telescopes can be expensive. If you lose interest in the hobby, you will have a sizable investment tied up in a telescope that may be tough to sell. Lose interest? Never, you say. Lots of people do. The reasons can often be traced right back to big telescopes that are awkward to set up. We have seen just as many people drop out of the hobby because they bought too much telescope as we have people who became disenchanted because they bought too small a telescope.

Our advice? The beginner should resist any temptation to buy a first telescope larger than 8 inches in aperture. Rule #3 for telescope buying: The best telescope for you is the telescope you will use most often. A small well-made instrument that is convenient to use will provide a lifetime of enjoyment.

## MATCHING SCOPE TO SITE

An important consideration when picking the best telescope is the observing site. Can you observe from your home? If so, are the skies dark, or are they heavily light-polluted? Are views restricted by trees, houses or streetlights? How far will the telescope have to be carried?

If you are plagued with light-polluted skies at home, f/6 to f/15 telescopes are preferred. As a general rule, they have better optical quality than most ultrafast f/4 to f/5 telescopes, while yielding higher powers with a given set of eyepieces. Both characteristics will provide more pleasing images of the Moon and the planets, the best targets in light-polluted skies.

If there is little possibility your telescope will be used at home, then base your decision on portability and ease of transportation. A 4-to-5-inch Maksutov-Cassegrain, a 5-to-8-inch Schmidt-Cassegrain, a 6-inch Newtonian or a 3-to-4-inch refractor on an

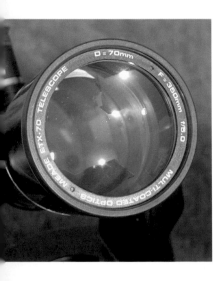

**Minimum Aperture** ▲
A 70mm-aperture telescope, like this Meade ETX70-AT, is the smallest telescope that aspiring backyard astronomers should consider.

equatorial mount will probably be used far more than will a bulkier instrument.

From firsthand experience, we suggest Rule #4: Telescopes that require more than 5 to 10 minutes to load into a vehicle or to set up suffer a steady decline in use after the first year of ownership. Instead of enjoying your telescope, you feel guilty for not using it. And a year or two later, you will probably sell it.

If, on the other hand, you are fortunate enough to live under dark rural skies and can conveniently store a large telescope for quick setup, then by all means consider a larger-aperture reflector, perhaps a 10- to-15-inch Dobsonian or a 10-to-12-inch Schmidt-Cassegrain. Even short-focus refractors, such as Meade's little ETX-AT models, scopes of limited use in the city,

perform well under dark skies, excelling at wide-angle views of deep-sky star fields.

## PHOTOGRAPHIC FEVER

A common demand of first-time telescope buyers is a big one: "I want to take pictures through my new telescope." Usually, this means long exposures of colorful nebulas and galaxies. This supremely difficult type of astrophotography requires a minimum investment of $3,000 in telescope gear alone, including a top-quality mount, plus the cost of specialized accessories.

Many types of astrophotos can be taken easily and cheaply, but close-up shots of nebulas and galaxies aren't among them. As enticing as the photographs look in the telescope brochures and in books like ours,

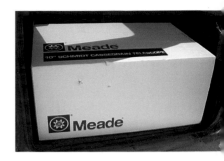

▲ **Will It Fit?**
Will the packed telescope fit into the family vehicle? This Meade 10-inch just makes it! Then how easy will it be to set up in the field in the dark?

# Types of Optics

### Achromatic Refractor
The classic refractor uses a doublet lens with elements made of crown and flint glass. In f/10 to f/15 focal ratios, chromatic aberration is negligible.

### Apochromatic Refractor
To eliminate false color, some apos use triplet lenses with elements of Super ED glass. Others use fluorite doublets or small corrector lenses near the focuser.

### Newtonian Reflector
Invented by Isaac Newton in 1668, this classic design uses a concave primary mirror (preferably with a parabolic curve) with a flat secondary mirror.

### Schmidt-Cassegrain
An aspherical corrector plate compensates for aberrations in the f/2 spherical mirror. A convex secondary folds the light path down the stubby tube.

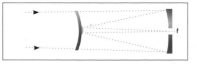

### Maksutov-Cassegrain
Based on a design invented by Dmitri Maksutov in 1941, the Mak-Cass uses a steeply curved corrector lens. The all-spherical surfaces are easy to mass-produce.

### Schmidt-Newtonian
This hybrid design, usually f/4 or f/5, combines a Schmidt corrector with Newtonian optics to reduce the off-axis coma inherent in fast Newtonians.

### Maksutov-Newtonian
Usually made in f/6 focal lengths, this design boasts a view free of aberrations across a wide field at low power and refractorlike images at high power.

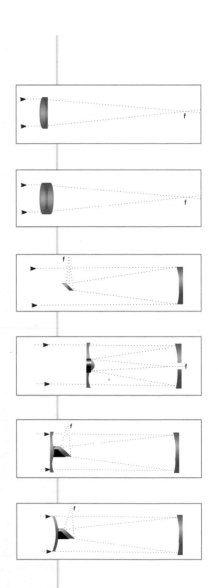

# Decoding Telescope Specs

The following terms, and not magnification, represent the most important optical specifications of any telescope.

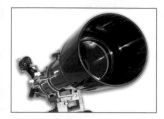

### Aperture and Light-Gathering Power

Telescopes are rated by their aperture. A 4-inch instrument has a main lens or mirror four inches in diameter. The larger the lens or mirror, the more light it collects, providing brighter and sharper images. An 8-inch telescope has four times the surface area, and therefore light-gathering power, of a 4-inch, making its images four times brighter.

### Resolution

In theory, an 8-inch telescope can resolve twice as much detail as can a 4-inch instrument. The resolving power of a telescope can be estimated with a simple formula: Resolving power (in arc seconds) = 4.56 ÷ aperture of telescope (inches); or 116 ÷ aperture of telescope (mm). This is the empirical rule devised by William Dawes in the 19th century. When manufacturers list a resolving power, they are merely stating the Dawes limit for that aperture of telescope, not a measured performance value for that specific model.

### Focal Length

The length of the light path from the main mirror or lens to the focal point (the location of the eyepiece) is the focal length. With Maksutov- and Schmidt-Cassegrains, the optical path is folded back on itself, making the tube shorter than the focal length.

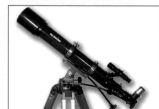

### Focal Ratio

The focal ratio is the focal length divided by the aperture. For example, a 100mm telescope with a focal length of 800mm has a focal ratio of f/8. For photography, faster f/4 to f/6 systems yield shorter exposure times (therefore, these are known as fast focal ratios). But when used visually, image brightness depends solely upon the aperture. Focal ratio has nothing to do with it.

### Diffraction-Limited

A promise of diffraction-limited optics means aberrations in the optics are small enough that image quality is affected primarily by the wave nature of light and not by errors in the optics. This is equivalent to stating that the optics provide a final error at the eyepiece of only one-quarter of a wavelength of light (the wavefront error), meeting the so-called Rayleigh criterion, a minimum standard for amateur telescopes. Anything worse, and planets will look soft, if not blurry. Contrary to some ad claims, diffraction-limited does not mean the optics cannot be improved upon. Premium telescopes can do better, with wavefront errors of 1/6 to 1/8 wave. Under good conditions, tests have proved that the difference is noticeable, but the performance edge over 1/4-wave optics comes at a high cost.

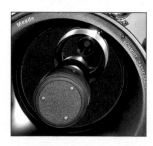

### Central Obstruction

While the secondary mirror in a reflector blocks some light, the loss is not significant. The noticeable effect is the smearing of image contrast caused by the added diffraction of light from the obstruction. This effect is proportional to the diameter of the secondary mirror. As such, central obstruction should be stated as a percentage of the diameter of the aperture. An 8-inch scope with a 2.75-inch-diameter secondary mirror has a central obstruction of 34 percent. To make the numbers seem smaller, some companies state obstruction as a percentage of area (12 percent in this example). In general, a central obstruction of 20 percent or lower *by diameter* produces a negligible effect.

..................................................................................................................

*The refractor telescope at top has an aperture of 70mm. Its focal length is 700mm, close to its actual tube length, as shown at center. This makes the entry-level scope an f/10 instrument, a focal ratio good for ensuring sharp views of the planets. Most reflector telescopes, such as the Schmidt-Cassegrain above, have a secondary mirror that obstructs the main aperture.*

we propose our Rule #5: Beginners should ignore the demands of astrophotography when making a telescope selection. Buy the sharpest, sturdiest telescope you can for the simple pleasure of exploring the sky by eye.

# Surveying the Telescope Market

No telescope design guarantees superb images. Only quality of craftsmanship can guarantee quality of images. A well-made Newtonian will outperform a poorly made refractor, and vice versa. Ignore Internet discussion-group gurus and ardent club members who tout one particular telescope design as "the best." Sooner or later, every telescope brand and type will be named, only adding to the confusion. Instead, seek out quality coupled with portability and convenience. Combine those traits, and you will have a great telescope no matter what its optical design.

That said, here is a rundown of what the current market has to offer in quality telescopes, sorted by optical design. We follow that with our recommendations, broken down by price category.

## ACHROMATIC REFRACTOR (70mm to 6-inch)

A 60mm refractor ($100 to $200) has long been a popular beginner's telescope. This instrument is in every big-box chain and camera store at Christmastime and is hawked on home-shopping channels. While there have been some excellent 60mm refractors in the past, such as legendary instruments from Unitron, we have seen few 60mm refractors in recent years that we could recommend.

The flaw of most 60mm refractors is not so much the main lens but everything else —poor eyepieces, dim finderscope, lack of slow-motion controls, flimsy mount and shaky tripod. Our advice to people who inquire about buying a low-cost 60mm telescope is to spend more to acquire an 80mm-to-90mm refractor or a 4-to-6-inch reflector,

or save money by purchasing binoculars.

With the jump to an 80mm or a 90mm refractor, the quality of the telescope improves greatly. Most f/10 to f/11 refractors in this aperture class are worthy beginners' telescopes. In our opinion, they represent the minimum for a serious buyer. Color correction of the crown/flint doublet lens is excellent, and the telescope is portable, durable and virtually maintenance-free for decades. Many companies and dealers offer a telescope in this category. Most are identical units made in China and branded by the local distributor. We've found their quality excellent for the low price of $300 for altazimuth-mounted versions. Look for units with slow-motion controls and one or two high-grade Kellner or Plössl eyepieces providing no more than 100x to 150x, rather than several eyepieces of dubious quality providing excessive magnification.

A just acceptable notch down is a smaller 70mm refractor ($150 to $200), but only if it is fitted to a sturdy mount, preferably an altazimuth mount with slow-motion controls. (Most equatorial mounts for telescopes of this size are too flimsy.) Even so, many 70mm refractors are still hampered by poor eyepieces (those marked H, HM or SR in the 0.965-inch barrel size, rather than the industry standard 1.25-inch size) and tiny 5x24 finderscopes with poor optics.

A step up in aperture to a 4-inch f/8 to

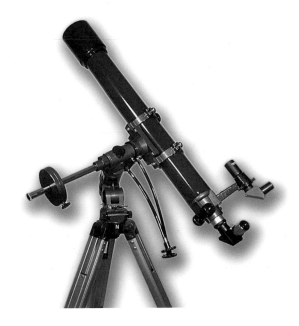

◀ **Classic 90mm Refractor** Achromatic refractors in the 80mm-to-90mm size and with f/10 to f/11 focal ratios have long been popular as entry-level telescopes, and rightly so. Made in China, this $400 SkyWatcher 90 on a sturdy EQ-3 mount provides stunningly sharp and contrasty planetary details, upholding the reputation of classic long-focus achromatic refractors. Unlike simpler altazimuth mounts, this German equatorial mount allows convenient tracking of celestial objects, either by hand or with an optional motor.

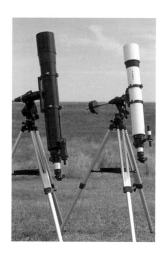

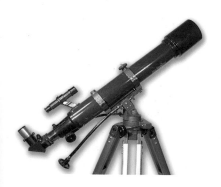

f/10 refractor is a fine choice for about $450. However, for this size telescope, an altazimuth mount doesn't work as well. The telescope hits the mount when aimed high, and the constant adjustments required to keep objects centered at higher powers become annoying. An equatorial mount to solve these problems is essential for this size scope.

Achromatic refractors larger than a 4-inch aperture used to be rare and expensive, but the line of achromats made in China by Synta Technology Corp. and carried by dealers under brand names like SkyWatcher now includes 4.7- and 6-inch refractors. For the breakthrough price of $550 for the 4.7-inch and $900 for the 6-inch, you get a refractor in a size that once cost thousands. To compete with Synta, Meade offers 5- and 6-inch achromats on its LXD55 German equatorial "Go To" mount.

There are trade-offs, of course. To keep the tube length manageable, the focal ratios have been reduced to f/8 or f/9. In an achromat, this inevitably introduces aggressive false color around bright stars and planets. (The faster the focal ratio of the refractor, the worse the false color becomes.) Jupiter appears yellow-green surrounded by a violet haze. Image sharpness of these large achromats varies, but views through units we've used have proved perfectly acceptable for such a low-price refractor, despite the false color. Some observers prefer to use a Sirius Optics minus-violet filter, which eliminates a good chunk of the false color. In doing so, it adds to the overall pale green of the planet, but the sharpness seems improved.

We particularly like the Synta 120mm (4.7-inch) f/8.3 refractor sold under a variety of brand names around the world. However, Synta's 6-inch f/8 achromat is a big instrument on a shaky mount that is awkward to use, with an eyepiece which ends up too close to the ground to be comfortable to look through. It fails to meet our criteria for a friendly telescope. But as big light-bucket reflectors did in the early 1980s, these low-cost light-bucket refractors have attracted a loyal following.

## Signs of a Good Starter Scope

Look for these features in entry-level telescopes. The smaller and less capable EQ-2 mount is usually supplied with this 90mm refractor.

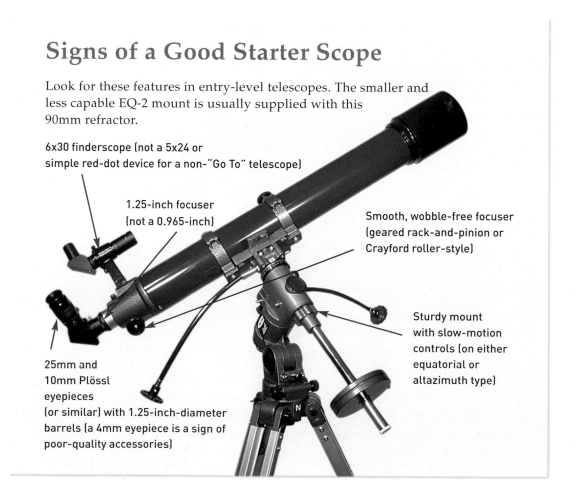

6x30 finderscope (not a 5x24 or simple red-dot device for a non-"Go To" telescope)

1.25-inch focuser (not a 0.965-inch)

Smooth, wobble-free focuser (geared rack-and-pinion or Crayford roller-style)

Sturdy mount with slow-motion controls (on either equatorial or altazimuth type)

25mm and 10mm Plössl eyepieces (or similar) with 1.25-inch-diameter barrels (a 4mm eyepiece is a sign of poor-quality accessories)

Many dealers stock so-called Short-Tube refractors, f/5 achromats with apertures ranging from 80mm to 120mm. These are fine, affordable instruments for low-power sweeps of the Milky Way under dark skies, but their optics are not sharp enough to make them suitable as general-purpose telescopes for high-power use.

At the other extreme in size and quality, lovers of classic achromats can select 5-to-11-inch models with f/12 to f/15 focal ratios made by D & G Optical. At $1,400 for a 5-inch or $1,700 for a 6-inch, prices are reasonable and well below that of the telescope which, in our opinion, sets the standard for optical quality, the apochromatic refractor.

## APOCHROMATIC REFRACTOR (3-inch to 8-inch)

Technically, the term apochromatic means the telescope brings three colors to the same focus, rather than just two, as in a standard achromat. In practice, however, any false color is reduced below the eye's threshold of detection. This is the level of performance achieved by two-element apos using fluorite glass in one element and by three- and four-element refractors using extra-low-dispersion (ED) glass. A 4-inch f/8 fluorite refractor less than three feet long can outperform a 4-inch f/15 achromat, long considered the minimum focal ratio for color-free performance from such a scope.

Apo refractors with triplet ED lenses provide color correction a notch better than do doublet fluorite refractors (according to theory), and in practice, these designs do provide superb images second to none in the telescope world.

Fluorite refractors were introduced by the Japanese firm Takahashi in 1974. In all sizes, from the diminutive 50mm to the giant 150mm refractor, Takahashi instruments provide outstanding images but command a premium price. A popular model, the FS-102 4-inch f/8 doublet fluorite with Takahashi's excellent EM-10 equatorial mount, for example, is about $6,600. Compared with other premium apos, however, a Takahashi refractor has a big advantage: Most models are available from stock. (We have yet to see any evidence that fluorite, an artificially grown crystal, degrades over

time, but the idea persists, perpetuated by makers of nonfluorite telescopes.)

More affordable are the Vixen apo refractors, sold for years in the United States by Orion. The VX102-FL, a 4-inch f/9 doublet fluorite, comes with the superb Great Polaris equatorial mount for about $2,900. The VX102-ED, an f/6.5 doublet ED, costs $700 less, but its level of color correction, though good, does not equal that of the fluorite.

Since the 1980s, Tele Vue Optics has offered consistently excellent apo refractors with color correction that has improved with every new model. Ultraportable 70mm, 76mm and 85mm models ($600 to $1,700) use doublet lenses for excellent color correction—only a tad below that of the best apos on the market. The top-of-the-line NP101, a 4-inch f/5.4 instrument ($3,400),

uses a four-element design for total elimination of false color and any other image-smearing aberrations. A front doublet is coupled to a rear-mounted, sub-diameter doublet that serves to flatten the field for pinpoint stars across a 35mm film frame. This telescope matches the best apos on the market. For $1,000 less, the doublet Tele Vue 102 offers similar levels of performance, but with a longer tube and f/8.6 focal length. For a primarily visual telescope, it is a great choice.

Some of the finest apo refractors and mounts sold today come from Astro-Physics.

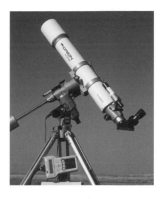

▲ **Vixen Value**
Made in Japan and now marketed in North America by Tele Vue, Vixen refractors like this f/9 fluorite offer consistently good value.

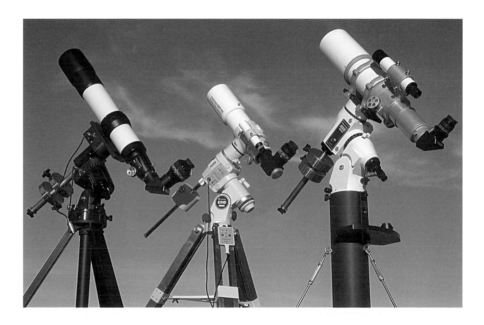

▲ **Apo Trio**
The cream of the optical crop (from left to right): the Tele Vue NP101 four-element apo on an Astro-Physics 400 mount; the Takahashi Sky90 fluorite doublet on Takahashi's EM-10 mount; and the hefty TMB Optical 105mm f/6.2 triplet with Super ED glass, carried on an Astro-Physics middleweight 600E mount. It doesn't get any better!

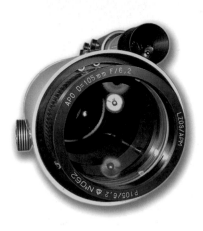

**Glorious Glass** ▲
The best-quality apos, like this Russian-made TMB triplet, use lens elements made of fluorite or extra-low-dispersion (ED or Super ED) glass for high performance but are costly.

**Dream Apo** ▶
The principal drawback to Astro-Physics telescopes, like this superb 5-inch f/6 EDF, is that they are in such demand, buyers must vie to get on a waiting list just to place an order for one. That wait alone can last for up to three years, followed by a wait of another year to receive the completed telescope. These really are dream scopes.

All its American-made refractors employ a three-element objective with a premium grade of ED glass—Super ED—forming the middle element. Images are completely color-free and textbook-perfect, even under extreme temperature conditions. Four models are offered: the wonderfully portable 4.1-inch f/5.8 Traveler 105 ($3,500), the 5.1-inch 130EDF f/6 ($4,800) and two versions of a 6.1-inch f/7, one with a standard 2.7-inch focuser ($6,500) and one with a giant 4-inch field-flattener lens and focuser for medium-format-film photography ($9,000). Over the years, Astro-Physics has produced larger and smaller refractors, as well as slower f/8 and f/12 versions, in limited

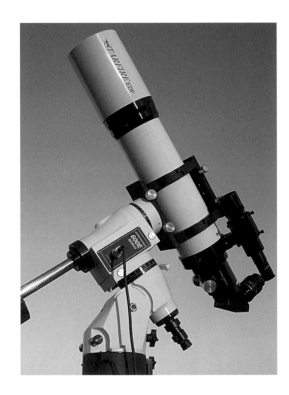

quantities. A used Astro-Physics refractor of any vintage is a prize.

To meet the demand for premium instruments, other manufacturers have introduced competitive telescopes that are more readily obtainable. Optical designer Thomas Back and his U.S. company TMB Optical offer triplet Super ED apos every bit the equal of Astro-Physics' optics. Lenses are designed by Back and manufactured in Russia by LZOS, a Zeiss subcontractor. The quality is exquisite. Models range from a jewel-like 80mm ($1,800) to huge 8-inch and larger refractors ($Call!). We've thor-

oughly tested the popular 4-inch f/6.2 and f/8 models ($3,300) and can attest to the first-class optics, mechanics and finish of the German-made tube assemblies. (The prices we're quoting for Tele Vue, Astro-Physics, TMB and Borg scopes are for optical tube assemblies only, without the mount.)

Long available in Japan, the Borg line of ED refractors is now enjoying wider distribution in North America. Models from 50mm to 150mm employ a modular construction, with a choice of tube diameters and focusers and a vast array of photo accessories and adapters. Prices are reasonable—$1,800 for a 4-inch doublet ED. From our testing, the apo models have color corrections slightly below that of top-of-the-line triplet EDs, but Borg telescopes offer wide, flat fields ideal for astrophotography in superportable packages. Premium models feature giant field flatteners for medium-format photography on 6x7 film and focal ratios as fast as f/2.8.

Other small companies such as Stellarvue and William Optics have introduced still more apos onto the market, usually in the 80mm-to-100mm-aperture range. In most cases, the optics are sourced from the Orient (China or Taiwan) or from Russia and are excellent.

## DOBSONIAN REFLECTOR
## (4.5-inch to 30-inch)

For our money, this telescope offers the most outstanding value in a beginner's scope. Originally popularized by the now defunct company Coulter Optical, today's low-cost Dobsonians provide generous aperture, good optical quality, excellent fittings and solid mounts for not a lot of money. What more could anyone want?

Apparently, an intangible sex appeal is also important. Dobsonians are dumpy, plain-Jane wooden scopes that are a tough sell on the sales floor of telescope shops. We have heard stories of youngsters bursting into tears after receiving a Dob as a Christmas gift. "That's not a telescope!" they blubber. It seems a long white tube sitting high atop a jiggly tripod is the only "real" telescope for them.

Although most knowledgeable amateur astronomers (including us) continue to praise and recommend these bargain telescopes,

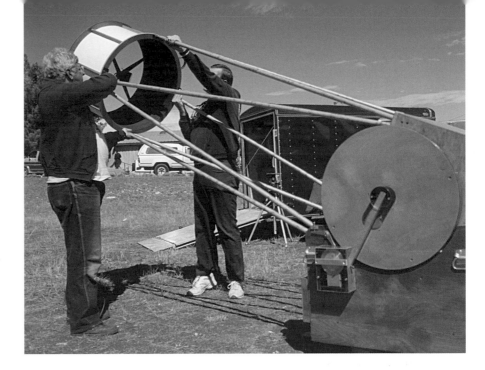

many buyers—especially first-time buyers —continue to bypass them for high-tech computerized scopes or reflectors on undersized equatorial mounts. We will continue to fight the tide and recommend a 6-inch f/8 Dobsonian reflector as the best telescope for a beginner. At $350, its value is unbeatable. Yes, to keep images sharp, all Newtonians require the occasional adjustment of the mirrors, a process called collimation (see www.backyardastronomy.com). But the procedure is not difficult and is a minor price to pay for the generous aperture of a Newtonian.

Various manufacturers have offered Dobsonians over the years, but many have dropped the lines due to sagging sales. Our favorites continue to be the metal-tubed models made in China by Synta and offered under the SkyWatcher brand name (and many others) and the Taiwanese-made Orion SkyQuest Dobs. The mirrors are well made and mounted, and the focusers and finderscopes are excellent for such low-cost scopes. The altitude tension control and Teflon bearings provide smooth motions with just the right amount of friction—not so tight that it binds and not so loose that it swings off target by accident. The 6-inch is a great starter scope capable of showing far more than any 70mm or 80mm refractor or 4.5-inch reflector of comparable price.

At $500, the 8-inch is perhaps the best telescope deal in the history of the hobby. We set up one (a Synta) right out of the box that was in perfect collimation after traveling all the way from China. In optical tests, it performed as well as scopes costing six times as much.

The 10-inch models ($650) are also an incredible bargain, providing even more aperture for hunting deep-sky denizens. All three sizes of introductory Dobsonians mentioned above have steel tubes and melamine (kitchen-cabinet-type wood) mounts that make them rugged performers in the field.

Other popular models are the series of American-made DHQ Dobs from Discovery Telescopes. Its 10-inch solid-tube DHQ offers fine craftsmanship and optical quality at an excellent price ($700).

Moving into 12-inch sizes and up, we recommend bypassing the low-end products and gravitating directly to the premium high-end Dob with its attention to detail of manufacture and superior optics. This class of telescope dispenses with the solid tube of the smaller scopes. Instead, the new-generation Dobs feature truss-tube designs that break apart into transportable components yet snap together in five minutes. In this league of telescope, prices vary but typically range from about $2,000 for a 12-inch telescope to $5,000 for 18- and 20-inch models. It is even possible to buy a giant 25- or 30-inch(!) Dob.

If the thought of a 25-inch telescope in your backyard is appealing, think again—the tube is typically 10 feet long. Observing over most of the sky means having to balance on a tall ladder. Although the tube may break down into small components, the ladder does not.

Leading manufacturers of premium Dobs are mostly small home-based operations run by dedicated amateur astronomers turned entrepreneurs, who exercise an uncommon attention to quality. Companies include Discovery Telescopes (12.5-to-24-inch), MAG1 Instruments (8-to-18-inch), Obsession (15-to-30-inch), Starmaster (11-to-30-inch) and Starsplitter (8-to-30-inch). We've used or owned telescopes from all these companies and love them for their craftsmanship, buttery smooth movement and superb mirrors from such top names as Galaxy, Nova, Pegasus, Swayze, Torus and Zambuto. Some of them even have com-

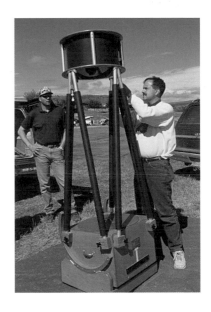

▲ **Land of the Giants**
It is fun to look through a giant Dobsonian at a star party, but do you really want to own a two-person telescope, like the one being assembled at top? A 12-to-15-inch Dobsonian, like the Obsession 15 above, is a much more comfortable size for personal use. Your feet stay close to or on the ground, yet the aperture is sufficient to present incredible views.

puterized drives that "Go To" and then track like equatorial mounts.

## EQUATORIAL NEWTONIAN (4.5-inch to 18-inch)

The standard instruments of the 1960s —6-inch f/8 or 8-inch f/7 equatorially mounted Newtonians—have attempted comebacks now and then but have never caught on. Dobsonian-mounted versions remain the more popular choice.

However, smaller equatorial Newtonians continue to populate the entry-level market. Tens of thousands of amateur astronomers started their astronomical lives with a Tasco 4.5-inch Newtonian, the 11TR, marketed for decades. It had several deficiencies: a small 5x24 finderscope, poor 0.965-inch eyepieces, a spherical rather than parabolic mirror and a shaky mount. Models similar to the old 11TR are available from Bushnell and as house-brand items from telescope and camera stores. We suggest bypassing these instruments altogether, no matter whose name is on them. (Tasco itself is now defunct.)

Largely replacing the Tasco-style 4.5-

inch is another ubiquitous telescope, the Synta 5.1-inch reflector offered in either f/7 or stubbier f/5 versions ($250 to $300). Sold under various brand names, this telescope's mount is the EQ-2 style that has been on the market for at least 40 years. The mount is just adequate, but the fittings and accessories of these 5-inch reflectors are fine, indeed. We recommend the shorter and sturdier f/5 instrument. The optical quality of its parabolic mirror is good.

Larger 6- and 8-inch Newtonians on Taiwanese-made mounts (such as Orion's SkyView Deluxe models for $400 and $550) also provide good-quality choices, though we suggest staying with the smaller scope. A larger telescope offered on the same mount is inevitably shaky.

For 25 years, a popular beginner's instrument has been the Edmund Astroscan (about $350), a 4.1-inch f/4.3 Newtonian in a teardrop-shaped housing that forms a ball for an ingenious ball-and-socket mount arrangement. The rugged telescope is highly portable and easy to use, and it has launched thousands of budding astronomers. It works best under dark skies for deep-sky scanning.

At the other end of the price spectrum, the finest big-aperture Newtonians on equatorial mounts are the NGTs (Next Generation Telescopes) sold by Jim's Mobile, Inc. (JMI). Both the 12.5-inch ($5,000) and 18-inch ($13,000) sizes are carried on a well-engineered, split-ring mount that is surprisingly compact considering the size of telescope it is carrying.

❖ A Variation:
    Schmidt-Newtonian
Meade introduced this hybrid design in the early 1990s. It consists of a primary and secondary mirror, as in a Newtonian, coupled with a Schmidt corrector plate for reducing some of the aberrations inherent in a fast Newtonian, notably off-axis coma. Models suffered from poor optics, and the design never caught on. In 2002, Meade reintroduced 6-, 8- and 10-inch Schmidt-Newtonians (all f/4 or f/5) on computer-controlled equatorial mounts as the LXD55 Series, with prices as low as $600 for the 6-inch.

With a large secondary-mirror obstruction and fast focal ratios, these Schmidt-Newtonians are best for wide-field scanning under dark skies, rather than high-

**Classic Small Newtonian ▼**
One of the most popular beginners' telescopes, the 4.5-inch Newtonian is being supplanted by the 5.1-inch Newtonian from Synta, such as this f/5 model. Units like this one, with parabolic, not spherical, mirrors, will provide the sharpest images.

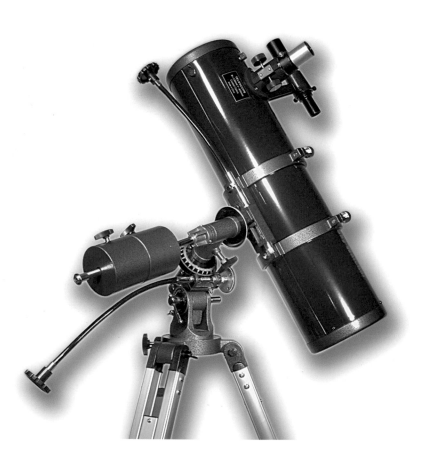

power planetary views in urban backyards.

## MAKSUTOV
### (3.5-inch to 10-inch)

The Maksutov telescope has been in and out of style during the history of amateur astronomy. Currently, it is enjoying a renaissance because of the demand for high-performance optics in a compact package.

The Maksutov is similar to the Schmidt-Cassegrain in that it employs a front-element corrector lens to eliminate the aberrations introduced by the primary mirror. Most Maksutovs are f/12 to f/14 Cassegrain systems, in which the light is directed out the back of the telescope through a hole in the primary.

The legendary Maksutov of this style is the 3.5-inch Questar. An f/14 Cassegrain, the Questar was introduced in 1954 as a top-of-the-line compact telescope. Fifty years later, it still is. Everything viewed through a Questar is wonderfully crisp and totally free of any aberrations, though contrary to mythology, a Questar cannot outperform larger instruments of good quality.

Nonetheless, the American-made Questar is a superb 3.5-inch instrument. Complete with mount, drive, leather case and tabletop tripod, it costs more than $4,500. Is it worth it? Judging by the thousands of owners and its half-century on the market, it is the right telescope for many people.

Larger Maksutov-Cassegrains are now available as tube assemblies or as fully mounted telescopes, with optics made in China or Russia. Prices are attractive ($300 for a 4-inch tube assembly, $400 for a 5-inch, $900 for a Russian 6-inch). All are f/12 instruments. The Russian telescopes by INTES and LOMO are more expensive but have better optical quality. Nevertheless, the bargain Chinese Maksutovs, such as Orion's StarMax units, are good general-purpose instruments that provide all the portability we suggest is so important to happy telescope use.

For a large Maksutov, Meade includes a 7-inch f/15 model ($3,000) in its LX200 GPS Series. Meade's big Mak has earned a reputation for fine optics, but its longer tube does not swing down through the fork, which makes this scope more difficult to

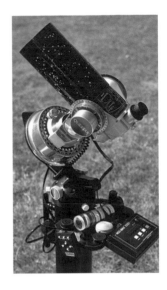

▲ **Legendary Telescope**
A rare sighting of a fully outfitted Questar 3.5 like this beauty is bound to attract attention among telescope fans at any star party.

# Do You Need an Equatorial Mount?

Many beginners are surprised to hear us and other advisers suggest a Dobsonian telescope. Older guidebooks give the impression that an equatorial mount (as in the traditional German-style design) is essential for a serious telescope. Bristling with setting-circle dials, chrome knobs and cable controls, an equatorial mount looks more scientific and professional. But it can be confusing to use. Polar alignment, though easy when you get used to it, continues to puzzle newcomers, and the telescope tubes can collide with the mount or drive-motor casing in some orientations. The other drawback is that low-cost units can be shaky. Yes, they can track objects, but the image still bounces around far more than with a solid Dobsonian mount, even with the nudging a Dob requires every few moments to keep objects centered.

We feel that an equatorial mount is simply not necessary for a beginner's telescope. Nevertheless, if an equatorial mount is appealing, then buy the best one you can, even if it means stepping up a notch in price.

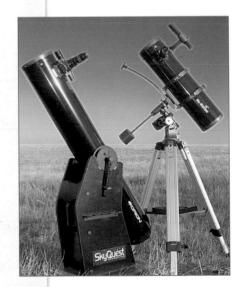

 *With similar optics and price, which of these telescopes has the sturdiest mount? The Dobsonian on the left. Dobs are also easier to set up and swing around the sky.*

# Pros and Cons of Telescope Types

| Type | Advantages | Disadvantages | Starting Price With Mount |
|------|-----------|---------------|---------------------------|
| Achromatic refractor | Low cost; sharp, contrasty images; rugged and durable design | Chromatic aberration, especially in fast f/5 to f/8 designs | $200 |
| Apochromatic refractor | Freedom from most aberrations | Highest cost per inch of aperture | $2,500 |
| Newtonian reflector | Lowest cost per inch of aperture; adaptable to Dobsonian mount | Coma at field edge; requires occasional collimation | $300 |
| Schmidt-Newtonian | Wide field of view; reduced coma | Large secondary obstruction | $600 |
| Schmidt-Cassegrain | Compact; good for astrophotography | Large secondary obstruction | $1,200 |
| Maksutov-Cassegrain | Compact; sharp optics; long focal length good for planetary viewing | Slow focal ratios and narrow field; long cooldown in large apertures | $400 |
| Maksutov-Newtonian | Wide, flat field; sharp, contrasty optics; small secondary obstruction | Long cooldown time; heavy; will not focus with 35mm camera | $2,500 |

**Little Meade Maks ▼**
The original 90mm (left) and newest 105mm ETX Maksutovs provide fine optics in a compact package. The computerized Autostar (on the 90mm here) is usually an optional extra.

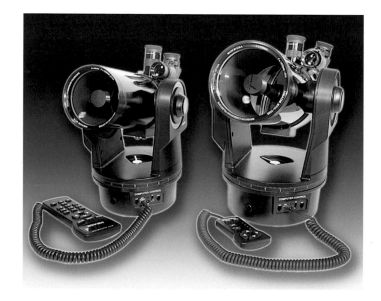

store than its Schmidt-Cassegrain cousins.

Favored by aficionados, the Telescope Engineering Company of Colorado sells 6-inch ($1,800), 8-inch ($3,700) and 10-inch ($7,000) Mak-Cass tube assemblies with first-class f/11 to f/20 optics optimized for planetary work. In the same league is the Astro-Physics 10-inch f/14.6 Mak-Cass, an $11,000 instrument designed to solve the long cooldown that plagues large Maks.

❖ The Trendsetting ETX

In 1996, Meade introduced a near clone of the little Questar, the ETX-90, short for Everyone's Telescope. The concept was to pack maximum performance into a small package for the lowest price possible. The introduction was an immediate success. And rightly so. The ETX-90s and the larger versions that followed consistently offered optics which matched the Questar for a fraction of the price. (The ETX-90 sold for $500 when it was introduced.) The ETX lacks the machining precision of the Questar. The fork mount and rear cell are ABS plastic, and the little 8x21 finderscope is marginal. But the brushed-aluminum tube is alluring, and the optics are tack-sharp. We've used several ETX-90s, and all have had fine optics.

A larger 5-inch ETX model followed ($900), and the line now includes a compact 4-inch Maksutov ($700). A smaller 70mm ETX model ($300) is not a Maksutov but a short-focus achromatic refractor that lacks the high-power sharpness of its larger Mak cousins. The ETXs are now offered with "Go To" Autostar computers (a must-have $150 option on the 90mm-to-125mm models), and it is safe to say that they have become the best-selling quality telescopes in the world. Celestron has a 4-inch "Go To" Maksutov, the NexStar 4, but its Chinese-made optics don't match the ETX optics for crispness.

The primary drawback to the Maksutov ETXs is the narrow field provided by the f/13.8 optics. The little focus knob is also hard to reach when the telescope is aimed up high. The Autostar system can be troublesome if not set up right (see Chapter 11), causing the telescope to bang against its hard stops as it tries to slew to a target. As with any "Go To" scope, you are dependent on battery power, which can drain to nothing in cold weather or with prolonged high-speed slewing. Unless placed on the optional Deluxe Field tripod, these telescopes tend to be bouncy. The plastic gears are noisy; the instruments sound as if something is not working right. But they do indeed slew to and track objects very well.

At lower cost, Meade's DS-2000 Series uses single-arm forks to carry a nice 70mm f/10 refractor or two sizes of short-focus Newtonians—of a type we usually avoid.

❖ A Variation:
Maksutov-Newtonian

In the early 1990s, Ceravolo Optical Systems—a small Canadian telescope company—introduced a type of telescope not seen before on the amateur market. Ceravolo's 5.7-inch f/6 Maksutov-Newtonian, the HD145, provided the high-resolution images normally associated only with an apochromatic refractor. This fine instrument is no longer made, but Maksutov-Newtonians are now available through U.S. dealers from the Russian companies INTES and INTES-Micro. Available in 5-to-8-inch sizes, these Russian Mak-Newts are superb telescopes. With f/6 focal ratios, they provide stunning wide-field, deep-sky views. Yet their small secondary mirrors and fine optics yield high-power planetary views that

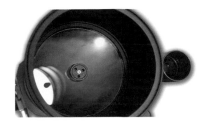

▲ **Refractorlike, but...**
It can take an hour or more for tube and mirror currents in a Mak-Newt to subside enough to allow the optics to perform at their best. The scope is heavy and demands a solid mount. Nevertheless, the small 18 percent obstruction helps it perform as the perfect, yet affordable, premium scope for all types of celestial viewing.

# Do You Need a "Go To" Telescope?

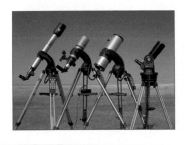

For the $300-to-$500 cost of a good beginner's scope, should you buy one with "Go To" capability, such as a Meade ETX-AT or DS-2000 model? They are seductive: Just press a button, and away they go. They make it fun to explore the sky with the kids. In the low-cost "Go To" scopes, what you sacrifice is image quality for any object you do find. For the same $300 to $400 that a small 70mm-to-80mm "Go To" telescope costs, you could have any of the following: a larger 90mm-to-105mm Maksutov, a 90mm long-focus refractor, a 130mm Newtonian reflector or a 6-inch (150mm) reflector on a Dobsonian mount. All will show brighter, sharper images, provided you do the work to find objects. Star-hopping guides and charts are the low-cost alternative to "Go To" computers. Using them to star-hop across the sky requires that you learn the sky, essential knowledge for cultivating a long-term love affair with the stars.

In higher-priced instruments, the trade-offs are not as clear-cut. Yes, for the same $1,000 to $1,800 it costs to buy a 5-to-8-inch "Go To" Maksutov- or Schmidt-Cassegrain, you could get a much larger reflector, such as a 10-to-12.5-inch Dob. But following our advice on portability, you might not want such a large instrument. In that case, why not get the compact high-tech model? We both use "Go To" scopes and now can't live without them. So it would be hypocritical for us to say you shouldn't have one. Our advice is to buy from a reputable dealer, preferably locally, in case repairs should be needed. (We had to exchange one "Go To" scope three times before getting a working unit.) Make sure the batteries are charged, and follow the instructions (see Appendix). Then have fun!

 *"Go To" technology is no longer just a high-end option. Low-cost computerized telescopes are now available with apertures starting at 70mm for small refractors and 114mm for small short-focus Newtonians.*

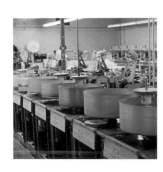

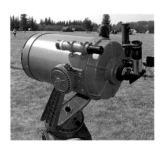

match the sharpness and contrast of a comparably sized (or slightly smaller) apo refractor at a fraction of the price. A 6-inch tube assembly sells for $1,200, a bargain for top-end optics. TMB Optical sells a 6.3-inch f/5.5 Mak-Newt for $2,300.

## THE REMARKABLE 8-INCH SCHMIDT-CASSEGRAIN

The telescope that for 30 years has been, and continues to be, at the technological forefront of amateur astronomy is the Schmidt-Cassegrain telescope, or SCT. Since 1970, the 8-inch model has been the top-selling serious recreational telescope. Its combination of generous aperture, portability, adaptability to astrophotography and all-round good performance has made the Schmidt-Cass the telescope of choice for backyard astronomers.

In the 1970s, Celestron had the field to itself. Its first serious competition came in 1980, when Meade Instruments brought out its Model 2040 4-inch f/10 and Model 2080 8-inch f/10 Schmidt-Cassegrains. Meade had started out in the early 1970s selling telescope accessories and good-quality Newtonians, but it soon realized that the future belonged to the Schmidt-Cassegrain. Throughout the 1980s and 1990s, Meade and Celestron battled with advertising wars, price-cutting and feature wars. Whatever one company did, the other soon copied or bettered.

The fierce competition continues to this day. Though models and features change, both Meade and Celestron have settled on three main product lines of 8-inch SCTs:

❖ **Entry-Level Models**
**($1,000 to $1,200)**
Both companies offer no-frills Schmidt-Cassegrains not much different from the models they started with a generation ago. These classic units can sometimes be purchased as deluxe packages with improved finderscopes or eyepieces, an added declination motor or heavier wedge and tripod. The upgrades are worth the modest extra cost. A wedge and tripod are essential.

In this price league, Celestron has its Celestar, while Meade has its LX10 and LX10 Deluxe, all battery-operated units with DC motor drives as standard. (The LX designation, used by Meade since the introduction of worm-gear drives in the mid-1980s, stands for "long exposure.") The $1,000+ price tag is the minimum cost for an 8-inch Schmidt-Cassegrain.

What are you giving up? The optics of these entry-level scopes are the same as those in the more costly Schmidt-Cassegrains. What they lack are the computers and sturdier mounts of the fancier models and features for astrophotography, such as periodic error correction. These models are best for visual use and limited astrophotography.

❖ **Midpriced Models**
**($1,500 to $1,800)**
The next jump up in price introduces "Go To" technology in two instruments that rep-

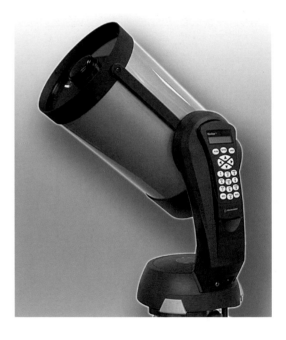

resent what are likely the most advanced telescopes the majority of buyers will need. Both the Celestron NexStar 8 and the Meade LX90 feature solid mounts and tripods that keep vibration to a minimum, especially when the telescopes are used as altazimuth platforms, driven by their computer-controlled motors. This arrangement is much more stable and vibration-free for any Schmidt- or Maksutov-Cassegrain than being tilted over on a wedge for traditional polar alignment, an unnecessary configuration unless you want to do long-exposure imaging.

The single-arm fork of the NexStar 5- and 8-inch models is not the flimsy structure it appears to be. It is hefty metal and holds even the 8-inch tube quite solidly. Yet the NexStar 8 is six pounds lighter than the LX90, making the Celestron model easier to cart around the backyard. The NexStar has a simple red-dot sighting device as a finder (good only for aligning on stars), while the LX90 has a proper, though poorly mounted, 8x50 finderscope (the plastic screws strip easily). Should batteries fail, the LX90 can be slewed manually; the NexStar can only slew electrically. The LX90 has more features for astrophotographers: a smoother worm-gear drive and the ability to accept an auto-guider (through an optional Accessory Port Module). The NexStar is designed more

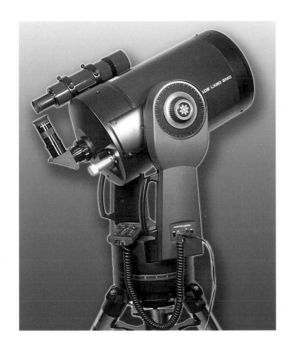

for uncomplicated visual observing. Both are battery-operated, but the LX90's C-cell pack lasts longer than the NexStar's penlight batteries. Neither has any hard stops to complicate initial setup.

❖ Top-End Models ($2,500)

Both companies revamped their high-end line in 2001-2002, bringing out models with the next logical step in "Go To" technology. The NexStar and LX200 GPS telescopes receive signals from the orbiting Global Positioning Satellites (GPS) that tell the telescopes where they are on Earth and the precise time. Additional on-board sensors detect north (more correctly, magnetic north) and level the telescope. Turn these

scopes on, and they dance to perform the initial alignment—no need to input latitude, longitude, date or time and no need to level the scope manually or aim it north, as with all other "Go To" models. These scopes then point automatically at the first alignment star. From then on, the process is the same as with other "Go To" scopes: You center that star, then a second one to calibrate the scope to the sky.

GPS technology adds another $1,000 or so to the cost of the scope. While it aids initial setup, it does nothing to improve the accuracy of finding targets or the view once you get there. However, the GPS models have other high-end features: Celestron's NexStars have carbon-fiber tubes for reducing focus shift in changing temperatures, a solid twin-arm fork, periodic error correction and auto-guider capability. Meade's LX200s add a more advanced version of the Autostar computer, as well as a sturdier fork, mirror lock-down and an external motorized focuser—excellent features for long-exposure imaging. If deep-sky imaging with film or CCD cameras is your intention, the GPS units are worth considering for these features alone. Celestron GPS scopes work only off external 12-volt power (i.e., a battery pack, a car cigarette lighter or an AC adapter); Meade's can operate off internal batteries, but expect only a night or two of use at best.

One caveat: Like any computer product, high-tech scopes quickly lose resale value. Five years from now, no one may want to buy, let alone pay much for, what is now the hottest thing on the market. For example, GPS and other top-end features will migrate down into lower-cost models, making the older units less desirable.

❖ Larger and Smaller SCTs

While the 8-inch models are the most popular, both Meade and Celestron manufacture other sizes of SCTs. Celestron has its NexStar 5, a wonderfully portable, compact instrument with consistently good optics and a rock-solid mount and tripod.

At the other end of the scale, Celestron has the excellent NexStar 11 GPS ($3,000), one of the finest big-aperture telescopes on the market. The optics are remarkably sharp, the mount and tripod vibration-free. For a semipermanent installation (the scope

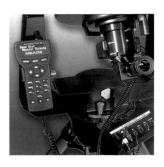

▲ **GPS Features**
Meade's GPS Series adds the improved Autostar II as well as an outboard motorized Crayford-style focuser, mirror lock-down and more control ports.

▼ **Big and Little Cats**
Celestron's NexStar 5 (below) is rock-solid, with a more accessible focuser than on Meade's ETX Series. Meade's 10-inch LX200 (bottom) is heavy but ideal for a permanent site.

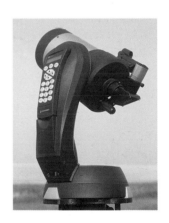

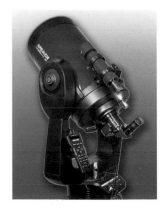

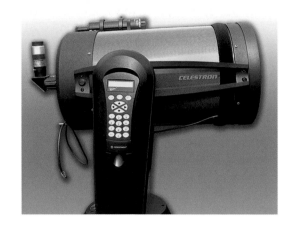

**Premium Celestron ▶**
The top of Celestron's line is the NexStar 11 GPS, a wonderful big-aperture telescope with fine optics. (It is shown here with a Kendrick Dew Remover System.) One of the authors uses this telescope regularly in a permanent observatory. For portable use, consider the smaller and much lighter NexStar 8 GPS.

weighs 63 pounds), this is as close to a dream telescope as most people will want.

In more conventional telescopes, Celestron also has, as of 2002, a 9.25-inch SCT for just $1,350 on its Chinese-made CG-5 mount and 11-inch ($3,200) and 14-inch ($5,300) models on its solid German equatorial mount, the CI-700 (a copy of the Losmandy G-11 mount). The C9.25 is a superb instrument optically, but the mount is a poor match. This scope deserves better. On the other hand, the CI-700 is a solid, well-

machined mount for the 11-inch, but the 14-inch strains its stability.

Meade has 10-inch ($3,000) and 12-inch ($4,000) SCTs in its top-end LX200 GPS Series. The 12-inch ranks as an observatory instrument. Its 80-pound heft means it will rarely be used as a portable telescope. However, the modest $500 price difference between the 8- and 10-inch models lures many buyers to the 10-inch SCT. But even here, Meade's 10-inch is much bulkier, and at 68 pounds, it is 60 percent heavier than the 8-inch. That's a lot of weight to lift to waist height to place precisely onto the tripod's center bolt. Unless you can leave the scope assembled and wheel it out on a dolly onto the driveway or patio, our advice to most people is to stay with the 8-inch.

In 2002, Meade's older 16-inch LX200, which had been a staple instrument for club and college observatories, was replaced by a new GPS-equipped model.

❖ Celestron vs. Meade Optics
A constant question in the minds of buyers is whether Meade or Celestron is better. We have examined images in dozens of Meades and Celestrons, in current versions and in telescopes that date back to their first years

# Picking a Schmidt-Cassegrain

Much has been written about the inherent quality of Schmidt-Cassegrain, or SCT, optics. Detractors maintain that the 35 to 38 percent obstruction of the secondary mirror degrades images unacceptably. In our experience, an SCT with well-made optics (as they have been for several years) provides images sharp enough to please the vast majority of users. We've seen stunning views of planets through SCTs (with its central obstruction, an 8-inch SCT can deliver contrast and sharpness that equal those of a 5-inch refractor). We're convinced the reason many SCTs don't perform well is that their optics are not collimated. With these telescopes, the slightest miscollimation of the critical secondary mirror softens planetary detail and degrades contrast. Once properly adjusted, collimation should be fine for years.

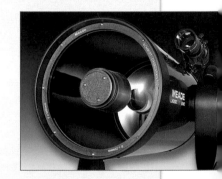

 *Meade offers Ultra-High Transmission Coatings as a worthwhile factory-specified option. They increase image brightness but, contrary to claims, do not increase resolution.*

of production, and have seen good and bad instruments from both companies. In the mid- to late 1980s, in particular, the marketplace was flooded with hundreds of telescopes that were unable to form clean star images. The reputation of the Schmidt-Cassegrain took a beating. Both companies instituted sweeping quality-control measures, and the units we have tested and owned in recent years have all contained excellent optics. We have seen no consistent difference in optical quality between Meade and Celestron. It's a toss-up. Meade used to offer f/6.3 versions of its 8- and 10-inch SCTs, but these were dropped in 2002 in favor of consistent and better-quality f/10 systems in all apertures.

In hardware design, Celestron models tend toward elegant simplicity and ease of use, including just those features that most users will actually need. Meade telescopes tend to impress buyers with a long list of features. Some are genuinely useful, others are not; but some buyers like to have them all "just in case."

❖ Meade Autostar vs.
   Celestron NexStar

The same can be said of computer software. The Meade Autostar software generally features a larger database of useful objects, with various catalogs of deep-sky targets nicely cross-referenced (for instance, you learn that Caldwell 1 is also NGC188). Meade also provides more scrolling information about more targets and handy utilities like Sun and Moon rise/set times. The Autostar software offers an excellent "Park" feature that, provided the telescope is not moved between sessions, allows the scope to be switched on and aimed at targets without the need to realign on stars—great for finding stars and planets in the daytime sky or for an observatory installation. Meade's programmed "Tours" are also more extensive and creative than Celestron's.

Meade's internal firmware can be updated via Internet downloads, but that in itself is a process which has tripped up even computer-savvy users. Plus the latest versions of firmware can be plagued with new bugs not present in the old versions, requiring the download of yet another patch or bug fix. (The main improvement provided by the Autostar II is quick access to catalogs from the keypad, rather than having to page down through menus.)

In our tests, the NexStar software for Celestron's larger 5-, 8- and 11-inch telescopes has proved reliable and bug-free from the start and has not needed to be upgraded. The Celestron software offers all the databases most users would want, quickly accessible through single keypad buttons rather than a hierarchy of menus, as in the original Autostar software. This makes the NexStar software simpler to learn, with little fussing over motor training and calibration, as is re-

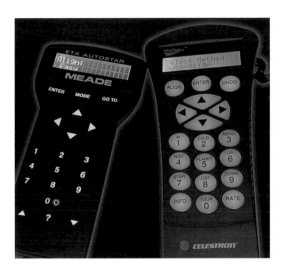

◀ **Push-Button Astronomy** Celestron's NexStar (the controller on the right) is easy to use and has all the features most observers would want. Meade's Autostar has more object information and extra features that can prove useful or perhaps just confusing. Both brands of "Go To" computers can guide users through a lifetime of sky exploration.

quired with the Meade units. Later models also have a feature identical to Autostar's "Park," called "Hibernate," which allows the telescope to be turned off, then turned on at a later date and immediately pointed to any object (as long as the scope was not moved during hibernation, of course).

❖ Do You Need GPS?

In our opinion, GPS receivers aren't necessary. After all, in most cases, your scope will never be used at more than one or two sites. It's easy to plug in those site coordinates by hand; the computer then remembers them. For time and date, other "Go To" telescopes we use have lithium batteries and internal clocks that keep time. We just turn them on, and away we go. If you can find north and "eyeball level" a scope, aligning a non-GPS scope is actually faster. The technology was added because many users of "Go To" scopes, unfamiliar with the sky, are unable to locate north at night. GPS also adds marketing value and high-tech appeal. Even so, GPS is fun and has its uses, especially if you do observe from many different locations. The GPS models are worth the extra cost for their other high-end features.

# Recommended Telescopes

Our list of favorite telescopes is biased toward models with several key characteristics: sharp optics, a steady, jitter-free mount, convenient portability and no-fuss ease of use, all at a price that is the best value in its class. Everyone nods in agreement that these are the important traits of a good tele-

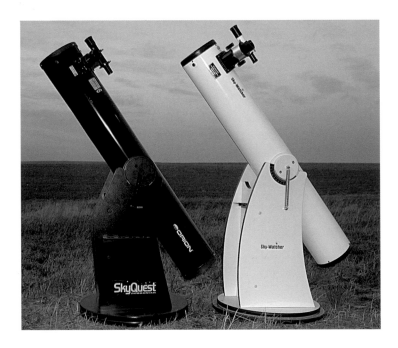

scope. Yet all too often, prospective buyers we talk to elect to ignore our advice, making their purchase based on other reasons: what was on sale at the local "Super-Mart"; which had the most convincing magazine ad; which has the biggest database in its handheld computer; or what might impress their friends and family the most.

Our aim is to outfit you with a telescope that will be easy to use and will show you the most for the money. The more you think "Wow!" when you look through the eyepiece, the more you'll want to continue to explore the sky. Unlike telescope companies and stores, we aren't trying to sell you a

telescope. But we would like to sell you on the wonder and enjoyment the hobby can provide. Of all the telescopes we've used of late, the following models are our personal favorites for doing just that.

## Getting Started ($150 to $450)

These models represent our first choices for your first telescope. None require a major outlay, and all will retain good resale value.

❖ 6-inch f/8 Dobsonian (Orion SkyQuest or Synta SkyWatcher)

These well-made scopes provide great optics on a stable mount. They represent our first choice for the best value in a starter scope. In side-by-side tests with off-the-shelf, inherently shaky and cumbersome 6-inch achromatic refractors, an elegantly simple 6-inch Orion SkyQuest won hands down for revealing crisp planetary detail.

❖ Orion SkyQuest 4.5

This little Dobsonian is physically too short to be used comfortably by grown-ups but is perfectly kid-sized. With its no-fuss setup, ease of use, erect-image finder, quality eyepieces and solid metal and wood construction, we can think of no better telescope to encourage the interest of a young astronomer.

❖ 70mm f/10 Refractor on Alt-Az Mount (Celestron, Meade, Orion, Synta SkyWatcher)

On a mount with slow-motion controls, small 70mm f/10 refractors provide good optics and ease of use. At $150 to $200, they are the lowest-priced telescopes of quality on the market. No other telescopes under $200 are worth considering. With its emphasis on high-tech, Meade has a nice "Go To" version on a solid mount, the DS-2070 AT.

❖ 90mm f/10 Refractor (Celestron, Orion, Synta SkyWatcher)

The Chinese-made 90mm long-focus refractors we've used have amazed us with the sharpness of their optics at any price, let alone the $300 these models cost. An altazimuth version is a fine starter telescope—most dealers carry a model under some brand name. For an equatorially mounted model, we prefer units on the larger EQ-3 mount. The added stability over lesser mounts is worth the extra cost.

❖ **130mm f/5 Parabolic Reflector (Orion, Synta SkyWatcher)**

The optics and fittings of these Chinese Newtonians are great, but the EQ-2 mount is just marginal. The tube tends to hit the optional single-axis clock drive. As with the 90mm refractor, a better though seldom-sold configuration is this tube assembly on the larger EQ-3 mount. The EQ-3 is used on Celestron's 6-inch f/5 Newtonian, the C150-HD, a step up in aperture and price.

❖ **Meade ETX70-AT**

The "Go To" computer finds and tracks targets just as advertised. Lots of people buy and enjoy these telescopes. The f/5 short-focus refractor optics work best under dark skies for crisp, low-power, deep-sky views. Compared with f/10 refractors, high-power images of the planets (you'll need a Barlow lens) are soft and colored by chromatic aberration.

## GETTING SERIOUS ($500 to $1,200)

If you're willing to invest more in a first telescope, these models provide either better op-

tical quality or more features. Note that we do not list the entry-level, non-"Go To" Schmidt-Cassegrains from Meade and Celestron. Though they're fine scopes, we suspect most buyers will soon long for a "Go To" version described in the next price category.

❖ **Meade ETX-105**

Of the popular Meade ETX line, this stands out as our favorite. It is compact, has good aperture and sharp optics and is vibration-free when placed on the sturdy but optional Deluxe Field tripod. And the "Go To" works. It may be all the telescope many newcomers to the hobby will need.

❖ **Celestron NexStar 5**

For the next step up in aperture in a "Go To" scope, we have been impressed with the NexStar 5. Its solidness, crisp optics, smooth, accessible focuser and reliable software make it a pleasure to use. We recommend it over the competitive Meade ETX-125.

❖ **4.7-inch f/8 Achromat Refractor (Orion AstroView or Synta SkyWatcher)**

If a refractor appeals to you, this is a terrific buy. The optics are fine and the fittings excellent for such an affordable scope. On the

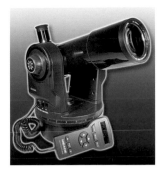

▲ **The Littlest ETX**
Though Meade's ETX telescopes, like this little 70mm AT, can be used on a tabletop, they are best placed on a tripod. The standard lightweight tripods tend to be shaky, so we recommend Meade's optional Deluxe Field tripod.

# Four Entry-Level Mounts

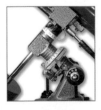

**EQ-1** This mount is too small and flimsy for most scopes it is asked to carry. An exception is Short-Tube 80mm refractors, but those wide-field telescopes are better placed on altazimuth mounts to make low-power scanning easier.

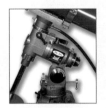

**EQ-3 (aka SkyScan)** We prefer this mount for 90mm refractors and 5-inch reflectors. It is too light for the 4.7-inch refractors often mated to it, taking seven to eight seconds to dampen vibrations.

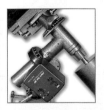

**EQ-2** This is a good mount for 60mm and 70mm refractors and Short-Tube 80mm and 90mm scopes but is strained by anything larger, such as the 4- and 5-inch reflectors with which it is commonly supplied.

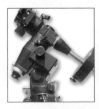

**EQ-4 (aka Celestron CG-5)** This clone of the Vixen Super Polaris mount is not as solid as the original but is suitable for casual visual use with 4-to-5-inch lightweight refractors and 6-inch Newtonians.

*These are the Chinese-made mounts you see most often on beginners' telescopes. The mounts aren't bad in themselves, but in most cases, the telescope is supplied on a mount that is one size too small for the weight and size of the scope. If you have the option to outfit the telescope with the next larger-sized mount, we highly recommend doing so.*

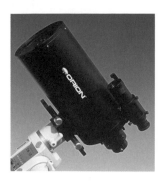

EQ-3 mount, images can bounce for several seconds before damping down; a better mount is the EQ-4. There are two versions of this refractor: long tube (f/8.3) and short (f/5). Get the f/8.3 model. It produces much better high-power views of everything, especially the Moon and planets.

❖ Orion StarMax 5-inch
   Maksutov-Cassegrain

The optics of this f/13.8 Mak-Cass are good, and the short tube mated to the EQ-3 mount is a solid, easy-to-use combination. The extra aperture and better mount justify the price jump over the 4-inch StarMax.

❖ INTES 6-inch f/12 Maksutov

The fine optics of this Russian Mak-Cass make it the equal of 4-inch apo refractors for sharpness and far superior to any low-cost achromatic refractor, all in a short tube that is easy to mount.

❖ 8- and 10-inch Dobsonians
   (Orion SkyQuest or
   Synta SkyWatcher)

These telescopes give you serious aperture for deep-sky viewing without costing a bundle. They are well made, with excellent

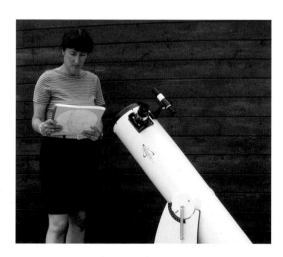

optical quality for the price. The 8-inch is light enough to be handled without difficulty by most teens and adults.

## GETTING MORE SERIOUS
($1,300 to $2,900)

This price range represents the top dollar that most buyers likely want to spend, though there is a soaring stratosphere of instruments above this.

❖ Celestron NexStar 8
   or Meade LX90

In this price range, the first choice has to be an 8-inch Schmidt-Cassegrain. All models feature the same optics regardless of price. With the luxury of "Go To" technology available for such a reasonable price in the

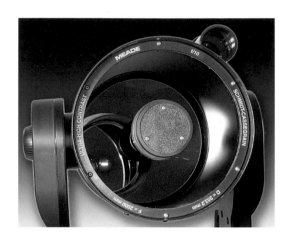

midrange Celestron NexStar 8 and Meade LX90 models—these units cost just $500 more than their no-frills counterparts —why not have it? The "Go To" hardware makes setup simpler (no polar alignment), the telescope is lighter and more stable (no tilting over on a wedge), and you have ready access to thousands of targets. These instruments are highly recommended for both serious beginners and intermediate astronomers.

❖ Celestron NexStar 8 GPS or
   Meade 8-inch LX200 GPS

The GPS option is a luxury, to be sure, but it is fun. More important, these models offer such premiums as improved mounts and tube assemblies and advanced astrophotography features: better drive gears, periodic error correction, auto-guider readiness and Meade's mirror lock-down and outboard focuser. Move up to GPS scopes only if you must have the latest gadgets or want to do serious photography or CCD imaging. In that case, be sure to buy the optional equatorial wedge.

❖ MAG1 Instruments
   PortaBall 8 or 10

What a neat telescope! What fun to waltz around the sky using Carl Zambuto's superb optics, which provide good deep-sky views as well as terrific planetary views

# Telescope Performance Limits

| Aperture (inches) | Aperture (mm) (approx.) | Faintest Stellar Magnitude Visible | Theoretical Resolution (arc seconds) | Highest Usable Power |
|---|---|---|---|---|
| 2.4 | 60 | 11.6 | 2.00 | 120x |
| 3.1 | 80 | 12.2 | 1.50 | 160x |
| 4 | 100 | 12.7 | 1.20 | 200x |
| 5 | 125 | 13.2 | 0.95 | 250x |
| 6 | 150 | 13.6 | 0.80 | 300x |
| 8 | 200 | 14.2 | 0.65 | 400x |
| 10 | 250 | 14.7 | 0.50 | 500x |
| 12.5 | 320 | 15.2 | 0.40 | 600x |
| 14 | 355 | 15.4 | 0.34 | 600x |
| 16 | 400 | 15.7 | 0.30 | 600x |
| 17.5 | 445 | 15.9 | 0.27 | 600x |
| 20 | 500 | 16.2 | 0.24 | 600x |

Even large scopes can rarely use magnifications over 600x or resolve better than 0.4 to 0.5 arc second.

(the central obstruction is only 20 percent). Yet these scopes are amazingly compact, especially when broken down and nested together. MAG1 owner Peter Smitka has designed one of the world's great scopes.

## NOTHING BUT THE BEST ($3,000 and up)

Here, we emphasize telescopes we have both used extensively and can recommend from firsthand experience. Other top-class brands exist and have well-deserved reputations for quality, but we simply haven't used them—yet. We'd be happy to receive samples for testing!

❖ **Tele Vue NP101 Refractor**
Optics don't get any better than this. From wide-field panoramas to high-power planetary views, this scope does it all. Couple it to Tele Vue's Gibraltar altazimuth mount and Star Tour digital setting-circle box, and you have computerized finding. For astrophotography on an equatorial mount, we suggest either a Vixen Great Polaris or a Losmandy GM-8, both first-class small mounts and an ideal match for the NP101.

❖ **TMB 80mm and 105mm Triplet Refractors**

These triplet Super ED apos by Thomas Back provide yet more tempting choices in the premium-refractor class. They are well machined and crafted, and their deeply coated optics provide maximum light transmission. Lovers of fine optics will want at least one! Although hefty, the German-made tube assemblies are a joy to use.

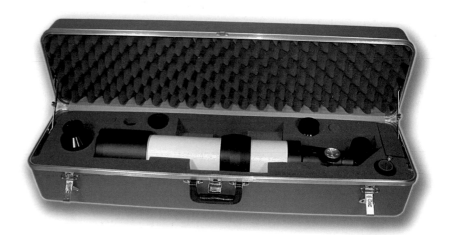

❖ **Starmaster Dobsonian Reflectors**
This ultimate premium Dobsonian has it all. Company owner Rick Singmaster spent

▼ **Top Tele Vue**
The Nagler-Petzval NP101's 26-inch-long tube comes in a foam-fitted case with precut compartments for all those eyepieces you are sure to collect.

**Top-Class Performer** ▶
TMB refractors use German-made tube assemblies that are heavy but beautifully engineered.

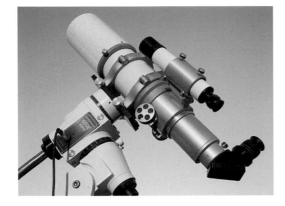

**A Unique Design** ▶
The MAG1 PortaBall truss-tube reflectors use a large fiberglass ball that rests on three Teflon pads. The scope is easy to swing around the sky and rotate to place the eyepiece at the right height and angle for ergonomic comfort. We love them!

**Cure for Aperture Fever** ▶
You want aperture without compromising quality of optics? This 30-inch Starmaster Newtonian is perhaps the finest big telescope money can buy. The views of deep-sky objects and planets are consistently breathtaking.

**The Borg** ▼
The 100ED Borg apo tube assembly splits apart into lens, tube and focuser for ease of airline transport.

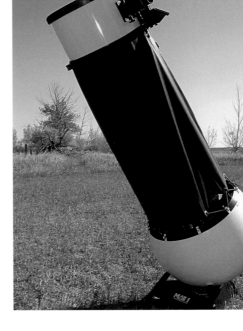

more than a decade perfecting the design of big but compact and highly portable Dobsonians with mounts motorized for "Go To" and tracking. Imagine an 18-inch Dob, only slightly taller than an adult of average height, that swings on command to any deep-sky object, then tracks like an equatorial mount. Now add unsurpassed

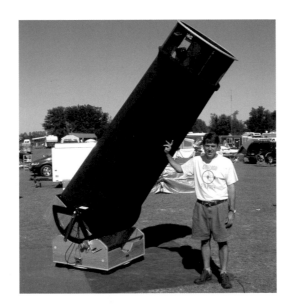

Zambuto optics, and wow! They are available with "Go To" in sizes from 14.5- to 30-inch. Although the bigger units strain the definition of portable, the 14.5-to-18-inch ones are wonderfully transportable.

❖ MAG1 PortaBall 12 Reflector
Ditto our comments on the 8- and 10-inch PortaBalls. The 12-inch f/5 provides the necessary aperture for serious deep-sky viewing while maintaining the superportability of the design. When broken down and nested into the ball, it sits on a car seat.

When set up, it swings around the sky with no awkwardness at the zenith, where traditional Dobs have a hard time aiming. The eyepiece rotates, so it's always at a convenient height and angle. Yes, it is expensive for a 12-inch scope, but the elegant design and Zambuto optics are worth every penny. The only drawbacks: Heavy eyepieces may cause the scope to sink, and digital setting circles or "Go To" motors cannot be added.

## ASTRO-IMAGING SPECIALIST ($3,000 minimum)

These instruments are designed first and foremost for deep-sky imaging, on film and with CCD cameras.

❖ Borg 100 or 125 Models
Borg refractors and fast astrographic lenses are optimized for photography. They project flat fields onto film formats as large as 6x7 centimeters yet are extremely light and break down into small components for airline travel. The 4-inch ED we used soon became a favorite scope.

❖ Takahashi FSQ106
Introduced in 1998, this telescope quickly gained a well-deserved reputation as one of the finest photo-visual refractors ever manufactured. Its two-element front objective (one fluorite element) forms half of the Petzval design. The other half is a two-element

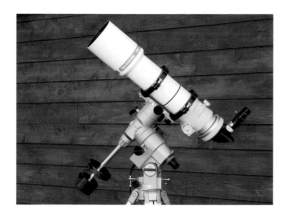

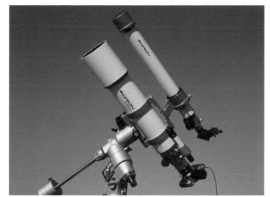

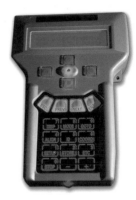

rear corrector/field flattener (again, one of the elements is fluorite). Takahashi calls it a Double Fluorite design, one capable of filling a medium-format film frame. The result is a superb visual and photographic no-compromise refractor for the aficionado.

❖ **Vixen ED Refractors**

The 102-ED and 114-ED Vixens coupled to the Great Polaris mount provide the flat-field optics and accurate tracking needed for astrophotography. At $3,000 to $3,600, these packages represent the minimum buy-in cost for a good astrophoto outfit. The

Great Polaris and heavier DX version can also be equipped with Vixen's SkySensor "Go To" computer, a $900 package (with motors) that we recommend—the standard dual-axis-motor package is $400, so adding "Go To" costs only another $500.

❖ **Premium Mounts**

Astrophoto systems are often mix-and-match affairs, with a tube assembly from one manufacturer mated to a mount from another. Astrophotography demands the best no-compromise mount. Two of the most popular models include the GM-8 and

# Making a Dob Track

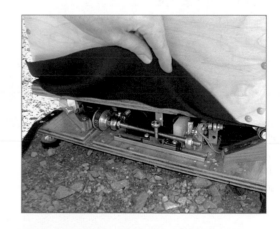

To allow tracking at high power, an increasing number of Dobsonian owners are investing in some form of tracking system. The traditional method uses a Poncet, or equatorial, platform, a stubby table-like device whose top swivels around an axis aimed at the celestial pole. The telescope simply sits on the platform. A Poncet platform must be manufactured for a limited latitude range (one made for Alberta, for example, won't work well in Arizona). Also, it provides tracking for only an hour or so before the threaded drive rod must be reset back to the start of its travel.

A higher-tech solution employs servomotors on each axis of the Dobsonian and a small computer to drive them, the same way other computerized "Go To" telescopes work. For example, Starmaster sells Dobs with "Go To" tracking hardware as an option, while JMI sells a system for retrofitting Obsession telescopes that is built around the well-designed Vixen SkySensor computer. Add-on systems like these provide not only tracking but also "Go To" capability for Dobsonians, amazing features for what were once considered no-frills telescopes.

 *Whether commercially made (from companies such as Equatorial Platforms) or homemade, as here, Poncet platforms provide a low-tech but reliable way to add tracking capability to any altazimuth telescope.*

G-11 models made by Scott Losmandy's company, Hollywood General Machining. These solid mounts provide superaccurate tracking and precision-machined compo-

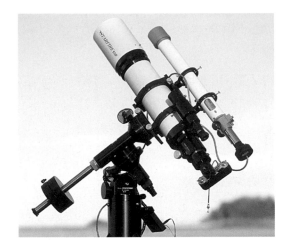

nents, all for a reasonable price of $1,400 for the GM-8 and $1,900 for the larger G-11. Losmandy's optional Gemini "Go To" system adds another $1,100 to $1,300. As an astrophoto platform, the GM-8 is suitable for scopes up to 4-inch refractors or 8-inch Schmidt-Cassegrains, while the G-11 handles 5-inch and lighter-weight 6-inch refractors and 9-to-11-inch SCTs. For the best values in a top-class mount available from stock, the Losmandy products can't be beaten. We highly recommend them.

Astro-Physics' premium "Go To" mounts (there are four in its catalog: the 400, 600E, 900 and 1200) have become the mounts of choice among many astrophotographers, including us. They are beautifully made and come with one of the best "Go To" systems we've used. The 400 mount easily carries a 4-inch apo refractor, the 600E is the choice for a 5-inch, and the 900 is a superbly solid mount for a 6-inch apo. The 1200 mount solidly carries just about anything an astrophotographer would want to attach to it. Unfortunately, these mounts require long waits just to place an order.

Other small companies, such as Parallax Instruments, Software Bisque (with its monster Paramount) and William Optics, also make top-grade mounts worth checking out if you are in the market. Unlike the products mentioned above, however, we have no experience with these brands.

## DEEP-SKY EXPLORERS ($3,000 and up)

If aperture fever has you in its grip, the only antidote is a big Dobsonian. The following are among the best.

❖ **Obsession Dobs**
These telescopes have long ruled the big-aperture universe. Designed by Dobsonian guru Dave Kriege, Obsession Dobs feature woodworking and finish of fine-grade-furniture quality. Though large, they break down into components that can readily be carried or wheeled by one person. However, setup and takedown of the largest instruments are usually best handled by two people. We particularly like the 15-inch model as a manageable and satisfying one-person telescope.

❖ **Starsplitter Dobs**
Just a notch down in the grade of woodworking from Obsessions, Jim Brunkella's line of high-quality Starsplitters has ex-

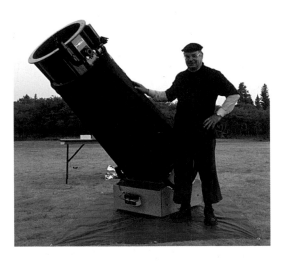

panded over the years to include light-weight designs and compact dual-truss models. As with other Dobs, we suggest a model in the 14-to-18-inch range for maximum convenience of eyepiece height.

## PLANETARY PERFORMERS ($2,000 and up)

Any telescope whose optics are good enough to reveal the finest planetary detail will work well on all types of viewing. But the following instruments are good choices if the planets are your first line of interest.

❖ Orion/INTES 6-inch
Maksutov-Newtonian

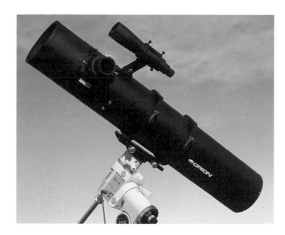

We are amazed at how good the images are in this telescope, matching or beating the views through a 5-inch apo that costs four times the price and blowing away any achromat of comparable size. Its 20-pound tube assembly needs a good mount.

❖ Astro-Physics 130EDF and
155EDF Apo Refractors

As with the TMB apos below, we list these premium refractors under the planetary category, but they could also be considered as among the finest astrophoto instruments on the market or simply as telescopes coveted by anyone looking for "nothing but the best." We love these scopes. Optics don't get any better—which is why these refractors are in such demand but low supply.

❖ TMB 130 and Larger Apos

Anyone looking for first-class performance and construction has lots to choose from in the extensive line of TMB triplet refractors. These heavy, battleship-grade tube assemblies demand a sizable mount, but there's no better choice for the well-heeled planetary enthusiast.

## PORTABILITY PLUS
## ($400 and up)

These instruments are durable, airline-portable telescopes that make traveling to observe under foreign skies easier.

❖ Orion AstroMax 90mm
and 102mm

With decent optics, these Chinese-made

Maks are great little scopes to stuff into carry-on bags for eclipse trips. Just don't forget the mount and tripod.

❖ William Optics Megrez 80
Semi-Apo Refractor

This doublet f/6 ED refractor has good color correction. At 4.8 pounds and only 14 inches long, it may be the lightest and most com-

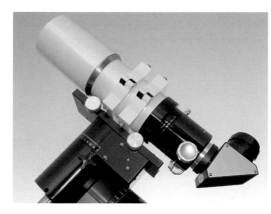

pact high-performance scope on the market. It provided super views of the eclipse, Mars and southern-sky wonders in Africa in 2001. With tube ring mounts and 2-inch diagonal, it's a bargain at $800.

❖ Tele Vue 85mm ED
Refractor

At eight pounds, this f/7 refractor is 21 inches long, approaching the limit of convenient travel-scope size (again, remember that all these scopes need a mount). The ED lens gives good color correction. It comes with 2-inch diagonal and tube ring but needs a field flattener for photography. It is priced at about $1,800.

❖ Takahashi Sky90 Fluorite Apo

About the same weight as the Tele Vue 85mm but the length of the Megrez 80, this doublet fluorite has dead-sharp optics with just a trace of false color. The optional Extender-Q reduces color at high power, while the optional reducer/field flattener is essential for deep-sky imaging. Fully outfitted, it is unmatched ($3,000).

❖ Astro-Physics Traveler
105EDF

The next step up in size is the Astro-Physics triplet 105mm f/6 Traveler. Color correction is better than the Sky90's, but at 12 pounds, its weight is significantly greater. Nevertheless, this rugged refractor has traveled the

**Supersharp Mak-Newt**
Insert a 35mm Panoptic eyepiece, and enjoy a stunning 2.7-degree field, sharp edge to edge. Put in a 3mm Radian, and you have unsurpassed planetary views.
◀ Mak-Newts can do it all.

◀ **Superportable
Megrez 80**
The smallest and lightest high-quality scope on the market is this William Optics 80mm semi-apo, with first-class tube finish and fittings.

▼ **Big-Sized Astro-Physics**
Planet observers lust after this telescope: the Astro-Physics 6-inch f/7. Images are tack-sharp, with very little bother from cooldown time or tube currents.

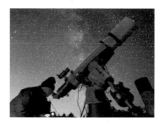

▼ **Portable Takahashi**
A high-performance travel scope, the Sky90 has fine 90mm f/5.6 optics and fittings typical of all Japanese-made Takahashis.

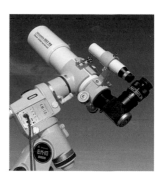

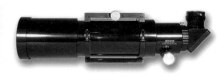

world with us and provided outstanding views. It is a joy to use, with its only downside being its curved focal plane, which prevents edge-to-edge sharpness on 35mm film even with the reducer/field flattener. Be prepared to line up to get one.

**Built to Travel ▲**
One of the finest portable telescopes ever made, the solid 105mm Traveler provides razor-sharp views.

## TELESCOPES TO AVOID

We have already warned against most 60mm refractors sold at Christmastime as "high-powered professional models." They are what we call Christmas trash scopes.

These are other common models to avoid.
 ❖ Short-Tube Newtonians
A number of companies sell 4.5- or 5-inch

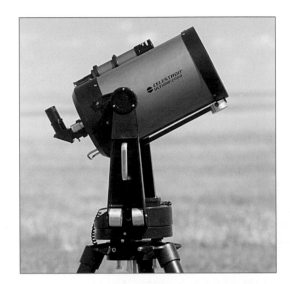

# The Used-Scope Lot

If well cared for, a telescope can last a lifetime and can work as well now as it did years ago. A used telescope from someone who has lost interest in the hobby or has traded up to a new high-tech model can be an excellent buy. Check the classifieds on the web at sites such as www.astromart.com.
Here are some of our favorite, and not so favorite, telescopes of years past that often appear on the used-scope market.

GREAT TELESCOPES OF THE PAST

**Celestron Ultima 8**
This was Celestron's premium instrument of the early 1990s, prior to the introduction of "Go To" scopes. Its solid fork mount and accurate periodic-error-correction (PEC) drive made it a fine scope for photography. Models that use a standard 9-volt battery as a power source are better choices than the early Ultimas, which used a rechargeable battery.

**Celestron Ultima 2000**
The successor to the Ultima 8 in the late 1990s, this battery-operated "Go To" scope featured light weight, quiet motors and great portability, as well as a mount that was sturdy enough for photography.

**Celestron C5+**
This battery-powered, one-arm forked unit produced in the early 1990s provided great optics in a compact package.

**Celestron/Vixen Fluorite Apo Refractors**
In the early 1990s, Celestron offered a series of supersharp Vixen 70mm, 80mm, 90mm and 100mm fluorite apo refractors, often on the superb Vixen Super Polaris mounts. They were labeled SP-70F, SP-80F, and so on, and are among the finest small-aperture refractors ever made. Even secondhand, these are far superior to the Chinese-made achromats and clone mounts being produced today. Fluorite models that were sold for a time under the Orion name are equally good.

**Meade ETX-90**
The original late-1990s ETX-90 was a plain non-"Go To" scope (offered for a time as the "RA" model). For a handy second scope with fine optics, a used "classic" ETX-90 is a great choice.

**Quantum 4**
This limited-production Maksutov-Cassegrain (a 6-inch was also offered) was produced in the early 1980s to compete with the Questar. Quantums are still prized for their fine optics.

**Any Early-Model Questar 3.5**
The quality of the Questar 3.5 has been consistent over the decades, with the only significant change being the option of a more convenient DC-powered drive (PowerGuide) in models made in the 1990s.

**Any Early-Model Takahashi Refractor**
This is another brand name you can trust for fine optics, no matter the era. A small Tak makes a great second scope.

Newtonians with f/4.5 mirrors but final focal ratios of f/9. A Barlow lens in the focuser doubles the effective focal length. Advertised as an innovative design, these short-tube models usually have soft optics—it is hard to make a quality f/4.5 mirror cheaply—and the Barlow merely amplifies the fuzziness.

❖ Small 3-inch Newtonians

Their small aperture and flimsy altazimuth or equatorial mounts make them poor choices compared with a 4.5-inch Dob.

❖ Undermounted Models

If you see a series of telescopes all offered on the same mount (6-, 8- and 10-inch reflectors like the Meade LXD55 Series, for example), chances are the small scope is solid, but the big scope will be shaky. The middle scope will be just acceptable.

❖ Spotting Scopes on Camera Tripods and Pan Heads

These are fine bird-watching instruments but are unsuited to astronomy. The straight-through optical configuration that works for birding is all wrong for astronomy. Also, the camera tripod doesn't have the fine motion control needed to track celestial targets.

**Any Early-Model Tele Vue Refractor**

The 4-inch models, in particular, were not as well color-corrected as the most recent models (the older the model, the worse the color), but Tele Vue optics have been consistently sharp and well made.

**Any Early-Model Astro-Physics Refractor**

Ditto on the color correction, but used Astro-Physics refractors are in such demand, they get snapped up largely by word of mouth and command as-new, if not higher-than-new, prices. These scopes actually increase in value.

USED TELESCOPES TO AVOID

**Any Early-Model Criterion or Bausch & Lomb Schmidt-Cassegrain**

The optical and mechanical quality of these 4-, 6- and 8-inch SCTs from the 1970s and 1980s left much to be desired.

**Almost Any German Equatorially Mounted Schmidt-Cassegrain**

With rare exceptions, these models by Celestron and Meade were (and are) consistently undermounted. The mounts may be fine for scopes a notch down in aperture.

**Almost Any Schmidt-Cassegrain Produced in the Mid- to Late 1980s**

During the Comet Halley era, optical quality slipped terribly, giving Schmidt-Cassegrains a bad reputation they have yet to shake off completely. Other older "Go To" SCTs can be an excellent buy, but test the optics first.

**Any Celestron C90**

Yes, it's a Maksutov, but its fuzzy optics and crude rotating-tube focusing are reasons to avoid the various incarnations of this 90mm telescope.

**Any Coulter Blue- and Red-Tube Dob**

They started the Dobsonian revolution, but today's entry-level Dobs are far better in all respects.

**Any of Meade's MTS Series Fork-Mounted Schmidt-Newtonians**

Produced in the early 1990s, these bargain 6- and 8-inch scopes suffered from shaky mounts and fuzzy optics.

**Any of the Early-Model Meade DS Series of "Go To" Scopes (not the current DS-2000 Series)**

These small refractors and reflectors with add-on motors had numerous problems with balky drive mechanisms.

**Meade 16-inch Starfinder Equatorial**

Bought by the aperture-fever-afflicted, this scope is way too big and shaky to be usable. Buy a truss-tube Dob.

**Meade Non-"Go To" 10-inch Schmidt-Cassegrains**

Ten-inch models from the 1980s, like the LX3, LX5 and LX6 units, have fork mounts and wedges too light and bouncy for the heavy tube assembly. Avoid the temptation of large aperture for low cost.

 *The Celestron Ultima 2000, facing page, is a fine "Go To" telescope from the late 1990s. A unique feature shared by few "Go To" scopes even today is its the ability to be moved by hand and still keep track of where it was pointed.*

# Making the Purchase

Now, the final hurdle. After making your selection, or at least narrowing down the choices, the question is, Where do you buy the telescope?

## WHERE TO LEARN MORE
### (Clubs, Internet, Reviews)

First, you may want additional real-world information. Observing nights and star parties hosted by local astronomy clubs are great places to see telescopes in action—especially scopes you may be considering

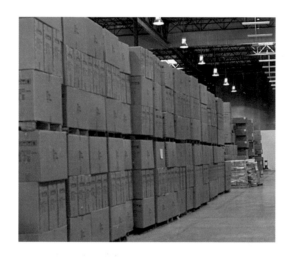

**Tons o' Scopes ▶**
Today, you can choose to buy from a major manufacturer that makes and stocks scopes by the thousands, as here, or from a "boutique" supplier that builds telescopes to order. The choice of models is now wider than ever.

purchasing. Seeing and handling a telescope provides a valuable reality check of your expectations of ease of use and of what a telescope can actually show you.

Magazine reviews (we regularly write them ourselves) also provide valuable details on how a telescope performs. Magazine websites usually have reviews archived for download, though perhaps for a fee.

The Internet is a source of still more opinions—lots more. But we caution you: Many of these opinions are highly biased, if not inflammatory, for and against specific brands or types of telescopes. Biased opinions that would never make it past the eyes of an astronomy-magazine editor go through no such filter when posted on a private web page or forum. They are not ob-

jective reports, though this may not be at all obvious to the novice looking for guidance. Many are written by owners who want to tell the world how wonderful their telescope is and what a great decision they made in buying it. Some are no more than a retelling of what they looked at on their first night out with their new telescope. Many reviews are written by people who have little experience with optics and lack any standards for judging performance. Unqualified statements such as "This is the best telescope I've ever used" beg the question, How many has the reviewer actually used?

A simple rule applies here: No matter how bad the telescope, someone will defend it as the best value on the planet! Yet even some of the poor reviews can provide an impression of what a telescope might be like. And amid the web dross are thorough reviews written by observers who know telescopes and whose expert opinions have value and can be trusted.

## DIRECT PURCHASE FROM THE MANUFACTURER

So where to buy? Many companies sell solely through dealer networks. Some small manufacturers have select dealers and also sell via mail or web-based orders. Others sell only direct from the factory.

In some cases, smaller-volume producers construct the telescope only after the order is received. Demand is often so great, they can never produce enough units to maintain a stock anyway. Such operations usually require a payment of one-third to one-half of the total cost of the instrument when the order is placed. The balance is due when the equipment is ready to be shipped.

Ninety-nine percent of the time, there is no risk in this procedure. But, as in any field of manufacturing, companies can go out of business. If this happens while a company has your money, you may be out of luck. To guard against such a problem, seek recommendations about companies from knowledgeable amateur astronomers. If there are waiting lists for a company's products, ask whether the company has shipped one of its products recently. If someone you know has been waiting for more than six months past the stated delivery date, then exercise caution. However, if the reports reveal that

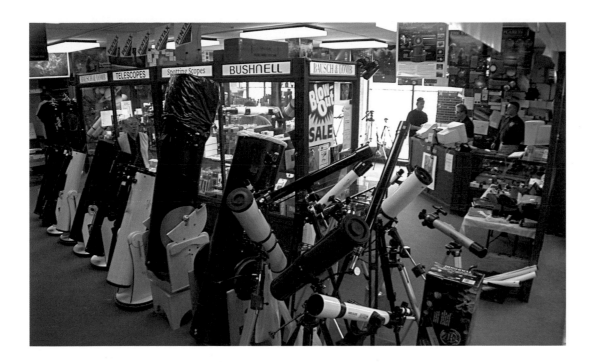

three-to-six-month delivery dates have been adhered to, it is probably safe to proceed. Although one of us waited 20 months for a telescope, it arrived only a few weeks past the originally stated deadline. Long waits are not in themselves cause for alarm, but repeated broken promises are.

## ORDERING BY PHONE OR VIA THE INTERNET

The major astronomy magazines have dozens of advertisements from dealers selling nationally by mail, phone and the Internet. Some, such as Astronomics and Orion Telescopes and Binoculars, market only telescopes, binoculars and accessories. Other outlets, such as Adorama and Focus Camera, are deep-discount warehouses that retail mostly cameras, VCRs and other consumer goods but also sell telescopes, often at appealing prices. The savings offered by out-of-town dealers may be attractive, but read the fine print about shipping costs, so-called crating charges and "restocking" charges if something is returned. Also consider the differences in personal service between local and mail-order dealers.

If the price differential or a lack of local outlets makes a mail-order dealer your choice, try to find someone who has dealt with that company. Discussion groups such as sci.astro.amateur on the web are filled with comments about users' dealer experiences. Ordering via the web is a well-established and reliable procedure. But if you are unsure about a company, call and talk to a real person. Telephone etiquette is an important tip-off. Any sign of impatience or lack of product knowledge by the salesperson at the other end of the line should be your signal to try elsewhere.

## PURCHASE FROM A LOCAL DEALER

The safest and usually the most convenient method of buying a telescope is from a dealer within driving distance of your home. You can get expert advice, see what you are getting before you pay and load the goods into your car and drive away.

If a local dealer does not stock the item you want, chances are, it can be ordered for you. We generally recommend this option over ordering it yourself direct from the manufacturer, even if it costs a few dollars more. Why? Consider this: If the telescope subsequently does not perform as advertised, it can be returned to the store where it was purchased. That's a tremendous convenience compared with packing up a telescope for reshipment back to the manufacturer, often at your own expense, then trying to deal with the manufacturer's customer-service personnel by phone.

# CHAPTER
# FOUR

# Essential Accessories: Eyepie

Eyepieces and filters
have improved
markedly since the
1980s, providing
today's backyard
astronomer with a
bewildering but
wonderful array of
accessories that, if
well chosen, can
enhance the perfor-
mance of any tele-
scope, large or small.

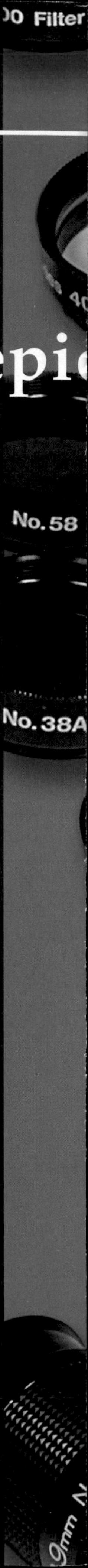

Since the 1970s, both of us have taught introductory courses for recreational astronomers. During each course, one or two class members come forward after the first or second session to ask why they are having trouble using their telescopes. Invariably, the instruments are the standard "450-power" beginner's model received as a Christmas or birthday present—the same type we unwittingly purchased as our own first telescopes.

Apart from their rickety mounts and vague instruction manuals, these telescopes (usually 60mm refractors) are notorious for their poor-quality eyepieces, filters and Barlow lenses. Typically, only one eyepiece of the two or three included —the one offering the lowest magnification—is usable. Our suggestion to the disappointed owner is to toss the other eyepieces in the trash, use the low-power eyepiece and forget the filters and Barlow lens. Those

items are added simply to give the appearance of fancy accessories. Upgrading to a better class of eyepieces is the best improvement a new owner can make to a starter scope.

Even owners of more expensive telescopes need to buy eyepieces, as upscale scopes rarely come with more than one eyepiece. Just as photographers soon add wide-angle and telephoto lenses to their camera bags, astronomers add eyepieces to enable their telescopes to show more. A well-chosen set of filters can also enhance the view through any telescope.

Together, eyepieces and filters represent the first and foremost accessories every telescope owner needs to consider. In addition, top-quality eyepieces tend to hold at least two-thirds of their value when sold as used equipment. This chapter is a guide to these essential accessories. We either own or have used virtually everything mentioned on the following pages.

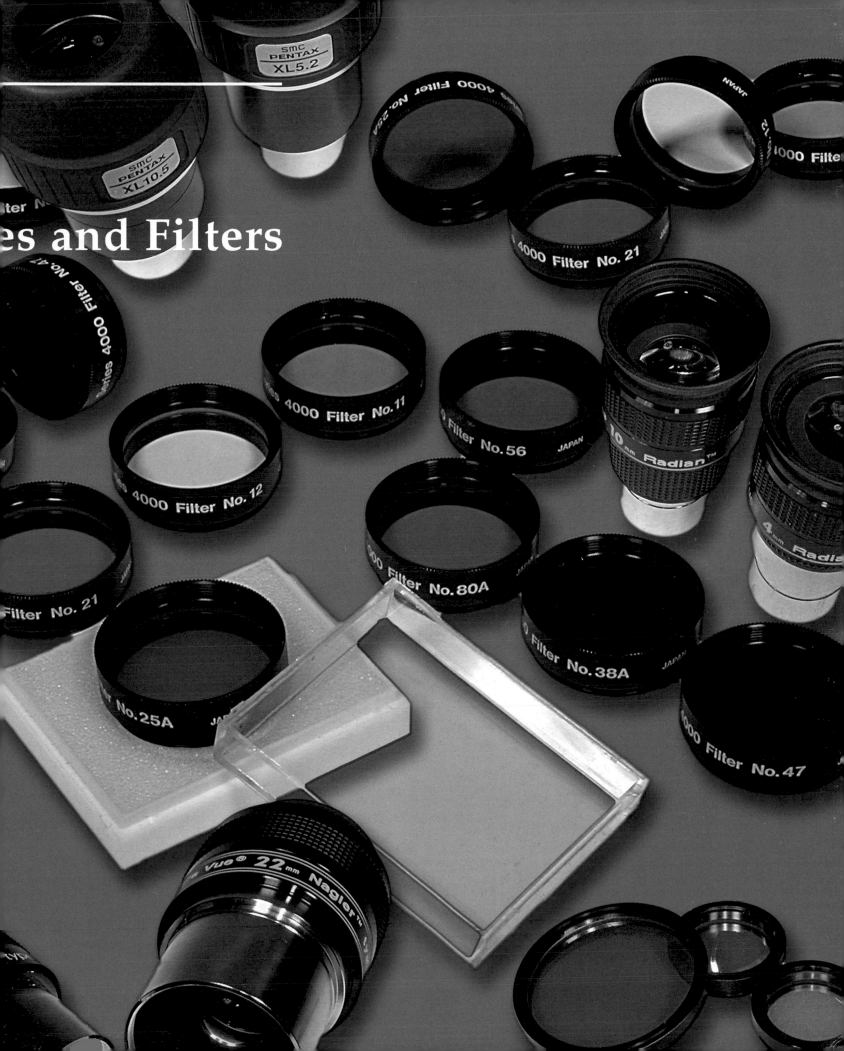

es and Filters

# Eyepieces

High-quality eyepieces, sometimes known as oculars, are as essential to sharp views as is a good primary mirror or objective lens. The telescope's main mirror or lens gathers the light and forms the image. The eyepiece

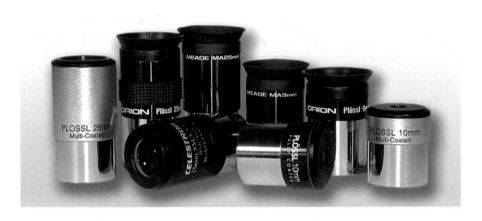

magnifies that image. Poor optics at either end of the telescope result in less-than-optimum performance.

On all astronomical telescopes, the eyepieces are interchangeable. Switching them is how you change magnifications. Selecting a set of eyepieces best for your telescope and budget requires understanding the merits of various eyepiece designs. However, the most important specification of any eyepiece is simply its focal length.

## FOCAL LENGTH

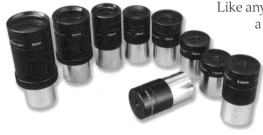

Like any lens or mirror, an eyepiece has a focal length, indicated in millimeters and marked on the top or side of the unit. A long focal length (55mm to 28mm) provides low power and shows a large region of sky. A medium focal length (26mm to 13mm) offers medium power and takes in a smaller area of sky (typically less than a Moon diameter). A short focal length (12mm to 3mm) produces high power and shows only a tiny region of sky.

As tempting as it might be, it is not necessary to collect an entire series of eyepieces, from the shortest to the longest focal lengths. A starter set of three (one from each eyepiece group) will provide enough magnifications to handle most astronomical targets, from large, faint nebulas to small, bright planets.

## FIELD OF VIEW

How much sky is seen through the eyepiece depends on the magnification it provides and on its apparent field of view. The apparent field of view depends on the optical design of the eyepiece. If you hold an eyepiece up to the light and look through it, you will see a circle of light. The apparent diameter of that circle (measured in degrees) is the eyepiece's apparent field of view, a figure usually given in the manufacturer's specifications.

Standard eyepieces, such as Orthoscopics and Plössls, have apparent fields of view

**Eyepiece Quality ▲**
Many manufacturers now supply at least one decent-quality eyepiece with their beginner telescopes. The units above are good examples. However, some telescopes still come with the same poor eyepieces that have plagued beginner scopes for decades.

**A Full Set ▲**
Eyepieces are available in a wide range of focal lengths. The shorter the focal length, the higher the power that eyepiece provides, a case of less is more.

# Calculating Power

To determine the magnification of a given eyepiece, divide the telescope's focal length in millimeters by the eyepiece's focal length. For an 8-inch Schmidt-Cassegrain telescope, for instance, with a 2,000mm focal length:

- a 40mm eyepiece yields $2000 \div 40 = 50x$ (low power)
- a 20mm eyepiece yields $2{,}000 \div 20 = 100x$ (medium power)
- a 10mm eyepiece yields $2{,}000 \div 10 = 200x$ (high power)

Those same three eyepieces on a telescope with a 1,000mm focal length would yield only 25x, 50x and 100x, respectively. As you can see, the power an eyepiece provides depends not only on its own focal length but also on the focal length of the scope with which it is used. That's why no astronomical eyepiece is marked with a magnification.

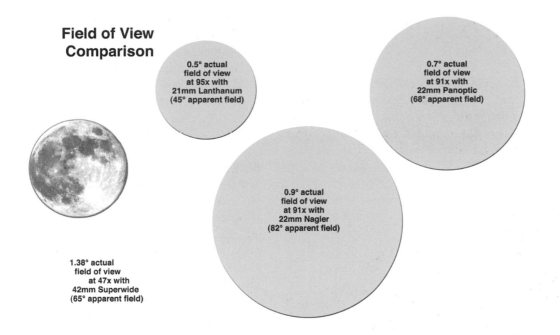

**Field of View Comparison**

0.5° actual field of view at 95x with 21mm Lanthanum (45° apparent field)

0.7° actual field of view at 91x with 22mm Panoptic (68° apparent field)

0.9° actual field of view at 91x with 22mm Nagler (82° apparent field)

1.38° actual field of view at 47x with 42mm Superwide (65° apparent field)

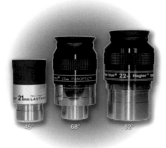

◀ **Normal, Wide and Wider** How much sky an eyepiece shows depends on its apparent field of view and its magnification with a particular telescope. In this illustration, several eyepiece fields are compared with the Moon's diameter as seen through a telescope of 2,000mm focal length. The three eyepieces— a 21mm Lanthanum, a 22mm Panoptic and a 22mm Nagler—have nearly the same focal length and yield almost identical magnifications (91x to 95x on this scope). Yet these three eyepieces show different amounts of sky. The standard-field (45-degree) eyepiece just takes in the whole Moon; the Panoptic (68 degrees) shows the whole Moon plus surrounding sky; and the Nagler (82 degrees) reveals an even wider field. The difference is in the eyepiece design—wide-angle eyepieces show more sky at a given power. Finally, the largest circle displays the maximum field that is possible on this telescope using a 2-inch-barrel eyepiece (a 42mm wide-angle model).

of 45 to 55 degrees. Wide-angle eyepieces have 60-to-70-degree fields. Extreme wide-angle eyepieces, such as Tele Vue Naglers and Meade Ultra Wides, have 82-to-84-degree fields.

To find the actual or true field of view (in degrees) that an eyepiece gives on your telescope, divide the eyepiece's apparent field by its magnification. For example, take a 20mm Plössl eyepiece with a 50-degree apparent field. On an 8-inch f/10 Schmidt-Cassegrain scope, it yields 100x. At that power, its actual field is about half a degree (50 degrees ÷ 100 = 0.5 degree), just wide enough to show the entire disk of the Moon. A typical 20mm wide-angle eyepiece (with a 65-degree apparent field) provides the same power but shows more sky, about 0.65 degree, enough to reveal black sky around the Moon.

For purists, another formula provides a more precise measure of actual field of view for wide-field eyepieces: (eyepiece field-stop diameter ÷ telescope focal length) x 57.3 degrees. The problem with using this formula is that almost no manufacturers, other than Tele Vue, supply field-stop-diameter specs for their eyepieces. An alternative and empirical method to determine actual field of view is to measure the time it takes for a star to drift across the eyepiece field. A star at the celestial equator takes four minutes to travel one degree.

Since they show a larger area of sky, eyepieces with wide fields are generally preferred for deep-sky observing. However, because of an aberration in the eyepiece optics called astigmatism, stars toward the edge of wide-angle fields can be distorted into lines or V-shaped blobs. Lateral color further spreads stars into little yellow-blue rainbows. The best (and usually the most costly) wide-angle eyepieces reduce these aberrations to a minimum, but none eliminate them entirely.

Although wide fields can provide exciting panoramic views of the Moon, they are not needed for planetary observing. Good views of the planets require high contrast and freedom from ghost images, characteristics compromised in some multielement, wide-angle eyepieces.

## BARREL DIAMETER

The lenses which constitute an eyepiece are mounted in a barrel that slips into the focuser or into the star diagonal of the telescope. Over the past century, eyepieces have evolved into three standard barrel diameters: 0.965-inch, 1.25-inch and 2-inch. The 1.25-inch size is by far the most common. Only the lowest-cost import telescopes still use the 0.965-inch standard. The selection and quality of eyepieces in this smaller barrel diameter are very poor. Indeed, the inclusion of eyepieces of this diameter as standard equipment with a telescope is a

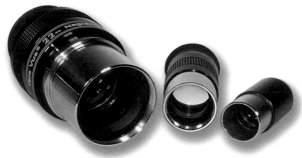

**Three Barrel Diameters ▶**
Eyepieces come in three barrel sizes: 2-inch, 1.25-inch and 0.965-inch. Today, all serious telescopes have standard 1.25-inch focusers and eyepieces. The smaller 0.965-inch barrel is now found only on the lowest-cost telescopes and should be avoided. The 2-inch size requires a telescope equipped with a 2-inch focuser and is generally reserved for low-power 22mm-to-55mm eyepieces.

**Dual-Barrel Convenience ▶**
Some premium eyepieces, like the trio at far right, are really 1.25-inch units whose barrels are equipped with outer sleeves that allow them to slip into 2-inch focusers, a convenient feature to minimize fumbling with adapter rings.

**Colorful Coatings ▶**
Fully multicoated lenses, critical in eyepieces with five or more elements, appear greenish, although some have a purple hue.

**Long Eye Relief ▶**
Eyepieces designed for long eye relief, such as the Orion Lanthanums (shown here) or the Pentax XL, Tele Vue Radian and Vixen LV models, have adjustable eyecups or extensions for comfortable viewing with or without eyeglasses.

sure sign of an inferior-quality instrument.

Achieving both long focal length and wide field in an eyepiece requires expanding the barrel diameter, hence the 2-inch eyepiece. On an 8-inch Schmidt-Cassegrain, for example, a 55mm Plössl or a

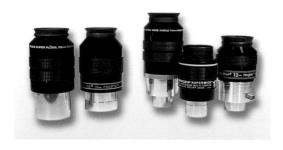

40mm König (both with 2-inch barrels) yields an actual field of view of about 1.3 degrees. But with 1.25-inch eyepieces, the widest field possible is about 0.8 degree. Many top-of-the-line telescopes are routinely equipped with focusers for 2-inch eyepieces. A step-down ring also allows 1.25-inch eyepieces to be used.

## EYE RELIEF

The distance the eye must be from the eyepiece in order to view the whole field is called the eyepiece's eye relief, an amount that depends on the eyepiece design. With most eyepieces, the higher the power, the shorter (and therefore less comfortable) the eye relief. Most Plössl eyepieces, for example, have eye-relief values of about 70 percent of their

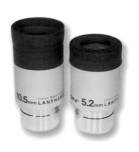

focal lengths. A typical 17mm Plössl has an eye relief of about 13mm, a comfortable value. However, 4mm-to-6mm eyepieces have eye reliefs of only a few millimeters, almost forcing you to touch your eye to the eyepiece to peer through it.

While a longer eye relief is usually desirable, some 30mm-to-55mm eyepieces have eye reliefs so high (more than 20mm) that it is difficult to position your eye for a proper view. However, eye relief is measured from the eyepiece's lens surface and can be reduced by extending the eyepiece housing or by adding a rubber eyecup.

While long eye relief allows the observer to wear glasses when viewing, only people with significant astigmatism need to keep their glasses on while observing. A quick refocus of the telescope corrects for normal near- or farsightedness. Observers who need their glasses, or prefer to keep them on, should select eyepieces with at least 15mm of eye relief. Anything less, and you won't see the full eyepiece field.

## COATINGS

Like camera lenses, all modern eyepiece lenses are coated to improve light transmission and to reduce flare and ghost images. The minimum coating is a single layer of magnesium fluoride applied to the eyepiece's two exterior lens surfaces, giving them a bluish tint. Top-quality eyepieces are multicoated, a complex process in which several layers of coating material are applied to the lenses to improve light transmission and contrast still further. In eyepieces

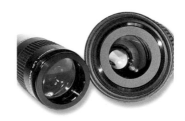

labeled as "fully multicoated," every air-to-glass surface is treated, a costly procedure limited to the finest oculars.

## MECHANICAL FEATURES

Some eyepiece brands are parfocal, which means that every eyepiece in the series focuses at the same point. Provided you stay within that manufacturer's series, switching eyepieces does not require significant refocusing, a convenient feature.

Ideally, the inside fittings should be blackened as a precaution against lens flares from bright objects outside the field. Better eyepieces also have a cleanly machined field-stop ring inside the barrel for defining the edge of the field. Low-cost eyepieces lack such a ring (yielding an ill-defined field edge) or have field stops with rough edges marred by metal burrs. On Orthoscopics and Plössls, the field stop is the metal ring just inside the bottom of the eyepiece in front of the first lens. On Naglers and many wide-field eyepieces, the field stop is inside the first lens near the bottom.

Many eyepieces are now supplied with rubber eyecups, which are good for blocking stray light and for keeping the eye prop-

# Upgrading to Larger Eyepieces

Upgrading a telescope to accept larger-diameter eyepieces can be as simple as adding the right accessory.

### Switching from 0.965-inch to 1.25-inch

Anyone with a telescope employing 0.965-inch eyepieces might consider switching to the larger 1.25-inch standard. For a refractor, this can be done in two ways. A step-up adapter tube that accepts 1.25-inch eyepieces can be inserted into the eyepiece holder. Alternatively, a hybrid star-diagonal prism that fits a 0.965-inch eyepiece holder at one end and accepts a 1.25-inch eyepiece at the other can be attached.

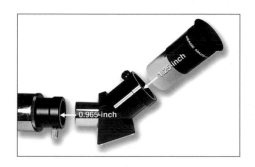

Neither a hybrid star diagonal nor an adapter tube works for a Newtonian with a 0.965-inch focuser, since both place the eyepiece too far from the tube to focus. Instead, the entire focuser on the side of the tube must be replaced with one that accepts 1.25-inch eyepieces. Check with local telescope dealers for the appropriate adapters and advice, or preferably, avoid buying an instrument with a 0.965-inch focuser in the first place, since it is definitely a liability.

But consider this: In many cases, an upgrade to 1.25-inch eyepieces costs as much as or more than the original purchase price of the telescope itself. Maybe selling the scope and starting over with a better, properly furnished instrument might be the best plan.

### Switching from 1.25-inch to 2-inch

To move up to the 2-inch world on a Schmidt-Cassegrain requires only an upgrade to a 2-inch star diagonal or visual back equipped with a lock ring that screws onto the back of the instrument in place of the standard 1.25-inch visual back.

Upgrading a refractor to accept 2-inch eyepieces may be impossible, while upgrading a Newtonian requires swapping the focuser for a 2-inch model. New focusers can be purchased from Meade, JMI and other specialty suppliers of Newtonian telescope parts. The best focusers for an upgrade are the low-profile type, which place the eyepieces close to the focal point of the telescope—a tall 2-inch focuser may place some eyepieces too far above the tube to reach focus.

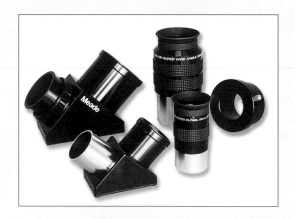

*A "hybrid star diagonal" allows 1.25-inch eyepieces to work on small refractors that accept only 0.965-inch accessories. Adapters to accept 2-inch accessories on Schmidt-Cassegrains are manufactured by Celestron, Meade, Tele Vue and Orion. A 1.25-inch adapter ring allows additional use of smaller eyepieces.*

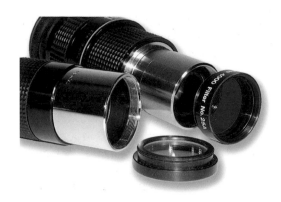

**Barrel Threads ▶**
Eyepiece barrels should be internally threaded for filters. Most 1.25- and 2-inch eyepieces have a standard thread that allows many kinds of filters to be screwed into the barrel. A few older eyepiece brands available second-hand have no threads.

**Classic Orthoscopics ▶**
The original legendary Zeiss Abbé Orthoscopics (shown here) had 0.965-inch barrels. Highly regarded low-cost Orthoscopics in 1.25-inch barrels are available from University Optics.

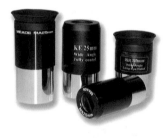

**Entry-Level Eyepieces ▲**
These entry-level Kellner-class eyepieces are good performers and are economical.

**Modern Plössl Roots**
The French-made Clavé and Tele Vue's original Plössls (two at far right) started the trend to this design and remain among the best available. Today, Plössls are the most popular eyepieces, available in a range of prices starting at $50, and are invariably made in Taiwan or China. Plössls work best in focal lengths of between 32mm and 15mm, where eye relief is good and aberrations are low. ▶

erly centered. In the best eyepieces, the eyecups are adjustable and integrated into the design, rather than being loose-fitting add-ons that can fall off and get lost.

## STANDARD-FIELD EYEPIECES ($30 and up)

An eyepiece design utilizes a particular combination of lenses (called elements) of a specific shape. The design determines the field of view and eye relief. With a few exceptions, manufacturers do not have exclusive license to a design. Plössl eyepieces, for example, are sold by nearly every telescope manufacturer. Many eyepiece designs are named for the pioneering telescope designers who invented them.

❖ Kellner and Modified
    Achromat ($30 to $50)

Since its invention by Carl Kellner in 1849, this was the standard workhorse eyepiece design until the 1980s. An economical three-element type, the Kellner produces average images in a fairly narrow field of view by today's standards—typically, 40 degrees. It works best on long-focal-ratio (f/10 or longer) telescopes and suffers from some chromatic aberration. Several manufacturers sell a variation called the Modified Achromat (MA), often included with entry-level scopes. Edmund Scientific has a version called the RKE (for David Rank modified Kellner for Edmund), in which the lens elements are reversed from the standard arrangement, giving it a wider field (45 degrees). Although not a true Kellner, the RKE is an improvement on the old design.

❖ Orthoscopic
    ($60 to $500)

In 1880, Ernst Abbé, a Zeiss optical designer, invented a four-element eyepiece with a 45-degree apparent field and less chromatic aberration and ghost imaging than a Kellner. The classic Abbé Orthoscopic is still considered by many amateur astronomers to be the best eyepiece for planetary observing.

Once the prime choice in premium eyepieces, Orthoscopics are now hard to find, having been largely supplanted by Plössls and their vari-ations. Takahashi of Japan sells a series of high-quality Orthos ($150 to $250) with 0.965-inch barrels, an odd holdover from post-

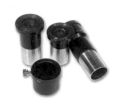

war days, when this barrel size, introduced by Zeiss in Germany, was the standard. Aficionados consider the current Zeiss series of Orthos (now with 1.25-inch barrels) the ultimate in high-contrast planetary eyepieces, though at premium prices ($200 to $500) and with limited availability.

❖ Plössls and Variations
    ($50 to $200)

This design, devised in 1860 by G.S. Plössl, enjoyed a resurgence in the 1980s as the result of intensive marketing by telescope companies stressing the advantages of the Plössl—not the least of which is that it is easier to manufacture than an Orthoscopic.

A true Plössl is a four-element design consisting of two nearly identical pairs of lenses. Compared with Orthoscopics, the Plössl has a slightly wider field (about 50 degrees), works better on f/6 or faster telescopes but has a shorter eye relief. The best Plössls are excellent for all observing tasks, particularly planetary viewing, although eye relief is poor in 13mm and shorter versions.

Some manufacturers market five-to-seven-element Plössl variations carrying various trade names. Examples are the superb and nearly identical Celestron Ultima, Orion Ultra-

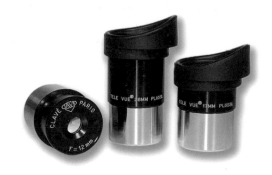

scopic and Baader Eudiascopic Series. Although closer in optical configuration to Wide-Field Erfles, they perform much like traditional standard-field Plössls. The highly regarded Vernonscope Brandon eyepieces ($200), designed by Chester Brandon in the 1940s, are another form of Plössl, with two unequal pairs of lenses. Takahashi's LE Series ($170 to $190 for 1.25-inch models) also falls into this category, though its "long eye relief" designation is a bit misleading—the shortest 12.5mm, 7.5mm and 5mm LE models have only 9mm of eye relief. Still, planet watchers praise Takahashi LEs for their contrast and clarity.

❖ Zoom Eyepieces
($180 to $400)

Why buy three or four eyepieces if one will do it all? That is the promise of zoom eyepieces, which use a sliding Barlow lens to vary their effective focal length. With improved glass types and coatings, today's best zooms offer superior performance to the cheap units of yesteryear, haunted as they were with ghost images. Commonly available with a focal-length range of 8mm to 24mm, zoom eyepieces ($180 to $220) afford a good range of magnifications. Unfortunately, their apparent field of view is restricted, narrowing to a tunnel-vision view of only 35 to 40 degrees at the lowest power. Inconveniently, most zoom eyepieces also require refocusing as you change powers.

Zoom eyepieces are great for public viewing sessions but have failed to win us over for serious observing—we prefer the higher quality and/or wider fields of view of separate fixed-focal-length eyepieces.

One premium zoom eyepiece that deserves mention is Tele Vue's $380 3mm-to-6mm Nagler Zoom. This is a

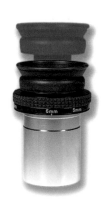

◀ **Optimum High Power**
An exceptional zoom eyepiece is the 3mm-to-6mm Nagler Zoom, a specialized high-power eyepiece for dialing in the optimum power for planetary viewing with short-focal-length telescopes, such as fast apo refractors. Despite its Nagler designation, the apparent field is a normal 50 degrees.

# Designs to Avoid

Older amateur-astronomy handbooks (particularly those written by British authors) regale the reader with references to exotic eyepiece variations. Some, such as the Hastings, Monocentric, Steinheil and Tolles, are praised by planetary observers but are so rare that few amateurs will ever encounter them. Others, like the Huygenian and Ramsden, are so poor, they deserve mention so that they can be avoided.

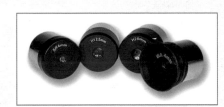

**Huygenian**

This ancient two-element design from the 17th century is commonly supplied with poor-quality telescopes. These eyepieces are marked with an H, AH (for Achromatic Huygenian) or HM (for Huygens Mittenzwey).

**Ramsden**

This primitive two-element design is of little value. Variations add more lens elements to produce an Achromatic Ramsden (AR) or a Symmetrical Ramsden (SR). All are junk, with fuzzy tunnel-vision views and little eye relief.

**War-Surplus Erfles**

Invented during World War II, the Erfle is a five-element, wide-angle design with a 60-degree apparent field. Ghost images from internal reflections make most models unsuitable for planetary observing, while astigmatism greatly distorts stars at the edge of the field. In longer focal lengths, however, Erfles can be decent, low-power, deep-sky eyepieces. Nevertheless, older war-surplus models (often filling swap tables at star parties) and new low-cost Erfles have been surpassed in all respects by modern wide-field designs.

 *These eyepieces give low-cost telescopes a bad name. They suffer from image-blurring aberrations, tunnel-like fields of view and lack of eye relief. Use the Huygenians and Ramsdens for dustcaps or solar projection.*

specialized eyepiece for high-powered planetary observing with apo refractors of 70mm aperture and larger. Imaging is superb, and—a first—the apparent field stays constant throughout the zoom. Click stops define the 6mm, 5mm, 4mm and 3mm positions, though the observer can select any value in between. Its 10mm eye relief is a bit cramped for those who wear their eyeglasses while observing.

## WIDE-FIELD EYEPIECES ($80 to $400)

In the 1980s, Tele Vue introduced its Wide-Field brand of eyepieces in focal lengths from 40mm to 15mm and with apparent fields of 65 degrees. The six-element design bore a family resemblance to the Erfle but with less ghosting and better imagery across the entire field, especially with fast focal-ratio telescopes. The Wide-Fields set the stage for a plethora of new eyepiece designs, although they have since been discontinued in favor of Tele Vue's even better Panoptics.

After Tele Vue's success with Wide-Fields, Meade came out with its Super Wide Angle six-element eyepieces, with 67-degree fields. Still offered in 13.8mm, 18mm, 24.5mm, 32mm and 40mm focal lengths, the Super Wides ($140 to $300) provide fine images, though softer toward the edge of the field than Panoptics.

In 1991, Tele Vue introduced its Panoptic Series (the name derives from "panorama optic"), now with 15mm, 19mm, 22mm, 24mm, 27mm and 35mm models ($210 to $365). What distinguishes the six-element Panoptics is their superior correction of aberrations over the entire 68-degree field. On well-corrected telescopes, stars are pinpoints almost to the edge, making Panop-

# Our Favorite Eyepieces I: Standard-Field Models

For a fine set of Plössls, we use models from either the Tele Vue or the Meade Series 4000 line. Both are top-class Plössls suitable for all types of viewing. At $80 to $100 each, however, a set of three or four eyepieces can cost as much as many beginner scopes.

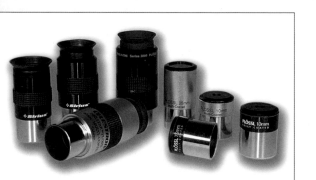

Brandon eyepieces are fine performers—we've both owned sets of them—and were once in the same price league ($80 each). But at their current $200 price point, we find it hard to justify their purchase over other top-end Plössls.

Nor have we felt a compelling need to spend $200 to $500 each for new Zeiss Orthoscopics, despite the collectible value of these prestigious eyepieces. We've tested older 0.965-inch Zeiss Orthos against modern top-of-the-line Plössls and found no performance edge for the legendary Zeiss oculars.

For general-purpose viewing, one of us (AD) recommends the Orion silver-barreled Lanthanum standard-field eyepieces ($110 to $120 each), preferred over the similar Celestron/Vixen models. The Orion's 20mm eye relief and flexible rubber eyecups make these eyepieces a pleasure to use. They provide crisp images in a compact, light package. For owners on a budget, we suggest Edmund Scientific's RKE line for its combination of quality and low price ($40 each). The Orion Sirius line and the Meade Series 3000 budget Plössls also provide excellent value at $50 per eyepiece. At that price, there's little reason to settle for anything less.

*Above: Generic Plössls, such as those included with many Chinese-made import scopes (right) or budget lines like Orion's Sirius Plössls (left), Meade's Series 3000 Plössls and Celestron's NexStar Series (center), provide good performance at an affordable price. We recommend any of these over Kellners and Modified Achromats because of their wider fields.*

tics the standard of comparison for Wide-Field eyepieces. Their drawback is some pincushion distortion, an aberration that gives the field the appearance of being projected onto the inside of a bowl. This warping of the field can be a little disturbing when you are panning across the sky.

Another eyepiece design similar to the Tele Vue Wide-Field and the Meade Super Wide is the König (König is a design, not a brand). This four- and five-element design provides a 60-to-70-degree field with fine edge performance. In the 1970s, these eyepieces were considered to be the best available. Today, with prices in the $80-to-$200 range, they represent outstanding values.

Other manufacturers followed suit with their own wide-angle-eyepiece designs, often carrying proprietary brand names, such as Celestron's Axiom Series (in focal lengths of 40mm to 15mm), Orion's Lanthanum Superwides (42mm to 3.5mm) and Pentax XLs (40mm to 5.2mm). The latter two series also fall into the category of long-eye-relief eyepieces.

## LONG-EYE-RELIEF EYEPIECES ($120 to $250)

An aging population of amateur astronomers has created a demand for eyepieces that are easy to look through, which translates into long eye relief and large eye lenses that don't require straining to see the whole field. Most eyepieces of all types in the 35mm-to-19mm focal-length range have excellent eye relief in any case. But finding comfortable viewing in shorter-focal-length eyepieces requires purchasing premium models designed specifically to provide long eye relief. Most achieve their short focal lengths and long eye relief by means of integrated Barlow lenses.

In the category of Plössl-class eyepieces,

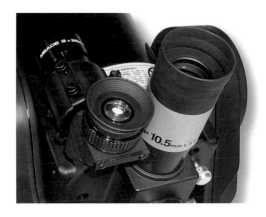

the Celestron/Vixen Lanthanum Series (20mm to 2.5mm) and the nearly identical Orion Lanthanum set (25mm to 2.3mm) offer a consistent 20mm of eye relief with standard fields of view of 45 to 50 degrees and no ghost images. On-axis images in the shortest-focal-length Lanthanum models tend to be a little softer than with conventional Plössl or Orthoscopic eyepieces, but these are comfortable eyepieces to look through, even when wearing eyeglasses.

A notch up in price and performance is Tele Vue's Radian Series, offered in 18mm-to-3mm focal lengths ($230 each). These six-element eyepieces have a uniform 20mm of eye relief with tack-sharp, high-contrast images across a 60-degree apparent field. Although this puts them just into the wide-angle-eyepiece class, the forte of Radians is high-power planetary viewing—they can compete against the best Orthos and Plössls. Focal lengths longer than 18mm (ideal for deep-sky viewing) are not possible with the existing Radian design.

Among the best values in quality wide-angle eyepieces are the Orion Lanthanum Superwides (42mm to 3.5mm; $220 to $300 each), made by Vixen of Japan. Each features 20mm of eye relief—an added bonus. Edge performance is very good, marred mostly by minor lateral color. All but the 42mm giant (2-inch only) have dual-sized barrels to fit both 2-inch and 1.25-inch focusers.

The more costly Pentax XL eyepieces (in focal lengths of 40mm to 5.2mm; $240 to $300) also provide a comfortable 20mm of eye relief and edge performance approaching pinpoint perfection in their 65 degree fields. (The 28mm model is barrel-limited to a 55-degree field.) Anyone seek-

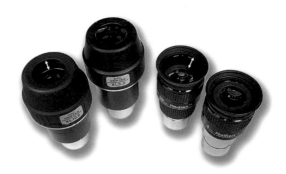

◀ **Lanthanum Advantage**
Half the price of other long-eye-relief models, the Vixen and Orion Lanthanum eyepieces ($110 to $120 each) are ideal on small scopes that can't handle hefty Wide-Field eyepieces, providing good eye relief and a tall eyepiece for comfortable viewing, such as with this Meade ETX-90 telescope. (With stubby eyepieces, your nose hits the scope!)

**Ahh...Eye Relief**
Pentax XLs (left pair) feature an adjustable eyecup that screws up and down as shown to place the eye at the right distance for optimum viewing. Tele Vue's sliding Instadjust eyecup, found on its Radians (right pair) and late-model Naglers, performs a similar function. Getting the Tele Vue eyecup positioned just right is important for minimizing blackout from kidney-bean darkening of the field (present on some Naglers and on most Radians). The Pentax design feels more comfortable on the eye socket and doesn't keep slipping out of position like ◀ the Tele Vue system.

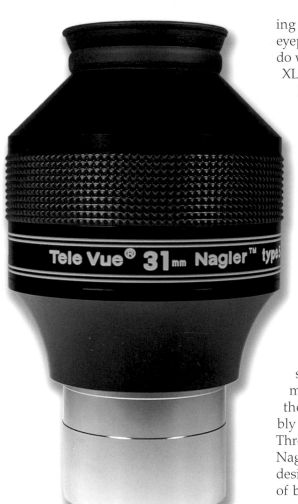

ing a matching set of the finest eyepieces money can buy would do well with a collection of Pentax XLs. Their principal disadvantage is that with the exception of the 40mm model, they fit only 1.25-inch focusers, despite the appearance of having 2-inch barrels. Changing between them and 2-inch eyepieces means swapping adapter rings in and out of the focuser, then refocusing over a wide range.

## NAGLER-CLASS EYEPIECES ($180 to $600)

Talk eyepieces with any seasoned amateur astronomer for more than a few minutes, and the name Al Nagler will inevitably come up in the conversation. Through his Tele Vue products, Nagler has revolutionized eyepiece design and raised the expectations of backyard astronomers with respect to eyepiece performance.

The flagship of the Tele Vue line is the series of eyepieces that bears Nagler's name. The original 13mm Nagler eyepiece caused a sensation when it was introduced in 1982. Other focal lengths soon followed, from 11mm to 4.8mm, all with seven steeply curved lens elements. Nagler eyepieces, sold exclusively by Tele Vue, were an instant hit, despite costing two to four times more than the best eyepieces of the day.

To create his revolutionary eyepiece, Nagler took an exotic, extremely wide-angle eyepiece design and placed a Barlow lens in front of it. The eyepiece and Barlow operate as a single unit—the aberrations of one cancel out the aberrations of the other—producing exquisitely sharp images edge to edge over a field of view wider than any other eyepiece commonly available at the time.

Other manufacturers were as impressed as the amateur astronomers using Nagler's new oculars. By 1985, Meade had introduced its competitive line, the eight-ele-

ment Ultra Wide Angle Series in 14mm, 8.8mm, 6.7mm and 4.7mm sizes, near clones of the original Nagler Type 1 eyepieces. Though unchanged since their introduction (as of 2002), the Ultra Wides remain a superb choice in the Nagler category of premium eyepieces.

The disadvantage of this class of eyepieces, besides price, is weight. At about 1.5 pounds each, the 13mm Nagler (now discontinued), the 22mm Nagler Type 4 and the Meade 14mm Ultra Wide are heavy, making them unsuitable for lightweight

**Biggest Eyepiece** ▲

The ultimate premium eyepiece is the six-element 31mm Nagler Type 5. Its long focal length coupled with a wide 82-degree apparent field produce a panoramic picture-window view unmatched by any other eyepiece. As attractive as this sounds, the giant eyepiece (shown here 80% of its actual size) has two distinctions that may deter backyard astronomers: It is the heaviest (2.2 pounds) commercially available eyepiece ever produced and is among the most expensive ($600). However, in its focal length, it is in a class by itself.

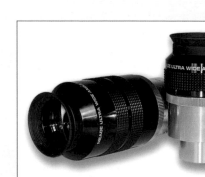

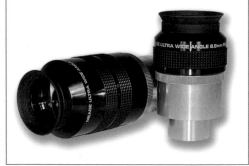

scopes. The original 13mm Nagler also had an unusual idiosyncrasy called the kidney-bean effect. If the eye was not properly centered over the eyepiece and the sky was not dark, an elliptical shadow appeared at the edge of the field and squirmed around as the eye moved.

In 1987, Tele Vue brought out a series of eight-element Type 2 Naglers that eliminated the kidney-bean effect, making it possible to use these eyepieces in the daytime. However, Type 2 Naglers (made in longer 20mm, 16mm and 12mm focal lengths) had

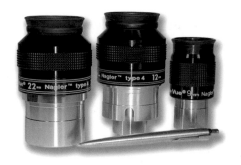

less eye relief than the originals. Now discontinued, Type 2 Naglers remain in demand on the used-equipment market.

# Our Favorite Eyepieces II: Wide-Field Deep-Sky

One of the most-used eyepieces in both our collections is Tele Vue's 22mm Panoptic. Its wide, flat field and its ability to fit both 2-inch and 1.25-inch focusers make it a favorite for all our telescopes. For higher-power, deep-sky views, venerable 14mm and 8.8mm Meade Ultra Wides are still excellent performers that see a lot of starlight.

Lately, we have both become fans of the Orion Lanthanum Superwide Series. They provide near-Panoptic performance at a reasonable price for a premium eyepiece with long eye relief. Their consistent dual-barrel design makes them convenient to use. In the lower midpower range (15mm to 22mm), two of our favorites are the compact Tele Vue 19mm Panoptic and the 16mm Nagler Type 5, with nearly identical actual fields of view.

From firsthand experience, author Dyer also recommends the Pentax XL Series of Wide-Field eyepieces. The 10.5mm and 5.2mm models provide razor-sharp, high-contrast views of both deep-sky objects and planets. "If I was restricted to owning one line of eyepieces," says Dyer, "it would be the Pentax XLs—though I'd still beg for Tele Vue Panoptics in the longer [22mm and greater] focal lengths!"

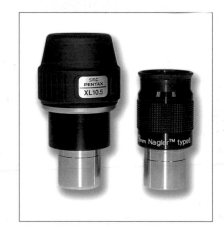

Author Dickinson says the Tele Vue 27mm Panoptic is among his standout eyepieces: "I notice this eyepiece is often overlooked by observers in favor of the 35mm Panoptic, but many of my most awesome deep-sky views have been with the 27mm Panoptic. It's an ideal combination of power, wide field and great optics."

The ultrawide fields and superb eye relief of the latest Nagler eyepieces are hard to resist. The 22mm and 12mm Type 4 models provide stunning deep-sky vistas at low to moderate powers, though their edge-of-field performance is inferior to that of other Naglers, past and present. However, the 9mm Nagler Type 6 quickly became a favorite for higher-power views of galaxy fields and globular clusters.

Be warned: The 22mm Nagler Type 4 (fitting only a 2-inch focuser) requires quite a bit of in-travel to reach focus, about half an inch more than a 35mm Panoptic, which itself demands a lot of racking in to focus. On one of our scopes, a MAG 1 PortaBall reflector, the focuser will not rack in far enough to allow a 22mm Nagler to reach focus. The same may be true of other Newtonians with limited focuser travel. We prefer the 22mm Panoptic in this power range.

*Some examples of outstanding Wide-Field eyepieces: Tele Vue 22mm and 19mm Panoptics (facing page, top), Orion 17mm, 8mm and 5mm Lanthanum Superwides (facing page, center), Meade 14mm and 8.8mm Ultra Wide Angles (facing page, bottom) and Pentax 10.5mm XL and Tele Vue 9mm Nagler Type 6 (above).*

**Guru at Work** ▶
Al Nagler has introduced
more innovations in tele-
scope eyepiece design than
anyone else in the history
of amateur astronomy.

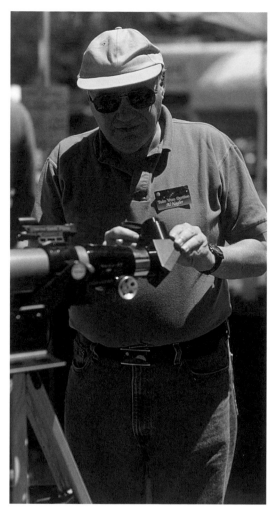

**Barlows: Power Boost** ▶
Barlows come in a wide
price range. The most
inexpensive (the long,
thin tubes shown top right)
make better doorstops
than they do lenses. The
Barlows included with the
lowest-cost import tele-
scopes fall into this category. We recommend the
top-end 1.25-inch Barlows
offered by Celestron,
Meade, Orion, Tele Vue
(bottom right) and Vernonscope, which sell for
$75 to $150 each.

In the late 1990s, Al Nagler began revamping his eyepiece lineup with new Nagler designs labeled Type 4, 5 and 6. (What the unseen Type 3 Nagler design was, let alone what lenses the Type 4 to 6 designs consist of, remains a corporate secret that in Al's words, "I could tell you all about as the cement hardens around your feet.")

A goal of the latest Type 4, 5 and 6 Naglers is to provide increased eye relief and contrast. They succeed in this. The Type 4 models, in particular, offer far greater eye relief than earlier Naglers of the same focal length, but we've found the edge-of-field performance of the Type 4s falls short of the pinpoint-edge sharpness of the old Type 1 and 2 models. Increased aberrations in the outer 25 to 40 percent of their fields distort stars into fuzzballs. The Type 5 and 6 Naglers, however, have the edge-to-edge sharpness for which these eyepiece designs are famous. The jumbo 31mm Nagler provides the most outstanding panoramic

views we've seen, while the wonderfully compact 16mm, 13mm and 9mm models have higher power and moderate eye relief.

If long eye relief is a requirement, the Pentax XL and Orion Lanthanum Super-wide Series provide excellent imaging across their entire, though smaller, fields of view. With a field as wide as a Nagler's, Meade's 14mm Ultra Wide presents a flat, sharp field with good eye relief. Since the 1980s, it has remained a favorite of ours. While Nagler eyepieces set the standard, other manufac-turers now offer tempting choices for those seeking the best eyepiece performance.

## BARLOW LENSES ($80 to $250)

Barlows are negative lenses that increase the effective focal length of a telescope and multiply the power of any eyepiece. (Meade Instruments calls Barlows "tele-negative amplifiers," eschewing the proper name that stems from Peter Barlow and George Dolland's 1834 scientific paper which first described this style of lens.) Bar-lows are available in magnifications from

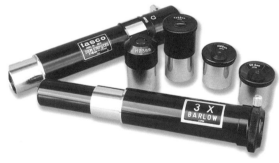

1.8x to 5x models, with 2x Barlows the most common. With a 2x Barlow, a 20mm eye-piece effectively becomes a 10mm one. Barlows can double your eyepiece set if

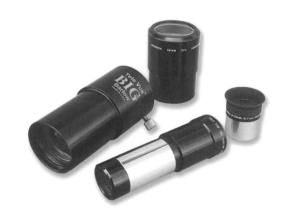

you plan carefully to avoid any unnecessary duplication of powers.

The advantage of Barlows is, they produce high power with longer-focal-length eyepieces that are easier to look through because of their better eye relief, compared with the eye-straining 8mm-to-4mm eyepieces. For owners of fast telescopes, such as f/4 and f/5 reflectors, Barlows are a recommended method of achieving high power. Their disadvantage is that they put more optics into the light path and potentially more aberrations and ghost images. Some people swear by them; others swear *at* them. Our tests show that the best multicoated Barlows introduce no detectable aberrations or light loss. Avoid Barlows with variable magnification, accomplished by a lens that slides up and down the barrel; Barlow lenses are designed to work best at one specific amplification.

Barlows come in various tube lengths, ranging from less than three inches to nearly six. Generally, the higher-amplification units are longer, but not always. It depends on

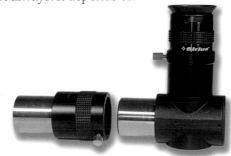

the specific lens design. Some manufacturers offer "shorty" Barlows, usually in 2x amplification, that are useful on refractors because the Barlow can be placed between the eyepiece and the diagonal to yield the normal 2x and can also be placed between the diagonal and the focuser, where 3x (approximately) is achieved—effectively two Barlows for the price of one. The Celestron

◀ **Adding Magnification**
Placing a Barlow lens ahead of a star diagonal (rather than inserting it into the star diagonal) adds approximately 50 percent to the magnification factor with a refractor. For example, a 1.8x Barlow becomes a 2.7x, a 3x becomes a 4.5x, and so on—an ideal arrangement for refractors in the 500mm-to-1,200mm focal-length range. Therefore, with one 25mm eyepiece and a 2.5x and 3x Barlow, a 1,000mm-focal-length refractor can be used at five magnifications: 40x, 100x, 120x, 150x and 180x.

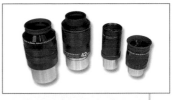

# How Low Can You Go?

A 40mm Wide-Field eyepiece (such as the Meade Super Wide, Orion Superwide 42mm, Pentax XL and University Optics MK-70 König) offers the maximum possible field of view for the 2-inch format. (Close runners-up are Tele Vue's 31mm Nagler Type 5 and 35mm Panoptic.) Advertising claims to the contrary, eyepieces with greater focal lengths do not provide wider fields of view than these 40mm designs. An eyepiece's field of view is limited by the aperture that receives the telescope light cone, not by the focal length. If the unobstructed diameter of the field lens is as wide as the inside of the 2-inch barrel, as it is with these 40mm eyepieces, the apparent field will be maximum as well.

On an 8-inch f/10 Schmidt-Cassegrain, for example, a 40mm Pentax XL produces 50x and an actual field of 1.3 degrees. On the same scope, a Tele Vue or Meade 55mm Plössl eyepiece, with its narrower apparent field of view, produces 36x but with the same 1.3-degree actual field.

Seekers of the widest, lowest-power field possible must also be aware of another limit to how low you can go. If an eyepiece yields more than a 7mm exit pupil in obstructed telescopes, such as reflectors and catadioptrics, you'll see a field with a dark hole in the center. Exceed the low-power limit with a refractor, however, and the only ill effect is that not all the light collected by the telescope enters your eye. Your eye "stops down" the telescope's aperture. Either way, there are only disadvantages to going lower than a 7mm exit pupil.

To determine the low-power limit of any telescope, multiply the telescope's focal ratio by 7mm (the diameter of the fully dark-adapted eye). For an f/4.5 reflector, for example, the longest eyepiece recommended is 31.5mm (4.5 x 7). A 40mm Wide-Field eyepiece would not work well.

*Above, the 55mm Meade Plössl (left) and the 42mm Orion Lanthanum Superwide (second from left) take in the maximum area of sky possible in a 2-inch-barrel eyepiece. For the maximum field in a 1.25-inch-barrel eyepiece, a 35mm Plössl, like the Orion Ultrascopic (second from right) and the Meade 24.5mm Super Wide Angle (right), provides nearly identical true fields of view.*

# Eyepieces: A Summary Comparison

| Type | | Apparent Field | Advantages | Disadvantages | Price |
|------|---|---------------|-----------|--------------|-------|
| Kellner | | 35° to 45° | Low cost. Good for long-focal-length scopes. | Narrow field. Chromatic aberration. | $30 to $50 |
| Orthoscopic | | 45° | Good eye relief. Freedom from ghost images and most aberrations. | Narrow field for deep-sky viewing. | $60 to $500 |
| Plössl[1] | | 50° | Excellent contrast and sharpness. Wider field than most Orthos. | Less eye relief than Orthos. Slight astigmatism at edge of field. | $50 to $200 |
| König and Erfle | | 60° to 70° | Wide field of view. Low cost for Wide-Field. | Astigmatism at edge of field. | $80 to $200 |
| Modern Wide-Field[2] | | 65° to 70° | Wide field of view. Minimal edge aberrations. | Moderate to high cost. | $140 to $400 |
| Tele Vue Radian | Ask Al Nagler! | 60° | Long eye relief with wide field and high contrast. | High cost. | $230 |
| Nagler[3] | | 82° to 84° | Extreme field of view with few edge aberrations. | Expensive. Low-power models are heavy and large. | $170 to $600 |

1. Also Meade Super Plössl, Orion Ultrascopic and Celestron Ultima
2. Includes Tele Vue Panoptic, Meade Super Wide Angle, Orion Superwides and Pentax XL
3. Also Meade Ultra Wide Angle

Ultima 2x unit is a first-rate Barlow in this shorty class. In our tests, it rated optically identical to the best "long" 2x Barlows, making us wonder why anybody makes a long 2x version. On the other hand, the couple of 3x shorty versions we are aware of did not compare favorably with the optical performance of the top longer-style 3x units.

Combined with the longer-focal-length Plössl eyepieces, a good Barlow (or two) provides a complete range of magnifications and high-quality images at modest cost. One veteran observer we know, who must use glasses at the eyepiece, employs 26mm and 20mm Plössls and one of three Barlows (1.8x, 2.5x and 3x) with an 8-inch Schmidt-Cassegrain to achieve 77x, 100x, 138x, 180x, 200x, 230x and 300x, all with good eye relief.

Some manufacturers have introduced 2-inch Barlows that double the power of giant eyepieces, such as a 40mm wide-angle or a 55mm Plössl. However, since 2-inch eyepieces are designed to yield low power, they are the ones you are least likely to want to amplify. Giant Barlows are of limited visual use, although they are excellent photographic accessories.

## EYEPIECE AND BARLOW PERFORMANCE

The purpose of a telescope is to collect light from a celestial object and bring it into focus at the eyepiece or camera lens. In effect, the parallel rays from the celestial source are forced into a converging cone by the main lens or mirror of the telescope. The higher the focal ratio, the longer and skinnier the cone. An f/15 system has a long, narrow cone; the rays at the edge are angled two degrees from those at the center. In an f/10 system, the edge rays are angled three degrees from the central rays, and in an f/5 system, they are angled six degrees. It is much easier for an eyepiece to accommodate rays at a two- or three-degree angle from the central axis than at a six-degree angle.

Try a typical eyepiece—say, a 25mm Plössl—on an f/15 refractor. The star field is perfectly sharp from edge to edge. Now use the same eyepiece on an f/5 telescope (regardless of the type); the stars toward the edge of the field are no longer perfect pinpoints but resemble seagulls or arcs, depending on the inherent aberrations of the telescope and eyepiece.

The steeper the light cone, the more difficult it is to design an eyepiece to suppress aberrations. At f/4, it is impossible. Even at f/5, the best eyepieces still display astigmatism in the field's periphery. It is here that a good Barlow can improve eyepiece performance. The negative lens increases the effective focal ratio by reducing the steepness of the light cone entering the eyepiece. A 2x Barlow reduces the cone's angle by half, so an f/5 telescope becomes, in effect, an f/10—a situation in which any eyepiece will show markedly improved performance.

## RECOMMENDED EYEPIECE SETS

Eyepieces are easy to collect. Buy one of a series, and you may long to own the whole set. In truth, five, six or seven eyepieces are not needed. A set of four will do well, as will just two or three to start. Here are a few recommendations.

❖ **Budget Set**
Many entry-level telescopes come with 25mm and 9mm Modified Achromat or economy Plössl eyepieces. These make an acceptable starter set of two eyepieces for most small scopes. For a set of three, 25mm, 12mm and 7mm Plössls provide low, medium and high powers. Or, better yet, consider 25mm and 17mm Plössls with

a 2x Barlow for a complementary range of four powers. Observers on a tight budget could select Edmund's RKE Series, Meade's Series 3000 Plössls or Orion's Sirius Plössls (nearly identical Plössls are sold by many other dealers). Most costly are Celestron's Ultimas (a Plössl variation) and units from Meade's Series 4000 and Tele Vue's Plössl line.

❖ **Expanded Budget Set**
For telescopes that accept only 1.25-inch eyepieces and already come with both a 25mm and a 10mm eyepiece, a top priority should be an ultra-low-power eyepiece. We recommend a 30mm-to-35mm Plössl. Regardless of brand, it is likely to have the

# Our Favorite Eyepieces III: Lowest-Power Panoramic

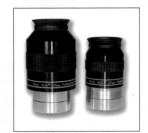

The 31mm Nagler Type 5 is arguably the best panoramic eyepiece on the market, but at a price nearly double that of the Tele Vue 35mm Panoptic. The Panoptic ($365) provides the sharpest, flattest field of any low-power eyepiece, with an apparent field only slightly smaller than that of the 31mm. Coupled with a 4-inch f/6 refractor, the Panoptic's 3.5-degree field provides stunning panoramas of Milky Way star fields.

Praised earlier, the 27mm Panoptic is a favorite 2-inch panoramic eyepiece, with slightly more power than the 35mm model. Its lighter weight (1.1 pounds versus 1.6 for the 35mm) makes it a good choice for scopes where balance is critical. The 27mm is the better choice of the two for f/4.5 or faster scopes, where eyepieces longer than 30mm are of no advantage.

The Orion 42mm Superwide shows moderate fuzzing of stars at the edges, even on f/10 telescopes, but provides the widest actual field of any eyepiece we've seen. Another fine performer is the University Optics 40mm MK-70 König. Priced at just over half the cost of a 35mm Panoptic, it's a bargain.

*Tele Vue's 35mm and 27mm Panoptic eyepieces count among our favorites for "big gulp" views of nebulas, galaxy clusters and Milky Way star fields.*

maximum possible field of view (for a 1.25-inch system) combined with excellent performance. (Plössls with a 1.25-inch barrel and 40mm focal length do not offer a wider field than the 30mm-to-35mm models.) Our second choice is the 28mm Edmund RKE. For adding magnifications at the high-power end, we suggest another Barlow, perhaps a 2.5x or 3x model, rather than adding 6mm or shorter Plössls.

❖ Adding a Single Premium Eyepiece

In our experience, the eyepiece you'll use the most for all types of viewing falls in the medium-power range. A rule of thumb says that the eyepiece which best matches

**Premium Choice ▶**
Compared with a generation ago, today's astronomy enthusiast has a plethora of premium-class eyepieces from which to select. Among them is the Orion Superwide Series, available in sizes ranging from 3.5mm to 42mm, all with 20mm of eye relief.

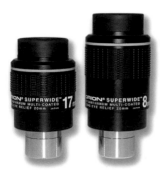

the eye's ability to resolve detail is the one that yields a 2mm exit pupil. To determine the focal length of this "optimum" eyepiece, simply multiply your telescope's focal ratio by 2. For example, the "best"eyepiece for an 8-inch f/10 Schmidt-Cassegrain would be a 20mm (10 x 2) eyepiece. It would provide 100x. Indeed, we did find that a 20mm eyepiece was the one we used most with the vintage C8 scopes we each owned for nearly a decade in the 1970s.

This rule isn't hard and fast, but it serves to point out which eyepiece you should invest in most heavily. If you can afford just one top-end eyepiece, make it the one with a focal length about twice your scope's focal ratio. A Meade Super Wide or Ultra Wide, an Orion Superwide, a Pentax XL or a Tele Vue Panoptic or Nagler would all be fine choices for that first premium eyepiece.

❖ Adding Deep-Sky Eyepieces
Another prime slot for a premium wide-angle eyepiece is in the 40x-to-80x range for general-purpose, deep-sky views. This might be a 24mm Meade Super Wide or a 22mm Panoptic (for 1.25-inch focusers); or for 2-inch focusers, a 22mm Nagler; or on longer-focal-length scopes, a 27mm Panoptic or a 32mm König or Meade Super Wide.

In the medium-to-medium-high range (120x to 200x), a Meade 14mm or 8.8mm Ultra Wide or a 9mm or 7mm TeleVue Nagler can provide enough power to resolve star clusters and reveal faint galaxies in a darkened sky without sacrificing field of view. These are perfect eyepieces for large Dobsonian reflectors.

❖ Adding a Low-Power "Panorama" Eyepiece
For the lowest-power eyepiece on a 2-inch

# Our Favorite Eyepieces IV: High-Power Planetary

For achieving high power and contrasty planet images, we routinely use longer-focal-length eyepieces in the 26mm-to-15mm range with a 2x-to-5x Barlow. Particular favorites of ours are Tele Vue's Powermate "image amplifiers," available in 2.5x and 5x models for 1.25-inch eyepieces. On our short-focal-length apo refractors, the 5x Powermate makes a great planet-viewing combination when coupled with a 26mm-to-20mm Plössl eyepiece (the Tele Vue and Meade Plössls are among the best).

Our other favorite high-power eyepieces include the Tele Vue Radians, which are, in effect, low-power eyepieces mated with a built-in Barlow. We use 6mm-to-3mm Radians on fast f/6 apo refractors and f/6 Maksutov-Newtonians. On longer-focal-length scopes, the 18mm-to-8mm Radians would be a fine choice. Other than cost, their main drawback is some kidney-bean field blackout if the eye isn't at just the right distance. As an alternative, we've found the Nagler 3mm-to-6mm Zoom to be a sharp performer on short-focal-length telescopes.

With a slightly wider field than a Radian, the 5.2mm Pentax XL also gets a lot of use on our fast focal-ratio scopes. Its wide field, high power and superb contrast are great for resolving clusters and picking out small planetary nebulas as well as providing crisp views of the planets. This is a heavy eyepiece for its focal length.

*The clean, sharp images of Tele Vue's Radian eyepieces combined with their generous field and comfortable eye relief at all focal lengths make high-power lunar and planetary viewing a real pleasure.*

system, consider the Meade 40mm Super Wide, the Orion Lanthanum 42mm Superwide or the University Optics 40mm MK-70 König. If a 40mm eyepiece yields more than a 7mm exit pupil on your telescope, then opt for a 32mm, such as the University Optics König or Tele Vue's 35mm or 27mm Panoptic.

Frankly, you can spend a lot on such an eyepiece and not use it that often. Don't forget to factor in the 2-inch nebula filter you'll also want. Costly, yes. But the views of big deep-sky objects, such as the entire Veil Nebula complex or the North America Nebula—possible only with such an eyepiece-and-filter combination—are unforgettable.

❖ Adding a Planetary Eyepiece
As mentioned, we use a good Barlow rather than a standard 4mm-to-6mm eyepiece. Eye relief is much better, and performance is not compromised. Three thoughtfully selected eyepieces and a Barlow make a versatile collection that will offer a magnification for every observing situation, including ultra-high-power views of planets and double stars when the seeing permits. A more costly high-power alternative is one or two Tele Vue Radian eyepieces.

❖ A Reminder:
Know Your Limits
When selecting eyepieces, remember the low- and high-power limits. Avoid eyepieces that yield a power less than four times the telescope's aperture in inches (to keep the exit pupil below 7mm) and more than 50 to 60 times the aperture (the maximum useful power). In practice, your most-used eyepieces will be in the magnification range of 7 to 25 times the aperture in inches.

## COMA CORRECTORS

While aberrations can be reduced to near zero in modern eyepieces, they are still present in the main optics. A premium eyepiece on a fast f/4.5 Newtonian telescope still shows some edge-of-field aberrations that make stars resemble tiny comets, the result of coma inherent in the parabolic primary mirror itself. Therefore, the next logical step is to optimize the eyepiece design to cancel out the flaws of the main optics, producing an entire telescope system that is aberration-free. A few coma-correcting eyepieces have been introduced over the years,

but none have caught on, simply because they are too specialized and are intended for use on only one type of telescope.

The more practical alternative is a coma-corrector lens inserted into the optical path

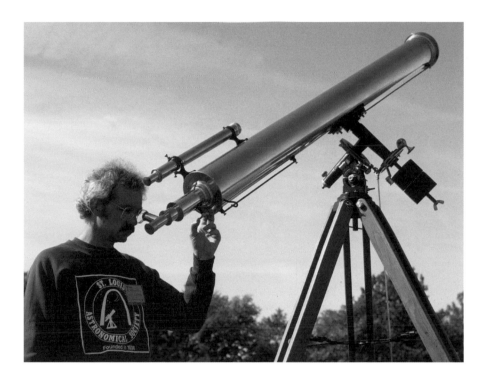

of a Newtonian in the manner of a 2-inch-barrel Barlow that then accepts any eyepiece. These accessories ($200 to $300) are highly recommended for f/4 to f/5 Newtonians to suppress off-axis aberrations. Anyone who spends serious money for a larger Dobsonian will want to have one.

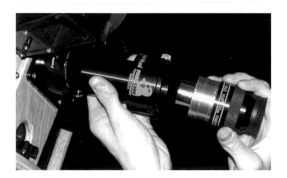

Coma correctors have their maximum effect at low power for panoramic viewing. The unit can be left installed for all but the most critical high-power planetary observing.

▲ **Vintage Observing**
Century-old refractors like Jon Slaton and Steve Sands' classic 6-inch Alvan Clark often have fine objective lenses, but most vintage eyepieces are vastly inferior to those of today.

◀ **Eradicating Coma**
Coma correctors are offered by Tele Vue and Lumicon. The Tele Vue Paracorr unit ($300) is available in both visual (shown here) and photographic versions.

# Filters

The best accessory for a telescope is a set of first-class eyepieces. The best accessory for the eyepieces is a set of good filters. The difference a filter makes is sometimes dramatic but more often subtle, requiring a trained eye to appreciate.

There are three types of filters for amateur astronomy: solar; lunar and planetary; and deep-sky. Although very different in construction, each has the same purpose: to reduce the amount of light reaching the eye. Considering that the goal of telescope owners seems to be to increase light-gathering power, this may sound strange.

It is easy to understand why a filter is required for observing the Sun, and because solar filters are integral to that task, they are dealt with separately in Chapter 9. But the planets? In large telescopes, planets such as Venus, Jupiter and Mars can sometimes be too bright; a filter cuts down glare without decreasing resolution. However, the main purpose of the filter is to enhance planetary markings by improving contrast between regions of different colors.

On the other hand, the motto for deep-sky observing is "Let there be light!" So how does a filter help? Light from deep-sky objects is usually accompanied by ambient light from sky glow and light pollution. Nebula filters block the unwanted wavelengths and admit those from deep-sky objects, improving the contrast between the signal (the target object) and the noise (the sky background).

## FILTER FEATURES

Most filters are mounted in cells that screw into the base of eyepiece barrels and are available for 0.965-, 1.25- and 2-inch eyepieces. All name-brand filters consist of optical glass with plane-parallel surfaces. Unlike for photographic applications, there are no gelatin filters for astronomy. Many current types of filter have the same anti-reflection coatings that are applied to lenses. With the exception of the Vernonscope Brandon models, all eyepieces use a standard filter thread, so any brand of filter can be used on any eyepiece. The Brandon eyepieces, however, require either Vernonscope or Questar filters. Some filters have threads on both sides, allowing them to be stacked, although combining two filters usually does not produce a more beneficial color than that achieved with a single filter.

## PLANETARY FILTERS

Beginners are attracted to planetary filters because they are inexpensive (about $15 each) and come in every color of the rainbow. (Planetary filters are labeled with the same Kodak Wratten numbers used in photography. A No. 80A blue filter for planetary observing is the same color as the photographer's No. 80A.) As with eyepieces, the temptation is to collect the whole set, but you do not need them all.

Of all the shades available, the most useful are No. 12 yellow, No. 23A light red (to increase the contrast between dark and light areas on Mars), No. 56 light green (to enhance features such as the Great Red Spot and the dark cloud bands on Jupiter) and No. 80A blue (for occasional glimpses of subtle features in Venus's clouds). These constitute a basic set for planetary viewing. A No. 8 light yellow can be substituted for the No. 12, and a No. 21 orange or No. 25 deep red for the No. 23A light red.

In all cases, the improvements offered by planetary filters are subtle—often too subtle for a beginner to notice. Knowing what to look for is the key. A general planetary filter designed to reduce the chromatic aberration inherent in achromatic refractors is described on pages 195-196.

## LUNAR FILTERS

The Moon can sometimes be too bright as well, especially in large telescopes. A yellow or neutral-density filter ($15) can cut glare and ease eyestrain. For refractors, a No. 8

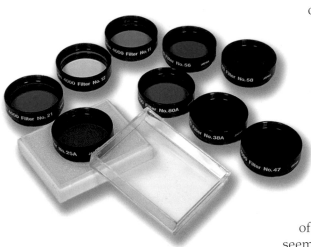

**Rainbow of Filters** ▲
Colorful filters that screw into eyepiece barrels enhance planetary details, but the improvement is subtle. First, you must ignore the overall tint the filter imparts and concentrate on the planetary shadings. If Jupiter's Great Red Spot is not visible without a filter, it will not snap into prominence when one is added.

# Planetary Filters: A Summary Comparison

| Wratten # | Color | Object | Comments |
|---|---|---|---|
| 1A | skylight | — | Haze penetration; mostly for photography |
| 8 | light yellow | Moon | Cancels blue chromatic aberration from refractors; reduces glare |
| 11 | yellow-green | Moon | Same as No. 8 but deeper color |
| 12 or 15 | deep yellow | Moon | Increases contrast; reduces glare |
| 21 | orange | Mars | Lightens reddish areas and accentuates dark surface markings; penetrates atmosphere |
| | | Saturn | May be helpful for revealing cloud bands |
| | | Sun | Cancels blue color of Mylar solar filters |
| 23A | light red | Venus | Darkens blue-sky background in daytime observations (for Mercury too) |
| | | Mars | Same as No. 21 but deeper in color |
| 25 | deep red | Mars | For surface details with large-aperture scopes |
| | | Venus | Reduces glare; may reveal cloud markings |
| 30 | magenta | Mars | Blocks green; transmits red and blue |
| 38A | blue-green | Mars | Reveals clouds and haze layers |
| 47 | deep violet | Venus | Reduces glare; may help reveal cloud markings; a very dark filter |
| 56 | light green | Jupiter | Accentuates reddish bands and Great Red Spot |
| | | Saturn | Accentuates cloud belts |
| 58 | green | Mars | Accentuates details around polar caps |
| | | Jupiter | Same as No. 56 but deeper color |
| 80A | light blue | Mars | Accentuates high clouds, particularly near limb |
| | | Jupiter | Accentuates details in belts and white ovals |
| 82A | very light blue | Mars | For Martian clouds and hazes |
| | | Jupiter | Similar to No. 80A but very light tint |
| 85 | salmon | Mars | Similar to No. 21; for surface details |
| 96 | neutral density | Moon & Venus | Reduces glare without adding color tint |
| — | polarizer | Moon | Darkens sky background during daytime for observations of quarter Moon |

See Chapter 10 for more about planetary filters.

light yellow or No. 11 yellow-green filter is also helpful in canceling the chromatic aberration—a bluish fringe most noticeable on the Moon, Jupiter and Venus—present in all but the finest models.

Polarizing filters are useful as simple neutral-density filters for viewing the Moon. Their ability to block light waves that vibrate in a particular direction makes them good for sunglasses but has a limited benefit in astronomy. Their main application is for observing the first- or last-quarter Moon in daylight or twilight. Light is most polarized in the region of the sky 90 degrees from the Sun, where the quarter Moon is found. With a polarizing filter (make sure it is rotated for the best effect), the sky background darkens, increasing contrast for daytime views of the Moon.

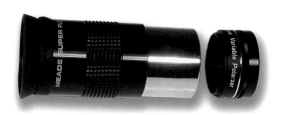

◀ **Dim the View**
Some manufacturers offer double polarizers containing two filters that rotate separately to create a variable neutral-density filter. Crossed at right angles, the filters admit only 5 percent of the light, which means that the Moon can be dimmed to a pleasing level no matter what its phase.

## DEEP-SKY OR NEBULA FILTERS

Nebula, or light-pollution-reduction (LPR), filters are rightly considered a significant advance in amateur-astronomy equipment. Much more than pieces of colored glass, nebula filters are, consequently, more expensive than lunar and planetary filters. Prices start at around $60 and run as high as $200 for a filter to fit eyepieces with 2-inch barrels.

These high-tech filters capitalize on the fact that nebulas emit light only at specific wavelengths—unlike stars, which emit across a broad spectrum of colors. Nebular light results mostly from hydrogen and oxygen atoms. Single gases have very well-defined emission lines, as do the gases contained in streetlights.

Mercury-vapor and sodium lights, which contribute heavily to the light pollution above and around cities, emit only in the yellow and blue ends of the spectrum. Since nebulas emit mostly in the red and green parts of the spectrum, one type of light can be blocked without interrupting the other. That's what nebula filters do.

Three kinds of deep-sky objects benefit from the use of a nebula filter: diffuse emission nebulas, planetary nebulas and supernova remnants. All emit their own type of light. Some nebulas (those seen as blue in long-exposure photographs) shine only by reflected starlight and therefore do not benefit from nebula filters. Nor do nebula filters help when viewing galaxies or star clusters. Filters will just make these objects and the sky look dimmer.

Nebula filters can transform a poor, light-polluted sky into a moderately good

### Nebula Emission Lines

Nebulas are bright in these wavelengths.

| Emission Line | Color | Wavelength |
|---|---|---|
| Nitrogen-II | red | 658nm |
| Hydrogen-alpha | red | 656nm |
| Oxygen-III | green | 501nm & 496nm |
| Hydrogen-beta | blue-green | 486nm |

location (at least for observing nebulas) and a good observing location into a great one. Contrary to popular belief, nebula filters are not just for city dwellers. This persistent urban myth discourages some telescope owners from laying out the money to buy one. Our advice: Don't delay—get one. Their effect is even more dramatic under dark skies, where there is always some sky glow caused by weak auroral and airglow activity that the filter can reduce.

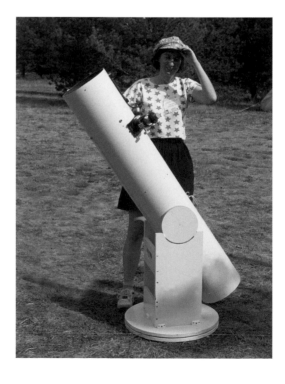

## TYPES OF NEBULA FILTERS

With nebula filters, the critical specification is the bandpass. All transmit a blue-green region of the spectrum, in addition to red wavelengths. However, some types let through a broad swath of light in the all-important green band. (The human eye is most sensitive to green, and the majority of nebulas have strong green emission lines.) These broadband types are intended for mild light-pollution reduction (LPR) when viewing all types of deep-sky objects. Examples are the Lumicon Deep-Sky filter, Orion SkyGlow and Thousand Oaks LP-1.

Another variety has a much narrower bandpass that blocks unwanted light more effectively, improving contrast still further, but only for emission nebulas. Other deep-sky objects just go dark. Examples are the Lumicon Ultra-High Contrast (UHC) filter, Orion UltraBlock and Thousand Oaks LP-2. Generic LPR filters from Celestron and Parks have bandpasses whose width falls somewhere between those of the broadband and the narrowband models.

There are also "line" filters (the Oxygen-III and Hydrogen-beta filters) that use ultra-narrow bandpasses adjusted specifically for wavelengths emitted by certain objects. O-III filters are excellent choices, but H-beta filters enhance so few objects (mostly the Horsehead Nebula), you may rarely use one. Examples are the Thousand Oaks LP-3 and LP-4 filters.

We use narrowband and O-III filters regularly but have found little use for the subtle improvement of broadband filters. If you are going to choose one filter, a UHC-class narrowband filter is the most useful for all telescopes. In our tests, even an 80mm f/12 refractor revealed the Orion, Veil and North America Nebulas much more distinctly with this filter. If you get hooked on nebulas, add an O-III filter later, though on slow f/10 to f/15 telescopes, these filters darken the sky so much that users can have difficulty seeing the field of view.

While spending as much as $200 on a filter may seem excessive, the improvement that a good filter can make to certain deep-sky views is like doubling the aperture of your scope. One view of the Veil Nebula both with and without a filter will convince you of a filter's effectiveness.

# Nebula Filters: A Summary Comparison

Nebula filters have become essential equipment for backyard astronomy because they improve the view of some of the best deep-sky objects regardless of light-pollution conditions. All these filters have a major transmission in the red portion of the spectrum (around the H-alpha emission line at 656nm). Their main difference is the transmission of wavelengths in the green band (around 500nm).

| Type | Bandpass* | Comments |
|---|---|---|
| Broadband | 90nm: 442nm–532nm | Widest bandwidth and brightest image but least amount of blocking of light pollution. Good for photography. Can be used to some degree on non-nebula deep-sky objects. Examples: Lumicon Deep-Sky, Orion Sky-Glow, Thousand Oaks LP-1. Good for slow focal-ratio scopes. |
| Narrowband | 24nm: 482nm–506nm | Narrow bandpass in green visual spectrum. Darkens sky further, with more dramatic contrast between sky and nebula. A good general-purpose filter for emission nebulas. Suitable for urban locations. Examples: Lumicon UHC, Orion UltraBlock, Thousand Oaks LP-2; filters sold by other manufacturers are similar in transmission characteristics. |
| Oxygen-III | 11nm, including 496nm & 501nm | A line filter—very narrow bandpass centered on green doubly ionized oxygen emission lines. Highest contrast and maximum blocking of light pollution. Good for planetary nebulas and supernova remnants. Best on fast focal-ratio telescopes. |
| H-beta | 9nm centered on 486nm | A line filter—very narrow bandpass centered on blue-green H-beta emission line. Useful for Horsehead and California Nebulas but little else. |

• Measured in nanometers. 1 nanometer – 10 angstroms = 1 millionth of a millimeter.

# The Backyard Guide 'Access

In the late 1950s, when the first Sputniks were flying and interest in space was keen, a major toy manufacturer introduced a unique Luminous Star Locater. The product was a series of clear plastic disks with constellation figures etched in them. Each disk could be placed in a holder shaped like a small tennis racket. A flashlight bulb in the handle illuminated the constellation etchings. Outside at night, the observer would hold the device up, look through the disk at the sky and use the illuminated outlines to identify the real stars.

Now fast-forward to the late 1990s and an $80 gadget billed as the Electronic Star Locater. A digital compass detects which way the user is looking and adjusts the glowing display accordingly. But the device was, in effect, the same gadget as its Sputnik-days predecessor. Both sounded like good ideas.

In reality, they were useless toys. Unlike the real sky, the illuminated stars all appeared about the same brightness, on maps that showed far too much sky at once. Both were clumsy to use. In short, neither device provided a better orientation to the sky than any $20 constellation guidebook.

Today, some more serious accessories for backyard astronomers are in the same category as these dubious Star Locaters. However, many other fine products add enjoyment to any session under the stars. Here is our catalog of common accessories, from the useful to the useless.

Attend any star party, and you'll see a wide range of eyepieces and accessories in use. The best stand out for the great views and convenience they provide. Others stand out for their gadget quality.

y Catalog'

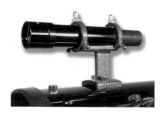

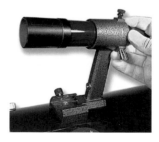

# Highly Recommended

Eyepieces and star charts top the list of must-have accessories, so much so that we cover these items in their own chapters. Accessories for photography are specialized products also discussed in later chapters. For the majority of observers, photographers or not, the following additional items deserve a place in website shopping baskets and on Christmas wish lists.

## UPGRADED FINDERSCOPE

One of the challenges (or frustrations) of observing, for both novices and veterans, is finding celestial objects. The better the instrument's finderscope, the easier this will

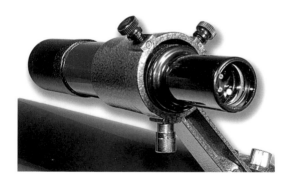

be. With finders, better usually means bigger.

Many manufacturers of entry-level telescopes continue to supply inadequate finderscopes with their instruments, because most first-time buyers do not realize how important that accessory is until they actually use the telescope at night.

Even more expensive telescopes (smaller reflectors and basic Schmidt-Cassegrains) often have only 6x30 finderscopes, meaning 6 power with a 30mm-aperture lens. These are just adequate. They are sufficient for locating bright targets but are limited when you are searching for deep-sky objects. For 8-inch and larger telescopes, a finderscope with a true 50mm aperture and 7x to 9x is a far better choice, able to show ninth-magnitude stars while maintaining a wide six-degree field.

All finderscopes come with eyepieces that have crosshairs. A few have a special reticle that indicates where the true north celestial pole is in relation to Polaris. While a polar-alignment reticle does not hurt, precise polar alignment is not essential unless you are doing astrophotography.

Some finderscopes have illuminated reticles, using a battery-operated light on the side of the eyepiece. However, the tiny camera batteries required are expensive, and we have found that it's easy to forget to turn off the light at the end of an observing session. As a result, the illuminator is dead most of the time.

A 50mm finderscope costs anywhere from $60 to $160, depending on extras such as right-angle prisms, illuminated reticles and dovetail brackets. An 11x80 finderscope, a good choice for large-aperture telescopes that can handle the weight, costs about $200 to $350.

## REFLEX SIGHTING DEVICES

They may seem like frivolous accessories, but most observers who use one soon can't do without it. Look through the angled window of a reflex, or unit-power, finder, and you'll see a red bull's-eye projected onto the naked-eye night sky. The most popular models are the original Telrad designed by Steve Kufeld and the Rigel Systems QuikFinder. Both attach easily to most telescopes using a plastic base that fastens with double-sided tape or screws. Both are light enough to pose few balance problems, and their LED lamps draw so little power that the batteries last for ages. Either model is the best $50 you can spend on improving your telescope.

Though not as versatile, an alternative is one of the gun-sight-style red-dot finders. Adapted from BB-gun sights, these reflex devices present a small circular window through which you see a red dot projected onto the sky. Examples are the Orion EZ Finder and Celestron's Star Pointer.

Their drawback is that the small window, sometimes dimmed with a coating, makes it hard to see enough of the sky for star-hopping to faint targets. However, for aiming at bright objects or just getting into the area, a red-dot-style finder works just fine. Compact and easy to attach to even

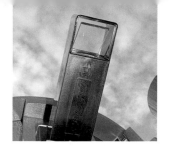

the smallest telescope, red-dot finders are good choices for owners of 60mm-to-90mm refractors and 100mm reflectors as supplements to a good optical finder. If you've ever groveled in the dirt to sight along the telescope tube only to discover you were still way off target, one of the reflex-style finders is the answer.

## RED FLASHLIGHT

You don't realize how essential this accessory is until you forget it one night. How do you read star charts? Equipped with thousands of dollars' worth of instrumentation, you are lost in the stars for want of a $15 flashlight.

Telescope stores sell astronomers' flashlights, but any pocket flashlight will do. Better yet, buy two or three. Paint the bulb red using deep red nail polish, or cover the face-

plate with red cellophane, red plastic, red paper (even brown wrapping paper) or red gel available from art-supply stores. The red illumination preserves night vision.

Buy a flashlight small enough to hold in your mouth so that both hands are free. The flashlight should be made of plastic; metal flashlights can become

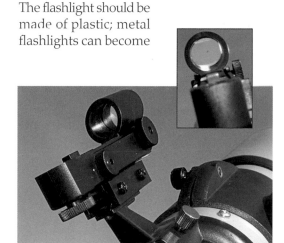

**Reflex Finders**
For locating objects, a reflex finder like the original Telrad, Rigel QuikFinder (above) or Celestron Star Pointer (left) may be all you need. One of our favorite telescopes, the 12.5-inch MAG1 PortaBall, uses nothing but a reflex finder, yet under dark skies and aided by star charts, it can be aimed precisely at faint deep-sky targets.

# Really Bad Finderscopes

Low-cost Christmas trash telescopes often have atrocious finderscopes. Sometimes, they are no more than hollow tubes with crosshairs. A few beginners' telescopes employ a flip-up mirror system that uses the main lens or mirror as a finderscope. It sounds good in theory, but in practice, the design fails miserably. The views are dim, and their orientation is confusing.

Inexpensive telescopes often come with a 5x24 finderscope. Most are junk. Many have aperture stops inside the tube, cut-

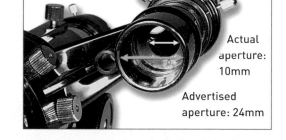

Actual aperture: 10mm

Advertised aperture: 24mm

ting the advertised 24mm aperture down to an actual working diameter of 10mm to 15mm. Such a small aperture makes it difficult to sight anything dimmer than the Moon. Upgrading such a junk finder to a good-quality 6x30 makes a vast improvement.

Another feature offered with some finderscopes, even larger ones, is a right-angle prism. While it is easier to look through a right-angle finderscope when viewing high in the sky, the views are mirror images. They do not match the real sky or printed star charts—all the constellation patterns appear flipped left to right. To match a star chart, the chart must be turned over and viewed from the back by shining a flashlight through the paper, awkward at best.

To make things worse, the observer does not look in the same direction as the telescope when using a right-angle finderscope. With a straight-through finderscope, the real sky can be observed with one eye while the other takes in the finderscope view. Even though a straight-through finderscope can be a literal pain in the neck near the zenith, we recommend you stay with the design when upgrading.

........................................................................................................................................

*Small 5x24 finders are often stopped down to 10mm or 15mm aperture by a stop just inside the main lens, because the finderscope lens is so bad that it cannot be operated at its full aperture of 24mm. If it were, its images would be completely fuzzy. The three-screw bracket of this economy model also makes it hard to align the finderscope precisely.*

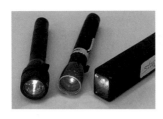

too cold to touch in winter. Flashlights with dual red and white LEDs are handy; a bright white light is good for assembling and dismantling equipment. Lights on flexible stalks are also useful for illuminating charts and tabletops.

## CLEANING AND TOOL KIT

A kit containing all the screwdrivers and wrenches your telescope may require can be an essential item. After a jostling road trip, parts come loose, optics need collimating and bolts require tightening. "No Tool" kits for some telescopes can replace screws and bolts with large knobs that are easy to adjust by hand.

A packet of lens-cleaning tissue, cotton swabs and lens-cleaning fluid should also be in every telescope-accessory case. Throw in some electrical tape as well—it is amazing what a roll of tape can fix.

Don't forget the insect repellent in your field kit. Mosquitoes love astronomers. But that insect repellent can be a nuisance, too, when it greases up the telescope's knobs and buttons. Worse, repellents with high DEET content can eat into optical coatings and the vinyl on binocular bodies and cases.

## CARRYING CASES

Eyepieces, filters and accessories require a proper case to carry them all. Check local camera stores. They offer all kinds of camera cases. Avoid those with several layers of compartments. The briefcase style is the best. Some have compartments with movable dividers. Others have foam that can be cut out as needed. Many have foam

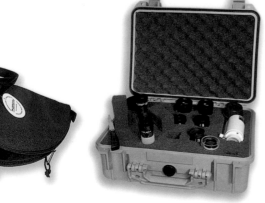

with precut squares that can be removed—these inserts tend to fall apart with use but can be replaced.

Then there's the telescope itself. Most telescopes don't come with their own case. Many telescope dealers sell padded or soft-sided bags for carrying almost every kind of small telescope. Some bags can even accommodate tripods and mounts. Hard-shell cases, often a manufacturer option, are better choices for larger scopes such as 8-inch Schmidt-Cassegrains.

## DEWCAPS

The first line of defense against dew or frost is a dewcap, a tube that extends beyond the front lens or corrector plate. Many refractors have a built-in dewcap. Yet the telescopes that most need dewcaps—Schmidt-Cassegrains and Maksutovs—rarely come with one. Exposed to the sky, the corrector lenses of these telescopes readily attract dew. Dewcaps can be purchased separately from many companies and dealers. Or a simple homemade device can be made out of cardboard, foam or plastic. Commercial or homemade, a dewcap that slides or folds back down the telescope tube for compact storage is the best.

Dewcaps function by maintaining a pocket of air in front of the objective or corrector lens that is slightly warmer than the outside air. A dewcap usually keeps moisture off the lens for about an hour longer than would be the case without the cap. But if dew is forming elsewhere, it will sooner or later cover all exposed optical surfaces. Then what?

## DEW GUNS

The first attack of dew can be repelled with a handheld hair dryer, blowing warm air (not hot) onto the affected lens. If you are running on 12 volts DC, use a heater gun sold in automotive stores for melting frost off windshields. It plugs into the socket of an automobile cigarette lighter.

As convenient as it is, a handheld hair dryer is only a stopgap measure. On humid nights, the optics always fog up again. Once the telescope has donated all its heat to the air and the air temperature stabilizes, dew keeps re-forming and hair dryers cease to be effective. Heat from a hair dryer also heats the air inside and around the telescope, temporarily blurring planetary views. Furthermore, after many dew attacks, a lens surface can accumulate a sticky, hard-to-remove residue from the heat-dried dust and moisture.

# Not Essential, but Nice to Have…

After the necessities of astronomical life have been met comes a list of items that aren't essential but can add enjoyment to any session under the stars.

## DEW-REMOVER COILS

The best way to eradicate dew and frost is to prevent them from forming in the first place. The trick: a low-voltage heater coil.

We recommend the Kendrick Dew Remover System, with heater coils and pads for all sizes of telescopes, finderscopes, Telrads, eyepieces, even laptop computers. The system uses a 12-volt, variable-intensity controller that can accommodate up to four low voltage heaters designed to wrap around the outside tube wall near the objective or corrector lens, the finderscope lens and eyepiece and the main eyepiece—any optical surface. The heaters provide just enough warmth to ward off the formation of dew. An eyepiece heater is not necessary if you make your own heated eyepiece-accessory box, because an eyepiece that dews over can be exchanged for a warm one.

Do heaters affect the telescopic image? We use heaters regularly on all exposed optical surfaces and find that the critical element is the objective- or corrector-lens heater. If there is too much heat, seeing is

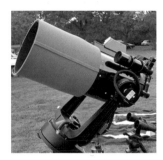

**Dewcaps**
A dewcap for a Meade ETX (left) or a larger Schmidt-Cassegrain telescope (below) may be the single most useful accessory a user can buy for this or any dew-prone telescope, especially in humid locations in eastern North America.

# Dew: The Vampire of Astronomy

Countless observing sessions are cut short by dew. Like a vampire, it silently sucks the life out of a telescope late at night. As the temperature falls during the evening, moisture condenses out of the air onto the optics, forming

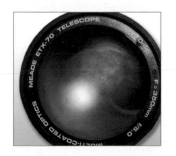

droplets of dew or, in cold weather, frost. Every backyard astronomer has cursed the plague of glazed optics that terminated a planet watch during an evening of excellent seeing or fogged an astrophoto midway through a 1-hour exposure.

Dew can be more than merely a nuisance—it can damage the coatings on mirrors and lenses. Our industrial civilization has, in many parts of the world, changed delicate dewdrops into acid dew, a cousin to acid rain. This stuff attacks coatings on lenses and mirrors and is especially brutal to enhanced silvered mirrors on reflectors. (The more common enhanced aluminum coatings are less susceptible.) In northeastern North America, where acid rain and acid dew are common, enhanced silver coatings are not recommended for Newtonians, where the optics are exposed to the air. (Such coatings are rarely offered today.) Owners of older Schmidt-Cassegrains or Maksutovs with enhanced silver coatings should ensure that the tube is always closed to outside air, with either an eyepiece or a plug in place.

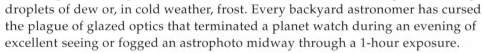

*Owners of standard Newtonian telescopes usually don't have to worry about dew, since the optics are at the bottom of a long tube that keeps dew off the main mirror. But for Schmidt-Cassegrains, Maksutovs and refractors (like this fogged-up Meade ETX70-AT), dew can be a drawback the salesperson neglected to mention.*

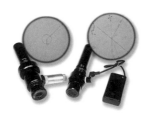

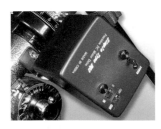

degraded; too little, and dew encroaches. When set to a happy medium, heater coils produce no ill effects.

## MOTOR DRIVES

Most first-time telescope buyers who purchase equatorial mounts do so because of the mount's ability to track the stars. Yet many never bother adding the optional motor drive needed to supply the very tracking the mount is capable of providing. Almost all import telescopes have optional battery-operated drives (110-volt AC motors have all but disappeared from the market). The lowest-cost models offer tracking in right ascension at a fixed speed. Better

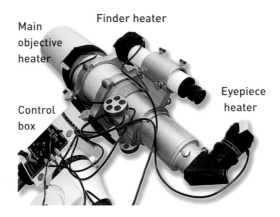

Finder heater

Main objective heater

Control box

Eyepiece heater

models have override buttons for an instant 8x, 16x or 32x speed change. This is useful for panning around the Moon and centering objects. Speed controls are worth having, because adding many of the optional drives to Chinese-import telescopes requires giving up the manual slow-motion control—there is no clutch assembly to allow both the motor and the slow-motion cable to turn the telescope, an inconvenience for centering targets.

Deluxe models of drives come with two motors, one for tracking in right ascension (east-west) and the other for centering objects in declination (north-south). If the mount has a good manual slow-motion control in declination (most models do), then a dual-axis drive isn't essential for casual viewing, but the push-button convenience is handy. However, a dual-axis drive is recommended if you wish to pursue guided deep-sky photography.

## POLAR-ALIGNMENT SCOPES

Equatorial mounts must be polar-aligned precisely for photography but only roughly

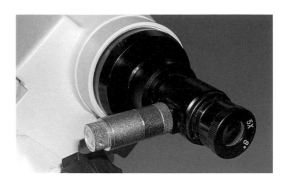

for casual viewing. No matter the technique and degree of precision, polar alignment requires some way of aiming the mount's polar axis at the pole. Almost all German equatorial mounts sold today come with suitable controls for making fine adjustments to the azimuth and altitude angle of the polar axis.

What may be optional is a small polar-alignment finderscope, located right in the mount, aimed up the polar axis. Looking through the polar scope makes it easy to sight Polaris and fine-tune the aim of the mount's polar axis. (See www.backyard-astronomy.com for detailed instructions on methods for precise polar alignment.) For visual observing, however, simply getting Polaris close to the center of the polar scope's field of view—within one degree of true celestial north—is good enough.

## POWER PACKS
## AND CAR CORDS

If the mount is driven by DC motors, it probably came with a battery pack that uses several penlight batteries or a single 9-volt transistor battery. These are not long-lived. D-cell batteries are really the practical minimum for drives. They hold up better in the cold through several long observing sessions. Another plan is to get the appropriate cable to plug the drive into the car battery. Or purchase a larger-capacity rechargeable pack that will be good for several nights' use before it needs to be recharged. A separate power pack allows the telescope to be set up anywhere and not tied to the car.

For powering just a telescope's dual-axis drive, a power pack with a 7-ampere-hour capacity is more than adequate for many nights' use. But the high-speed

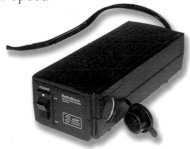

motors of some high-torque "Go To" telescopes can draw as much as 3 amps at 12 volts, quickly draining a small power pack before the night is over. Add antidew heater coils and CCD cameras, and the modern observer quickly encounters a mini energy crisis in the field. If everything has been hooked up to your car battery, you may not be able to start your car in the morning. Separate power packs with 15-to-30-ampere-hour capacities are a minimum for the high-tech observer. Rechargers that work off solar panels are a great idea for long stays at remote sites.

## HEAVY-DUTY TRIPODS AND WEDGES

Some models of telescopes offer an upgrade to a heavier tripod or, in the case of noncomputerized Schmidt-Cassegrains, a heavier wedge assembly. Tripods from Celestron often fit Meade scopes, and vice versa, so if one manufacturer does not have a tripod that meets your needs, chances are, the other might. The added stability of heavy-duty tripods makes them worth the extra cost, especially if you intend to do any imaging through the telescope.

## OBSERVING CHAIRS

Treat yourself to a place to sit while observing. If possible, lower the telescope so that the eyepiece is at eye height when you are seated. The increase in comfort is sheer luxury. A good observing chair is a stool whose height is adjustable. It should also

fold up for easy storage. The stools sold in music stores for drummers are very comfortable, although not tall enough for some telescopes. Instruments bigger than 15-inch Dobsonians require an observing ladder. A small kitchen stepladder may be just the thing for reaching the eyepiece of these big telescopes when aiming at the zenith.

## OBSERVING TABLES AND ACCESSORY TRAYS

Many observers have found it convenient to take a folding camp table to the observing site to hold star charts, books and other stargazing paraphernalia. Without one, observers are forced to work off tailgates or out of car trunks.

Some small suppliers make tripod-mounted trays for common telescopes, such as Schmidt-Cassegrains, that lack a place to put anything. These accessory trays are great for holding eyepieces, auto-guiders and computer hand controllers that would otherwise dangle with no place to clip. Certainly, any aid that reduces fumbling around in the dark is a welcome convenience, as long as it does not require an inordinate effort to set up in the first place.

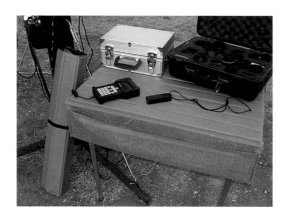

## WHEELEY BARS AND SCOPE COVERS

Borrowing an idea from large TV studio and movie-camera dollies, one innovative accessory company, Jim's Mobile, Inc. (JMI), designed a set of rolling casters that fit under the tripod legs of large scopes, such as 10-to-12-inch Schmidt-Cassegrains. If you can store such a telescope in a garage and

◀ **Supplying 12-Volt Power**
Users at home face the problem of powering an array of 12-volt devices from 110 volts AC. Small wall transformers, or "battery eliminators," are simply not sufficient for some power-hungry "Go To" scopes, even without the added draw of heater coils. A 12-volt power supply (or two!) capable of outputting at least 3 amps, like this $60 Radio Shack unit, is essential.

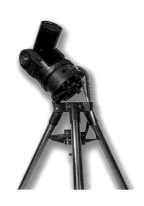

▲ **Deluxe Tripods**
Celestron's heavy-duty tripod (Meade has a similar model) provides a sturdier platform for Meade ETX telescopes than the barely adequate tripod supplied as standard equipment.

◀ **A Place for Everything**
This durable camp table, sold by Orion, folds up into the tight roll seen propped up at left, making it easy to stuff into a car for field trips to dark-sky sites.

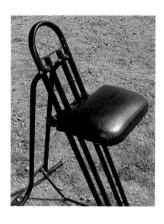

**A Place to Sit ▲**
This clever observing chair can be adjusted over a wide range of positions to suit different eyepiece heights.

**Shelter From the Storm ▶**
At star parties during the day, many scopes are cocooned under tarps or silvery Mylar "Desert Storm" covers; the reflective blankets help keep the scope cool. After prolonged use, however, the material in these high-tech covers can break down and tear apart.

**Aids for Alignment ▼**
A collimation tool, like a sighting eyepiece (left) or a laser collimator (right), is a useful accessory for owners of f/6 to f/8 Newtonian telescopes and is an essential tool for Newtonians with optics faster than f/6, which need tighter collimation. The slightest miscollimation in faster Newtonian systems can degrade image sharpness and contrast, blurring planetary and double-star views in particular.

if there's a clear access from there to your backyard viewing site, JMI's Wheeley Bars are a great solution for a no-fuss setup. Just wheel the scope outside as a single unit.

For garage storage or for star parties and backyards where you can leave your telescope set up for a few nights, simply protect the telescope with a weatherproof cover. Orion Telescopes and Binoculars offers an excellent set of waterproof nylon covers to fit many sizes and styles of telescopes. We use them regularly and can recommend them. They are far more durable than garbage bags!

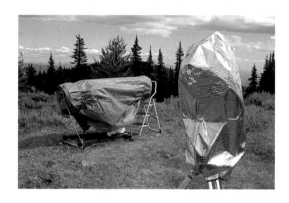

## COLLIMATION TOOLS

Every owner of a Newtonian telescope should have a collimating eyepiece, a simple tube with crosshairs and a diagonal surface to reflect light down into the telescope tube. Sighting through such an eyepiece (one type is called a Cheshire eyepiece) makes it much easier to center all the optical elements, especially if both the secondary and the primary mirrors are marked with a dot or ring at their centers (see www.backyardastronomy.com for collimation procedures). A collimating eyepiece costs no more than $40. Put one on your stocking-stuffer Christmas list.

A higher-tech solution is the laser collimator. This unit sells for $75 to $200 each. Placed in the focuser, it projects a beam of laser light down the Newtonian tube assembly. The user watches the beam of light and adjusts the tilt of the telescope's mirrors until the beam returns exactly onto itself.

# If You Can Afford Them…

For the observer who has collected all the required toys, there is another realm of wonderful accessories, but some of them come with hefty price tags.

## DIGITAL SETTING CIRCLES

As described in the Appendix, much of the computer-finding functionality of "Go To" telescopes can be added to any telescope. Celestron, JMI, Lumicon, Meade, Orion, Sky Commander, Tele Vue and other manufacturers all offer add-on boxes and axis encoders for a variety of telescopes that provide digital readouts of the telescope's position. Databases of objects (the more costly the model, the larger the database) allow you to find targets simply by punching in the object's catalog number (M31 or NGC4565, for example). You then watch the display as you manually swing the scope. When the display's coordinates read 00 00, you are on target.

Installed properly and after a nightly alignment on two known stars, digital setting circles work well. With the required optical encoders that must be installed on each scope axis, expect to pay $350 to $600 to equip a telescope with one of these high-tech finders.

With "Go To" telescopes now so affordable and prevalent, add-on digital setting circles are becoming less popular, but they are still excellent choices for adding computerized finding to Dobsonian reflectors.

## EQUATORIAL PLATFORMS

The major limitation of Dobsonian telescopes—no tracking—can be overcome with an ingenious device: the equatorial platform. The telescope rests on a small, ankle-high table that can swivel around one axis, driven by a motor. Platforms, also known as Poncet mounts, typically provide about an hour of tracking until the threaded drive rod needs to be reset back to the beginning of its travel. Having an object stay put in the high-power eyepiece of a massive

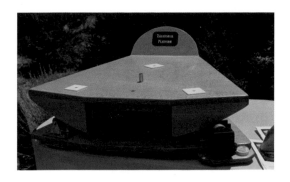

20-inch or larger telescope is a sight to behold. Platforms can transform Dobs with premium optics into the finest planetary telescopes available. Faint galaxies and nebulas that normally fly across the field of view in an undriven scope suddenly stand still for detailed inspection, and deep-sky observers wonder how they saw anything without a platform turning their giant Dob.

Expect to spend $750 to $2,500 for a commercially built platform, depending on the size of telescope it has to carry and sophisticated extras such as dual-axis motion.

## SUPERBRIGHT STAR DIAGONALS

If the focuser can accommodate it, a 2-inch star diagonal is a great addition to a refractor or a Cassegrain-style telescope. Most star diagonals in this size use mirrors. The best employ the same full-thickness diagonals used as secondary mirrors in Newtonians. The very best incorporate multilayer dielectric coatings that reflect 99 percent of the incident light, a sizable notch up from the 89 percent reflectivity of standard overcoated aluminum. Modern enhanced coatings are ultrasmooth and pose no durability or tarnishing problems. Though costly ($300

◀ **Poncet Platform**
Some amateur astronomers build their own tracking platforms, like this unit, complete with a built-in battery for the power source. However, small companies such as Equatorial Platforms supply well-crafted commercial products.

# Where to Keep Your Telescope

Telescopes are pretty rugged instruments that, if handled and stored properly, can last a lifetime. There are three basic storage situations for telescopes used at home: a portable telescope kept in the house; a portable but bulky instrument stored in a garage; and permanently mounted equipment in an unheated observatory.

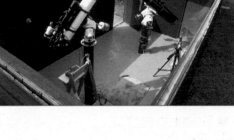

The key to effective storage is to avoid moisture, which attacks the surfaces of mirrors in reflector telescopes, the coatings on lenses in refractors and eyepieces and the metal parts and fittings of telescope mounts. In a dry climate, the problem is dust.

For storage inside the house, avoid a basement location prone to moisture problems, condensation or mildew. Because of space limitations, an unheated garage or shed is often a necessary storage facility. One advantage is that the telescope remains close to the outside temperature, reducing the wait for the optics to reach the temperature stability required for optimum performance. However, garages can be prone to fumes and oil that might coat the optics. Bring the telescope inside when it's likely to go unused for months at a time.

A backyard observatory is the luxury solution to telescope storage. One precaution is to have an enclosure that is as airtight as possible—to prevent dust, snow and other material from blowing in. A modest amount of electric heating from time to time, just to have the inside temperature slightly warmer than the outside, can greatly reduce moisture problems, though at the risk of introducing a mouse problem. An alternative to heating the whole building is to install permanent low-voltage heaters near the optical elements.

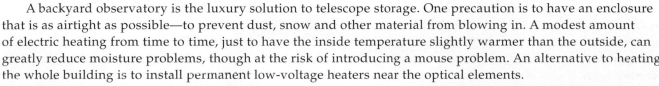

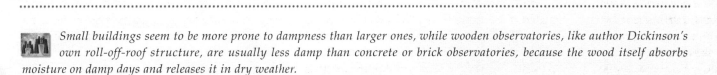

*Small buildings seem to be more prone to dampness than larger ones, while wooden observatories, like author Dickinson's own roll-off-roof structure, are usually less damp than concrete or brick observatories, because the wood itself absorbs moisture on damp days and releases it in dry weather.*

in a machined housing), such a superdiagonal extracts the ultimate performance from a high-end telescope. Newtonian owners

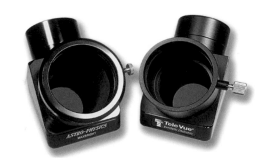

can replace their standard-coated secondary mirrors with mirrors featuring similar EnduroBrite coatings. Visit www.backyard-astronomy.com for lists of suppliers.

## BINOCULAR VIEWERS

Now we're talking expense. A high-quality binocular viewer designed specifically for telescopes costs upwards of $1,000. Then factor in the dual sets of eyepieces and filters, and the luxury of viewing through

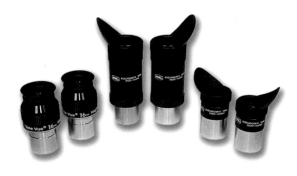

a telescope using two eyes proves to be a costly one, indeed. Is it worth it? Once you are satisfied that you have a good telescope and good eyepieces, there may be no other accessory you can add which will provide as significant an improvement to your viewing. Even your nonastronomy friends and relatives will be wowed.

Binocular viewers utilize a series of prisms to split the single beam from a telescope into twin collimated beams, one for each eye. Lower-cost units ($500 to $750) are adapted from microscopes. They suffer from two problems: As you change the interocular distance (the spacing between the two eyepieces) to accommodate differ-

ent people, the focus changes. The prisms are also undersized, resulting in a vignetted field with lower-power eyepieces, just the kind you might like to use the most.

More advanced designs, such as the Astro-Physics Baader Binocular Viewer and Tele Vue's Bino Vue, use larger prisms and a folding body, much like a normal binocular that maintains focus as you adjust the eyepiece spacing.

The added length of the light path through a binocular viewer requires as much as four to five inches of extra in-focus travel of the focuser. Some premium apo refractors can accommodate this, as do most Schmidt-Cassegrains. But Newtonian reflectors certainly cannot, unless the telescope tube or strut length is cut down specifically for binocular use.

To get around this, binocular viewers are usually supplied with a Barlow lens (often a 1.5x-to-2x model), which extends the focal point out far enough that the eyepieces, now perched atop the binocular viewer, will reach focus. However, the extra power of the Barlow and the restriction to 1.25-inch eyepieces conspire to make it impossible to achieve very low power and wide fields with binocular viewers. Don't expect three-degree-wide views of the North America Nebula in two-eyed splendor. But deep-sky objects at moderate power and planets at high power become all new experiences when viewed with eyes wide open—both of them!

## DOMES AND SHELTERS

The ultimate accessory for any telescope is a house to put it in. Observatories can be rotating domes or shelters with roofs that roll off, flip off or fold down. Most backyard

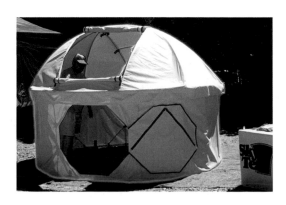

astronomers build their own structures, though commercial companies offer pre-fabricated domes for existing structures or complete observatories from the ground up. Even if you live under less-than-ideal sky conditions, having your scope permanently set up and ready to go may be the most convenient arrangement for you.

For a less permanent facility, companies such as Sky Tent offer nylon domes wrapped around a skeleton of strong plastic tent poles. Staked down, these domes can withstand high winds, allowing them to remain in a backyard as a temporary structure. (Visit www.backyardastronomy.com for links to manufacturers, or look in a current issue of *Sky & Telescope* for ads of the latest products from dome and shelter suppliers.)

# You Can Live Without These…

Some items are awfully tempting, but in an effort to "simplify, simplify," these are gadgets you can happily live without, keeping your nightly sessions uncluttered and trouble-free. The more add-on gadgets you have, the more time it takes to set up and the more that can go wrong.

## FOCUS MOTORS

Celestron, Meade and third-party suppliers, such as JMI, make battery-operated motors for virtually every focuser and telescope on the market. Hands-off focusing reduces vibrations and image shake, but despite thousands of motorized focusers being sold, we see few scopes equipped with them at star parties. Those that are often suffer from hesitation, backlash and run-on, making it hard to home in on precise focus. In fairness, some of the problems may reside in the focuser itself, not the motor. Nevertheless, in frustration, we just want to grab the focus knobs and turn them by hand, something not possible with some electric-focus add-ons. We find the extra batteries and cables a nuisance. Some people love focus motors; we have survived without them.

## VIBRATION DAMPENERS

These small pads are placed under the legs of a telescope tripod. A ring of rubber set into high-density metal absorbs vibrations and prevents their being transmitted up the tripod leg. These $60 pads work, cutting the time it takes vibration to dampen by a factor of two or more, but few observers use them. People prefer their telescopes to be firmly set into solid ground. This is especially true of telescopes that need precise alignment, such as "Go To" scopes and those intended for photography.

## LAPTOP/PALMTOP COMPUTER

"Go To" telescopes can be controlled by an external computer running astronomy software. Many programs offer telescope control as one of their functions. The advantage over using the telescope's own hand controller is that the computer (perhaps a laptop, a pocket PC or a palmtop) can display a star map showing where your telescope is pointed and the location of other nearby targets. This makes it easy to explore a region of the sky thoroughly, instead of randomly bouncing around the sky following the hand controller's lists of numbered targets. The disadvantage is the extra gear that needs to be set up, the tangle of connecting cables and adapter plugs required and the fuss about charged

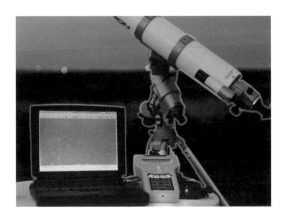

batteries to run it all. Computers make sense in a permanent observatory, but few mobile observers have taken to them as necessary gear.

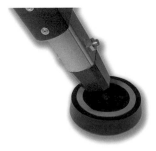

▲ **Footpads for Scopes**
A partial solution to shaky mounts are these vibration-suppression pads.

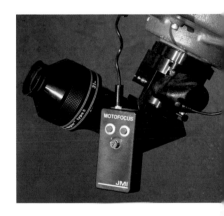

▲ **Motorized Focusing**
Focus motors, like this unit from JMI, reduce vibration introduced by touching the focuser. Of course, the ideal solution is a more rigid mount. Another application of motorized focusing is extremely precise focusing, as is often required in astro-imaging.

◀ **Computer-Aided Viewing**
"Go To" telescopes, such as this Vixen Great Polaris with SkySensor, can be connected to a computer. Click on an object on the computer's display and— whir—the telescope slews to find that target. Very neat, but not essential.

## VOICE-ACTIVATED TELESCOPES

Tell your telescope to "Find M31," and off it goes to center on the Andromeda Galaxy. When it gets there, a synthesized voice tells you all about M31. It may sound like science fiction, but voice-recognition software such as the Astro-Physics DigitalSky Voice does just that.

Hailed as a breakthrough when first released in the late 1990s, voice-activated programs simply haven't swept the hobby, despite the proliferation of "Go To" telescopes. The reasons are simple. First, most of us feel a little silly talking to our telescopes, alone in the dark. Will the neighbors hear you? Or your telescope? Second, these programs require another layer of technical complexity at the telescope—at least a laptop computer with an external microphone. A voice-commanded telescope can be a fun gadget for the home or club observatory, especially for impressing visitors and novices. But in the field, we suggest keeping it simple.

## ERECT-IMAGE FINDERS

A finderscope equipped with an Amici prism presents an erect image that matches the sky and star charts. The right-angle construction is convenient to look through when aimed high in the sky. This sounds ideal, yet Amici-prism finders still suffer from the problem inherent in all right-angle finders: They force you to look 90 degrees away from the direction the telescope is aimed, making it hard to relate the finder view to where the scope is pointed. Some dealers sell straight-through, erect-image finders, but models are generally 6x30 units

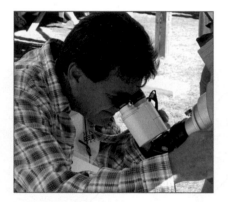

# The Two-Eyed Advantage

Looking through a telescope equipped with a binocular viewer is an impressive sight. Objects appear to float in three-dimensional space. Planets seem to take on extra detail. The Moon looks like a real landscape just outside your spaceship window. And the comfort and ease of viewing are far greater than with one-eyed observing.

The obvious question with binocular viewers is the loss of light. Each eye can receive, at best, only 50 percent of the light that a conventional single eyepiece would provide. However, when the brain merges the two dimmed images back together, the result is a view that is nearly as bright as the same view would be with a single eyepiece.

Where light loss enters the picture is from all the extra optical elements the light must traverse getting to your eyes. Users report about a 0.5-magnitude loss of light-gathering power compared with a single-eyepiece view. For smaller scopes, that's roughly the equivalent of dropping down one imperial-unit notch in aperture: an 8-inch scope performs like a 6-inch for image brightness, a 10-inch like an 8, a 12.5-inch like a 10-inch, and so on.

Binocular-viewer aficionados maintain that these drawbacks are offset by the additional detail which all objects seem to present. Our experience tends to confirm this conclusion. Images from both eyes present the brain with a view that has a greater signal-to-noise ratio. Views seem less grainy. Those dark and nasty eye floaters (dead blood cells in the eye) that drop in front of planets at high power are now largely suppressed, allowing continuous and unimpeded views of the planets.

Depending on the telescope and what you're looking at, the binocular-viewer experience can be unforgettable. But going double-eyed with a telescope is certainly costly: For one thing, you'll need two identical eyepieces at every magnification. Yet few who make the leap regret the expense. Tele Vue and Baader make the best viewers.

. . . . . . . . . . . . . . . . . . . . . . . . . . . . . . . . . . . . . . . . . . . . . . . . . . . . . . . . . . . . . . . . . . . . . . . . . . . . . . . . . . . . . . . . . . . . . . . . . . . . . . . . . . . . . . . . . . . . . . . . .

 *For use on a refractor or Schmidt-Cassegrain, the Tele Vue Bino Vue needs to be coupled to a star diagonal. However, the Baader Binocular Viewer, shown here, comes with a low-profile, prism-type star diagonal that helps minimize in-focus travel.*

—best for daytime use, not for astronomy. To those new to the sky, erect-image finders seem like an essential item, but most observers learn to live without them.

## BLINKING TRIPOD-LEG LIGHTS

Just like aircraft warning lights on the top of antennas, these little LEDs attached to tripod legs tell unsuspecting visitors: Don't trip here! They are strictly for the gadget lover, although we'll grant them one legitimate use: to mark the location of a camera atop a tripod placed off in the distance for an unattended long-exposure photo.

## EYE PATCHES AND GOGGLES

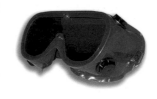

In theory, an eye patch makes it easier to view through your telescope with your uncovered eye. Perhaps, but no one wants to look like a geeky landlocked pirate, even in the dark. Another goofy accessory, red-filtered goggles, helps you retain your dark adaptation if you go inside a brightly lit building. Double ditto on the geek factor.

◀ **Astro-Goggles**
Yes, they work, preserving night vision under bright lights, but do you want to be seen wearing them?

# Astro-Travel and Touring

The best accessory for your telescope is a dark sky. But why not have a dark sky that comes already equipped with great equipment and comfortable accommodation? That's the attraction of a growing number of astro-resorts. Destinations such as StarHill Inn and New Mexico Skies offer an impressive array of telescopes and cameras for visitors to rent on a nightly basis, or you can set up your own gear next to your private cabin under pristine desert skies. At Jack and Alice Newton's Observatory B&B, visitors are offered sky tours and optional CCD-

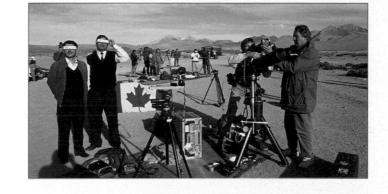

imaging tutorials as a bonus to first-class bed-and-breakfast accommodation. Even Kitt Peak Observatory near Tucson is in the business of astro-tourism: Visitors can book guided observing sessions or reserve a Meade 16-inch telescope for a night of personal viewing and imaging.

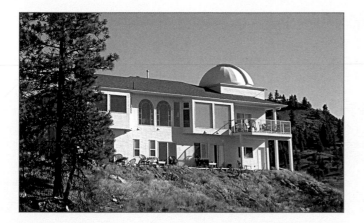

In the 1980s and 1990s, eclipse chasing grew from the pursuit of a dedicated few to a sizable segment of the ecotourism market. For each total eclipse of the Sun, dozens of tour companies offer to take you into the path of the Moon's shadow. Eclipses are a great way to combine the trip of a lifetime to an exotic locale with the experience of a lifetime: the sight of the Sun suddenly extinguished from the daytime sky. In the next few years, eclipse travelers will explore Turkey (2006), the Canadian High Arctic (2008), China (2009), Easter Island (2010) and the South Pacific (2012). For the addicted, maps of future eclipse paths serve as lifetime vacation planners.

*Jack and Alice Newton offer luxury accommodation at their Osoyoos, B.C., bed and breakfast, with telescopic tours in the dome as a unique bonus, left. Eclipse chasers, right, set up on the volcanic landscape of the Chilean altiplano in November 1994.*

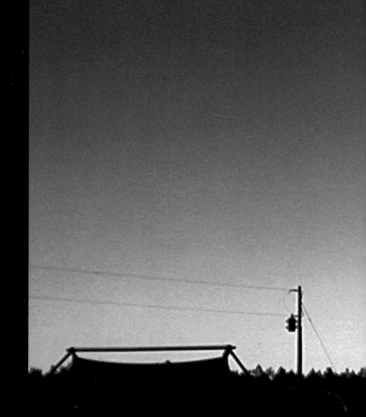

## CHAPTER SIX

# Using Your New Telescope

Whether you are unpacking your first purchase or the latest addition to a growing collection, setting up a new telescope is always a thrill. If you have never set up a telescope before, the anticipation of seeing your first celestial object through the sparkling new optics may make you toss the instruction manual aside.

Even if you are patient enough to read manuals, you may find the instructions for your foreign-made telescope more than a little baffling. We've seen many imported telescopes packed with misleading or, worse, missing manuals.

To aid first-time telescope owners, we've assembled this chapter as our own "guide to using your new telescope," with tips drawn from our own experiences and from the many questions we've handled over the years from puzzled owners.

First light through a new telescope is an experience charged with anticipation, whether the telescope is a high-end model, as here at a star party, or a modest starter scope set up in the backyard. Photo by Terence Dickinson.

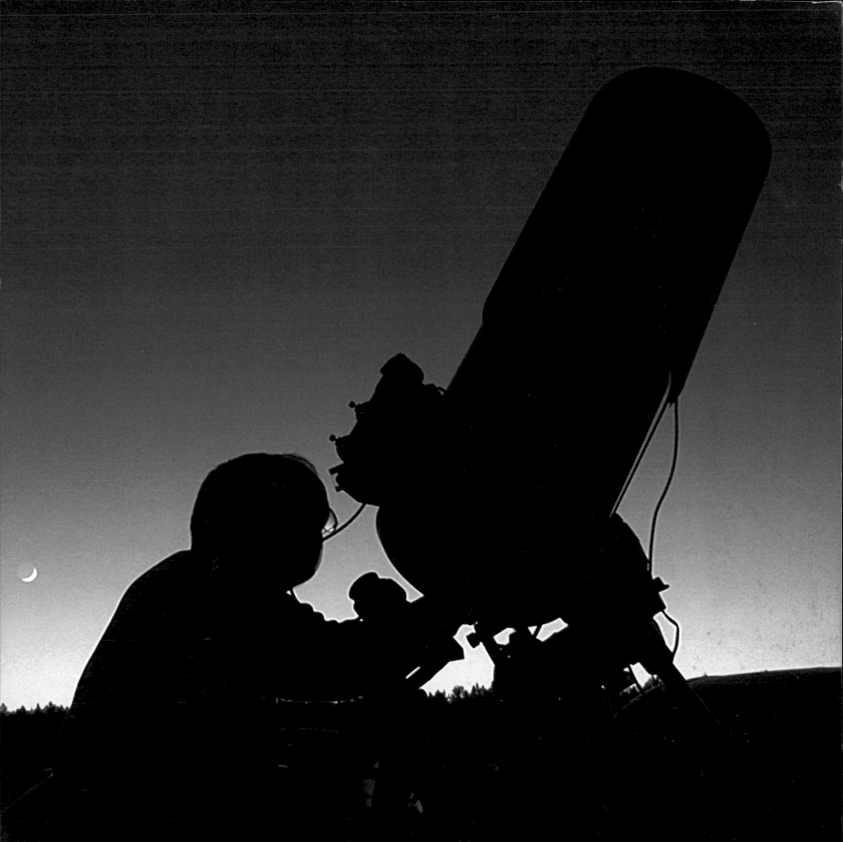

# Introducing the Beginner's Telescope

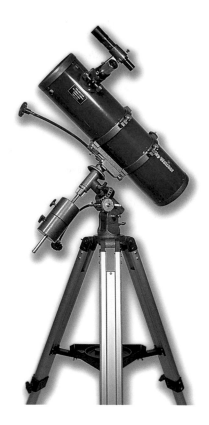

This chapter is a guide to setting up and using your first telescope. As examples of beginners' scopes, we have selected two Chinese-made import telescopes, since these represent the most common entry-level telescopes you are likely to encounter. Telescopes virtu-

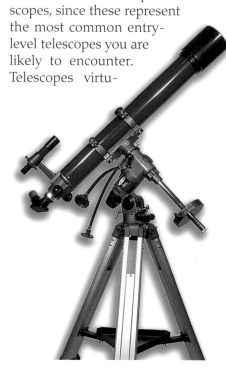

**What Is Best to Start?**
Though a Dobsonian-mounted telescope (in use below) remains our first recommendation for a beginner's telescope, models with German equatorial mounts (right) remain popular as starter scopes. This chapter addresses the common questions owners have about setting up and using these models.

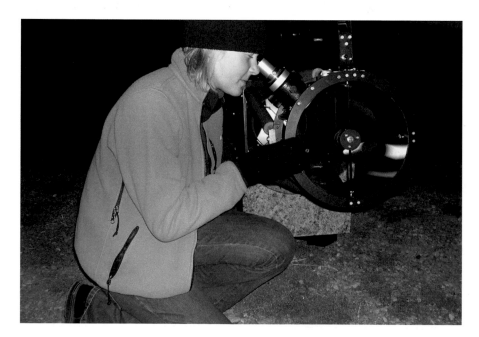

ally identical to the two shown throughout this chapter are sold worldwide by a variety of dealers under many national and house-brand names.

We're not saying that these specific telescopes are necessarily the ideal beginners' scopes; but the fact is, they are very successful products, and huge numbers of them are in the hands of novice astronomy buffs.

From our experience, these telescopes, although excellent buys, also elicit the most cries for help from new owners. The scientific appearance of the models, which tends to attract the novice in the first place, is also at the root of the "What's this for?" questions that emerge when it comes to using these instruments.

Consider this chapter a tutorial for the assembly and use of any beginner's scope, because many of our tips about finder-scopes, eyepieces and successful first-light views apply to any instrument. (For advice on using today's generation of computerized "Go To" telescopes, see our array of tips in the Appendix for accurate pointing with these electronic marvels.) Many beginners, though, go with a noncomputerized scope for starters, which is why we concentrate on that type here.

## A TOUR OF YOUR TELESCOPE

While some telescopes, such as Meade's ETX-AT series, come out of the box as one-piece items, most require some assembly. Tripods must be bolted together, mount heads need adjusting and tightening, and fittings have to be attached to the main optical tube.

Before diving into the assembly tutorial, we will tour our typical telescopes to learn what all the parts are and what they do. We will even point out parts that have no purpose other than to confuse new owners! Although our illustrations show specific models, keep in mind that the same components can be found on any equatorial mount and on most reflector and refractor telescopes.

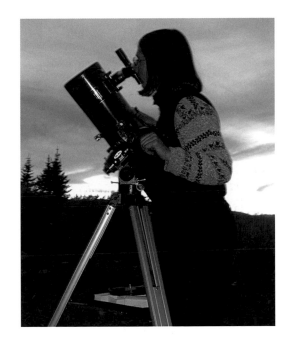

# Decoding Directions

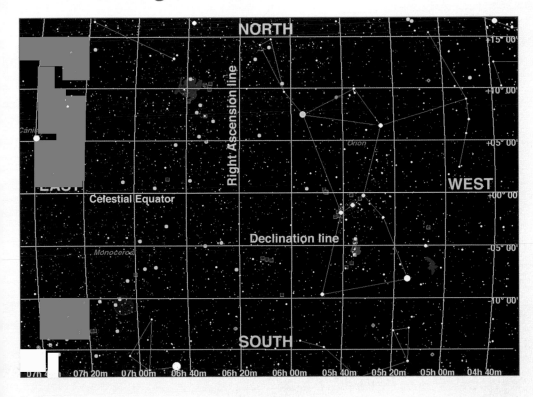

We cover celestial coordinates more thoroughly in Chapter 11, but assembling any equatorial mount introduces the owner to the concepts of right ascension (R.A.) and declination (Dec.). An equatorial mount moves through these two directions.

Declination is like latitude on Earth: Changing the declination moves the telescope north or south in the sky. Right ascension is like longitude: Changing the right ascension moves the telescope east or west. All equatorial mounts have two axes set at right angles—one for right ascension (east-west) motion and one for declination (north-south) motion.

*The sky is gridded into lines of declination and right ascension. While using celestial coordinates to locate objects is not the method we recommend for finding your way around the sky (see Chapter 11), learning which direction is which helps you understand how your equatorially mounted telescope moves on its two axes. (Map courtesy TheSky™/Software Bisque)*

# THE MOUNT

The standard beginner's telescope is often supplied with a German-style equatorial mount like this, commonly called the EQ-2 model. It's not as complicated as it appears, and there are some knobs and dials you will probably never use. But first, you need to know what the parts are called and what each does.

**Setting Circles and Index Pointers**
Graduated scales for dialing in the coordinates of celestial targets. You'll rarely use these. The right ascension (R.A.) setting circle (at bottom, marked 0 to 23 hours) can be turned independently of the mount; the declination (Dec.) setting circle (at top, marked 0 to 90 degrees) should be fixed.

**Declination Axis Lock**

**R.A. Setting Circle Lock Screw**
Tightening this little screw allows the R.A. setting circle to turn with the telescope when the R.A. lock is loose. This is unnecessary—leave this little screw loose.

**Motor-Attachment Bolt**
(under R.A. setting circle) This is where an optional fixed-speed R.A. motor drive attaches (the lowest-cost motor option).

**Slow-Motion Control**

**Latitude Bolt and Scale**
This is where the angle of tilt of the mount is adjusted for your latitude, a one-time adjustment. Turn this bolt with caution, as the entire mount will flop down once it is loose.

**Fine Latitude Adjustment**
(partially hidden at back) Turning this moves the mount up and down slowly for fine-tuning the tilt of the mount. This is for ease of polar alignment.

**Azimuth Lock**
Loosening this allows the entire mount head to swing in azimuth (i.e., parallel to the horizon). This lock and motion are for polar alignment only, not for turning the telescope to find targets.

**Slow-Motion Controls** (two of them)
Once the locks are tightened, the slow-motion controls engage and can be used to fine-tune the pointing and to follow objects.

**Right Ascension Axis Lock**
Loosen the R.A. and Dec. locks to allow the scope to turn freely to swing to a new target. Tighten them once you are close to a target.

**Mystery Wheel and Lever**
This is a gear-and-clutch mechanism for older-style AC motors and one type of variable-speed DC motor.

**Mystery Screw**
Does little. This is removed to attach the variable-speed motor. Can serve as a spare for the little setscrews on some focusers.

**Equatorial Head**
Comes as a single unit out of the box, but its angle of tilt needs to be set to your latitude, a onetime adjustment. Here, the mount is set for 45 degrees latitude.

**Counterweights**
The number and size vary with the telescope. These slide up and down the screw-in counter-weight shaft to balance the telescope around the R.A. axis.

# THE MOUNT

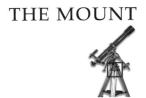

This EQ-3 model is a somewhat beefier German equatorial mount commonly supplied with an 80mm-to-100mm refractor telescope. The advantage with this stronger mount is increased stability, an important factor because refractors require taller tripods. The illustration highlights parts unique to this larger mount.

**Motor-Attachment Plate**
In optional dual-axis drives, this is where the declination motor (not shown) attaches.

**Locks** (two of them)
Lever-type knobs lock the scope in position in right ascension and declination. On this mount, these can be awkward to find in the dark. The R.A. lock must be engaged for the motor to drive the mount.

**Motor-Attachment Bolt**
This is where the optional variable-speed R.A. motor drive attaches.

**R.A. Setting Circle Lock Screw**
Leave this loose for the R.A. setting circle to work properly.

**Bubble Level**
(partly hidden on base)
To aid in leveling the tripod. You'll likely never use it.

**Fine Azimuth Adjustment**
Two more bolts push the scope from side to side for precise polar alignment. Very nice feature.

**Polar Scope Fitting**
This mount's polar axis is hollow and can accept an optional borescope for sighting Polaris for polar alignment.

**Fine Latitude Adjustment**
Two screws, in a push/pull arrangement, allow for fine aiming of the telescope's polar axis toward the celestial pole.

# THE OPTICAL TUBE

The Newtonian optical tube is the business section of the beginner's telescope. The finderscope and focuser are the most frequently adjusted parts of your telescope.

**Finderscope**
A low-power (usually 6x) scope for sighting and centering targets. Many now attach to the main tube with a dovetail bracket (shown here), aiding quick removal for packing and transport.

**Finder Adjustment Screws**
Two bolts at right angles move the finder so that it points at the same place as the main telescope. The third knob contains a spring-loaded bolt that holds the finder tube in place.

**Dustcap**
Obvious, but what's the off-axis hole for? For viewing the Sun through a smaller aperture, with unsafe and now largely unavailable eyepiece solar filters. This off-axis hole has little purpose today.

**Piggyback Bolt**
This 1/4-20 stud bolt and threaded ring allow you to attach a camera (or ball-and-socket tripod head) for piggyback photography. Very handy.

**Spider Bolts**
These three bolts hold the struts (called the spider), which in turn hold the secondary mirror in place. Don't loosen these.

**Focuser**
A rack-and-pinion type whose tension can be adjusted by the two or four screws between the focus knobs. The black screw-on ring at the top is essential (it's the adapter for 1.25-inch eyepieces). If it is missing, eyepieces have no place to sit. This ring is threaded for optional camera adapters.

**Eyepiece Setscrew**
One or two screws that hold the eyepiece in place. Don't lose them!

**Tube Rings**
Loosen the clamps to slide the tube up and down the rings for balancing in declination. In Newtonians, these also allow you to rotate the tube to place the eyepiece at a convenient angle and height.

# THE OPTICAL TUBE

The optical tube of a refractor is basically a simple spyglass with the main lens at the top and an eyepiece at the bottom for magnifying the object being viewed.

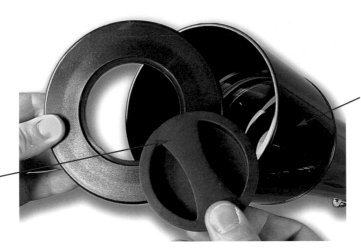

**Dustcap**
Yes, there's that small-aperture hole again (for solar viewing with dangerous filters and for stopped-down dim viewing of night-sky objects). Of little value. Tape the small filler cap in place so that it won't get lost.

**Dewcap**
A plastic extension tube helps prevent dew from forming on the main lens.

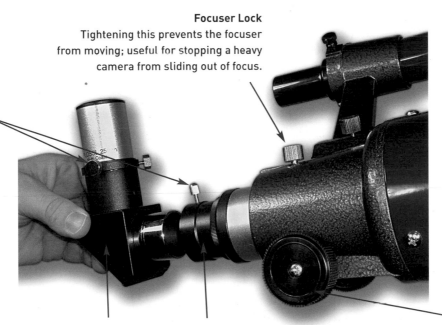

**Focuser Lock**
Tightening this prevents the focuser from moving; useful for stopping a heavy camera from sliding out of focus.

**Setscrews**
Little screws hold the diagonal and eyepiece in place. These "mission critical" components are easy to lose and almost impossible to replace.

**Focuser**
A rack-and-pinion gear type (usually) that slides the eyepiece back and forth to focus. Once focused, an eyepiece won't need refocusing as you move around the sky (all celestial objects are at infinity). However, changing eyepieces generally requires changing the focus.

**Star Diagonal**
A 90-degree prism for preventing neck and back strain. With it, you can comfortably look down into the telescope instead of having to stoop and crane to look straight up through the telescope. Essential for some refractors to reach focus.

**Adapter Tube**
Adapts from focuser drawtube's diameter to 1.25-inch eyepieces. Removed for some camera adapters. Don't lose this.

## THE TRIPOD

Many beginners' telescopes are supplied with lightweight but sturdy aluminum tripods. The legs and fittings are usually packed as separate components that must be bolted together.

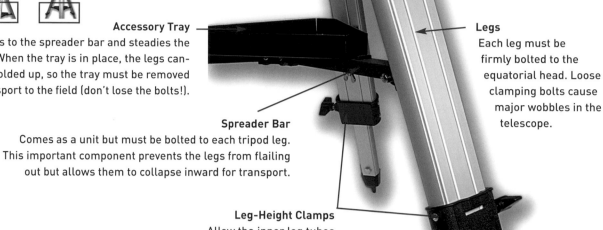

**Accessory Tray**
Bolts to the spreader bar and steadies the tripod. When the tray is in place, the legs cannot be folded up, so the tray must be removed for transport to the field (don't lose the bolts!).

**Spreader Bar**
Comes as a unit but must be bolted to each tripod leg. This important component prevents the legs from flailing out but allows them to collapse inward for transport.

**Legs**
Each leg must be firmly bolted to the equatorial head. Loose clamping bolts cause major wobbles in the telescope.

**Leg-Height Clamps**
Allow the inner leg tubes to slide down to extend the tripod's height. These should be tight to prevent a leg from collapsing.

**Foot Bolt**
Some tripods come with a short extension knob that screws onto this bolt. Even without it, though, this vestigial bolt can be used for its intended purpose, as a point to place your foot to press the tripod leg firmly into the ground.

## EYEPIECES

At least one eyepiece, if not two or three, is supplied with most telescopes. You change power by changing eyepieces. The lowest-power eyepiece is the one marked with the largest number (often 25mm).

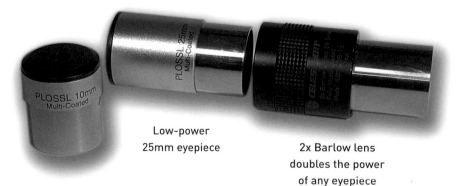

Low-power
25mm eyepiece

2x Barlow lens
doubles the power
of any eyepiece

High-power
10mm eyepiece

**Barlow Lens**
Inserted between the eyepiece and the telescope, a Barlow lens doubles or triples the power of any eyepiece. The premium-grade Celestron Ultima Barlow is one of our favorites. The low-cost Barlows included with many beginners' telescopes are often of just acceptable if not downright poor quality.

## HOW A TELESCOPE MOVES

To move across the sky, an equatorial mount swings around two axes. One of these, the polar axis, must be set to aim toward the sky's celestial pole (near the North Star for observers who live north of the equator).

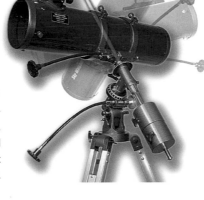

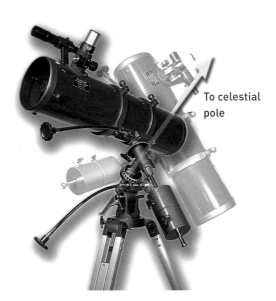

To celestial pole

**Declination Axis** ▶
The telescope moves north-south around the axis shown. You move the scope around this axis to find a new object in the sky.

**Right Ascension, or Polar, Axis** ▶
The telescope moves east-west around the axis shown. To follow an object across the sky as Earth spins, the scope turns around this axis. Because the axis must be aimed toward the celestial pole to polar-align the mount, it is commonly called the polar axis.

# How *NOT* to Set Up a Telescope

What's wrong with this picture? Lots! Yet this is how reflector telescopes are often depicted in discount-store catalog ads (we have the ads on file to prove it!) and set up in chain stores by staff who know little about telescopes.

1. The eyepiece is at the bottom of the telescope, where all eyepieces should be, right? Not on a Newtonian reflector. This scope has its tube pointed at the ground.
2. The finderscope is mounted backwards, here pointed at the sky, to be sure, but 180 degrees away from where the main optics are aimed, rendering it useless as a finder. Worse, we've seen the Barlow lens installed here as if it were the finder.
3. The equatorial mount is polar-aligned for a location near the North or South Pole, near 90 degrees latitude, an unlikely home for most buyers. It will be impossible to find and track objects.
4. The counterweight shaft and its weights are missing, making for an unbalanced and unstable telescope.
5. The 2x Barlow and 4mm eyepiece are in place, in the misguided belief that more power is better.

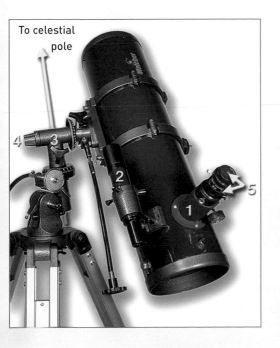

To celestial pole

# Some Assembly Required

The majority of entry-level telescopes require a similar assembly procedure for turning the box of parts into a working telescope. Be sure to do this first-time assembly indoors, in the light, warmth and comfort of your home.

## TELESCOPE ASSEMBLY, A 10-STEP PROGRAM

Here's the basic procedure for putting together the typical import reflector or refractor on an equatorial mount. The steps apply to most small telescopes and involve attaching the legs to the equatorial mount first, then assembling the rest of the tripod. Only after you have a secure and solid tripod (Steps 1 through 3) should you attach components such as the counterweights and tube to the mount. Fail to follow that order, and your telescope could collapse.

## STEP 1

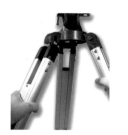

**Bolt Legs to Mount Head**
Notice there's a hinge halfway down each leg. The hinge goes on the inside and is for holding the spreader bar.

**Variation:** On some mounts, the legs attach to a separate top plate. The main mount head then attaches to this plate with one large bolt. This arrangement makes the scope easier to disassemble for transport.

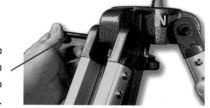

## STEP 2

**Attach Spreader Bar**
Stand the tripod and mount on a nonslip floor, and attach each arm of the spreader bar to a tripod leg by pressing the bolt through the leg's hinge and the spreader bar's arm. A Phillips head screwdriver (likely not supplied) is required for the final tightening. On some scopes, the legs come with the spreader bar already attached.

## STEP 3

**Attach Accessory Tray**
Use the small wing-nut bolts to attach the tray to the spreader bar. This helps stabilize the scope.

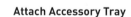
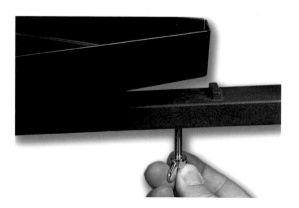

## STEP 4

**Attach the Counterweight Shaft**
This screws into the head (leave the weight off for now). The shaft acts as a good handle for the next step, which is likely necessary, since most mounts come packed in the wrong orientation.

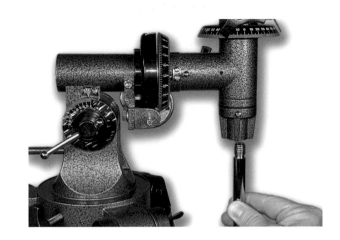

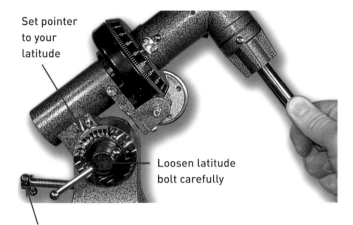

Set pointer to your latitude

Loosen latitude bolt carefully

Latitude fine-adjust screw

## STEP 5

**Adjust Mount Angle**
Loosen the large latitude bolt (carefully!) while holding the mount head. Tilt the mount to your latitude (use the scale on the side of the mount as a guide). Just get it close at this point. Then lock that large bolt down good and tight. You won't need to adjust it again. Turn the latitude fine-adjust screw in until it pushes against the mount. This will help hold it in place.

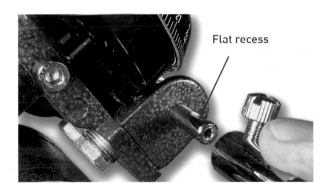

Flat recess

## STEP 6

**Attach Slow-Motion Cables**
Use a small flat-bladed screwdriver (or the little tool supplied) to attach each cable to its shaft on the mount. The setscrew in each cable's sleeve mates with the flat side of the D-shaped shaft. The long cable goes on the declination axis.

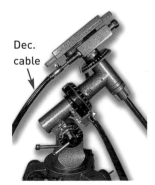

Dec. cable

**Refractor**
For a refractor, turn the head so that the declination slow-motion control extends to the bottom, to place it near the eyepiece.

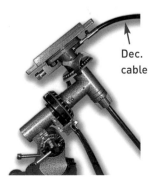

Dec. cable

**Reflector**
For a reflector, turn the head so that the declination slow-motion cable is at the top—closest to the eyepiece.

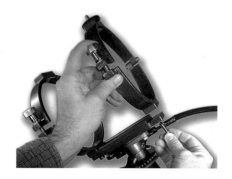

## STEP 7

### Bolt Tube Rings to the Head
Use the supplied bolts and lock washers to fasten each ring to the head. The ring with the piggyback camera bolt can go at the top of the mount. Do this without the telescope tube —just fasten the rings on first.

## STEP 8

### Slide on the Counterweight
Undo the screw at the bottom of the shaft so that the counterweight can slide onto the shaft. Put the weight about halfway up the shaft for now. Be sure to replace the screw at the bottom of the shaft—it prevents the weight from accidentally sliding off the shaft and onto your foot!

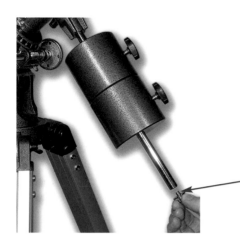

Install this

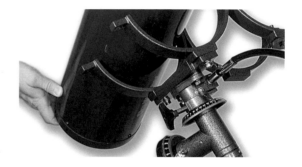

## STEP 9

### Place the Tube in the Cradle
Remember, if it's a reflector, the eyepiece goes at the top!

## STEP 10

### Install the Finderscope
First, insert the finderscope into the bracket. Many finders come with a small rubber O-ring (you may find it strung around the bracket). The O-ring goes into the recessed notch on the finder tube and acts as a spacer to hold it steady. To insert the finder, you may have to pull back the spring-loaded silver bolt (if your finder has that style of bracket). The bracket then clamps into the dovetail channel on the telescope tube.

Pull back spring-loaded bolt

O-ring

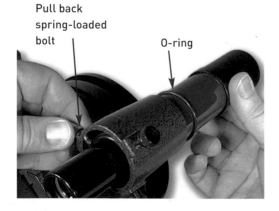

**Variation:** Some brackets must be bolted directly to the tube.

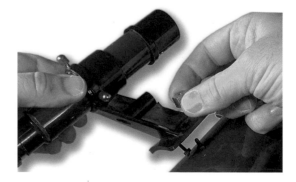

# MOTOR-DRIVEN

The most common option for this EQ-2 mount is the single-speed DC motor for the right ascension axis. It attaches to a bolt hole on the side of the polar axis and slides onto the shaft normally occupied by the R.A. slow-motion control. Attaching the motor forfeits the east-west slow-motion control and any ability to fine-tune the position of the scope manually in the east-west, or R.A., direction. But once on a target, the scope continues to track it automatically, a great convenience.

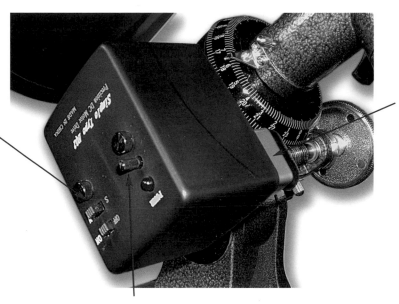

**Direction Switch**
Flicking the N-S switch to S makes the motor turn the opposite way for use in the southern hemisphere. If you live north of the equator, set it to N. If objects drift out of the field even faster with the motor turned on, chances are that it's on S.

**Battery**
All drives today operate with batteries rather than AC power. This one uses a 9-volt transistor-radio battery that lasts, at best, for only a few nights of use. A rechargeable battery is a good choice here.

**Speed Controls**
Turning this knob speeds up or slows down the motor. Getting it to turn at the right speed to track stars is trial and error on this model. Turn it to the highest setting as a start.

**Another Option**
A better, though more expensive, choice for the EQ-2-class mount—and the only choice for better-grade mounts, like this EQ-3 model—is a variable-speed motor drive. This unit has push-button controls for 2x and 8x speeds for fine centering of objects and for guiding long-exposure astrophotos taken with a camera piggybacked onto the telescope tube. Variable-speed motors also run off separate battery packs with multiple C or D cells for much longer battery life. Better dual-axis units can also control motors on both axes.

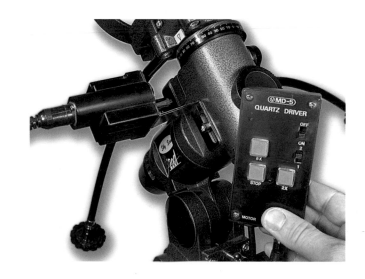

# Daytime Adjustments

Setting up at night is *not* the next step to using your new telescope. Take your scope out first during the day for some simple but crucial adjustments. This daytime checkout is a highly recommended procedure that many owners of new telescopes bypass, anxious to experience the thrill of first light with their telescope under a starry sky.

However, essential adjustments, such as lining up the finderscope and getting the telescope focused for the first time, are much easier to do during the day than at night. You can find suitable targets more easily, and you'll know what they are supposed to look like once you get them in view.

Attempting to make these initial adjustments at night "cold turkey," on targets you can't find (because the finder isn't lined up yet) or can't focus on (because the focuser is way off position), is inviting frustration and disappointment.

## UP TIGHT

First, ensure that all the tripod bolts are firmly cinched down. A loose tripod makes for a wobbly telescope and a jittery image. Then play with moving the telescope around. Learn where the locks are. Unlock the axes to swing the telescope close to a target. Once there, lock the axes and use the slow-motion controls to center a target. As you do, you'll notice a couple of things:

The slow-motion controls probably won't do much as long as the axes remain unlocked. Lock the axes (with the levers shown here) to engage the slow-motion controls.

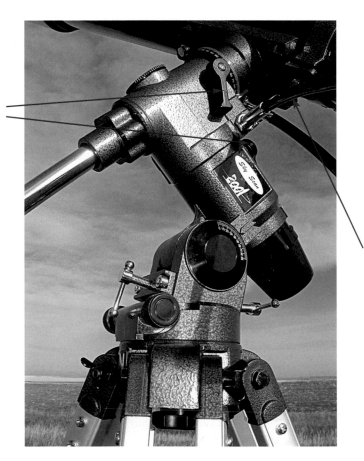

The telescope tube will collide with the slow-motion cables or with the tripod itself when aimed in certain directions (mostly straight up). There is a trick to avoid this to some extent. Read on!

## BALANCING ACT

Unlock the telescope, move it to point at a new area of the sky, then let it go. If the scope swings away on its own, it is out of balance. An unbalanced telescope tends to point wherever it wants to and makes it hard to find the targets you'd like to observe. Motor drives might not work well either, as they labor to turn the telescope. To fix this:

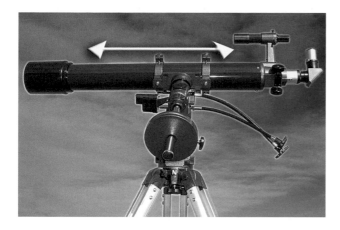

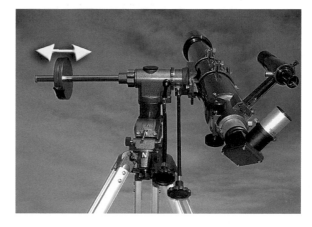

**Adjust Balance in Declination**
With the declination lock loose and the tube horizontal, loosen the cradle rings and slide the tube along the cradle until it is balanced in the north-south, or declination, direction. In most cases, the telescope balances with the cradle in the middle of the tube. Large refractors are top-heavy, however, and balance with the main lens nearer the cradle.

**Adjust Balance in Right Ascension**
With the R.A. lock loose and the counterweight shaft horizontal, slide the counterweight(s) along the shaft. The scope is balanced when, with the lock loose, it stays put when you let it go, no matter where it is pointed in the sky. When balancing any scope, be sure the finderscope and the eyepiece are in place and the lens cap is not.

## BECOME FOCUSED

It is much easier to focus a new telescope first during the day on a familiar landscape object, rather than at night on an unfamiliar stellar target. To do this:

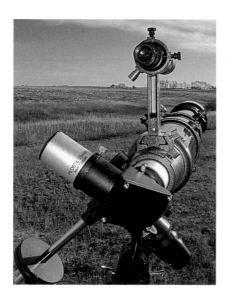

1. Insert the low-power eyepiece. With most entry-level telescopes, that's the one marked 25mm or 20mm, not the ones marked 12mm, 9mm or 6mm. If you have a refractor, insert the star diagonal first, then slide the eyepiece into it. Don't use any Barlow lens that might have been supplied.

2. Swing the telescope to a distant target on the horizon (sight along the tube to get it close). It should be several hundred meters away, not just across the street.

3. Rack the focuser back and forth until the image sharpens to a crisp focus. With a daytime object, it should be obvious when you are getting closer to focus.

4. Leave the focuser as is. You are now close to being perfectly focused for nighttime objects in the sky.

## GETTING LINED UP

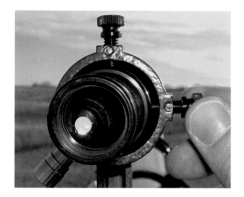

Use the adjustment bolts (often just two on many models now, but three or six on some finderscopes) to tilt the finder's tube until your target object is centered at the intersection of the finder's crosshairs.

Once the low-power finder is collimated in this fashion, any object placed on the crosshairs is automatically centered in the main high-power telescope. You may need to recollimate the finder from time to time.

The little low-power finderscope on the side of the tube is an essential aid to trouble-free telescope enjoyment. To be of value, the finderscope must be lined up so that it points to the same place as the main telescope. This is easiest to do during the day on a distant landscape target. To do this:

With the low-power eyepiece in place in the main telescope, find a recognizable object on the horizon (a power pole or an antenna, perhaps).

Now look through the small finderscope, and locate the same object; it will likely be off-center.

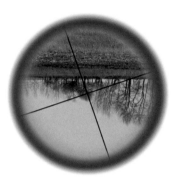 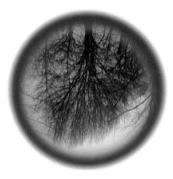

View through the finderscope          View through the telescope

## SHARPENING THE FINDER

If the view through the finderscope is fuzzy, you'll need to focus the optics of the finder. Most units allow this, though it is not always obvious how. Some finder eyepieces turn. As they twist, they slide in and out to focus. For other less obvious units:

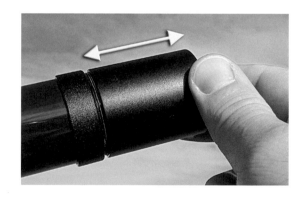

1. The finder view is focused by moving the main lens, not the eyepiece, up and down the finderscope's tube. This lens (usually a 25mm- or 30mm-diameter objective) is contained in a metal or plastic cell screwed onto the tube. It is not glued and is designed to turn. To adjust the focus, first grip this black lens cell.

2. Turn the lens cell counterclockwise to unscrew it. A short retaining ring behind the lens cell will also loosen at this point. You may need to turn this ring to back it away from the main cell. This will expose enough of the threads to give you room to move the main lens. The threads are fine, because adjusting for most users' eyesight requires only a minor focus change.

3. Turning the lens cell moves the main lens up and down the tube, changing the focus. Don't worry if it comes off the tube; the optics shouldn't fall out.

4. When images of distant objects look sharp, turn the retaining ring so that it locks back up against the lens cell. The finder is now focused for your eyes and shouldn't need to be adjusted again. Choose whether you want the finder to be focused for viewing with your glasses on or off—either way is fine.

# Nighttime Use

Many entry-level telescopes are compact and light enough that they can be left assembled and carried outside as a unit for impromptu viewing sessions. That is the best arrangement. But usually, removing the tube from the mount is required each night.

When reassembling your telescope, you may need to balance it for proper operation of the mount.

Once outside, all equatorial mounts must be aligned to the celestial pole, a procedure with an undeserved reputation for being complex and intimidating. Yet it can be accomplished in seconds.

Even when the mount is aligned, aiming at some areas of the sky may be awkward. This is one quirk of a German equatorial mount that is easy to work around.

## TELESCOPE BREAKDOWN

If you do need to break down your scope after each night's use, keep the procedure as simple as possible.

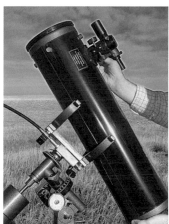

**1.** Remove the tube. Unless your telescope tube and rings attach to the mount via a quick-release dovetail mechanism, the easiest "no-tool" method of removing the tube is to open the cradle rings and simply lift out the tube. When replacing it, be sure the tube is balanced in declination and not sitting too far up or down the cradle rings.

**3.** If you need to break the scope down further, remove the tripod's accessory tray so that the tripod legs will fold in. Remove the tripod legs themselves only if you need to pack the scope into a small space for transport.

**2.** Remove the counterweights from the shaft. This makes the mount lighter and less cumbersome.

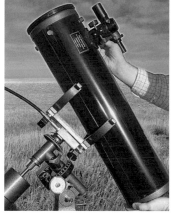

**4.** If you are packing the telescope in a car, remove the finder-scope, slow-motion controls and eyepiece. Protruding as they are, they can get damaged or lost. Keep the counterweight well packed so that it can't bang against the tube.

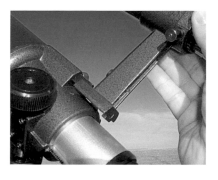

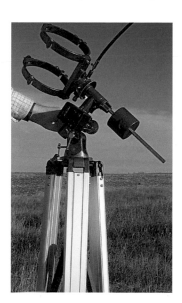

# ON THE LEVEL

Precise leveling of any telescope mount simply isn't necessary. However, if you have the angle of the equatorial head set for your latitude, then leveling the scope ensures that the polar axis of the mount is indeed pointed close to the celestial pole.

**1.** Use the height adjustment on the tripod legs to level the mount. Eyeballing it should be sufficient.

**2.** If the scope is out of level, just turn the fine latitude adjustment screws to shift the mount's angle up or down to aim at the celestial pole.

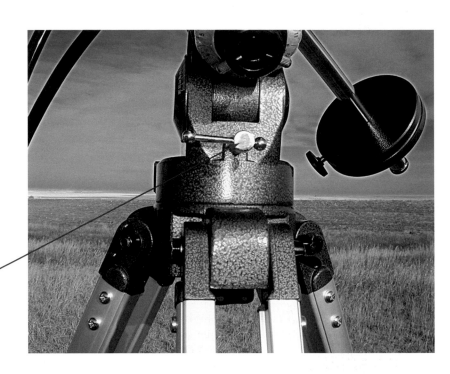

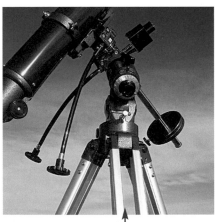

Adjust south leg

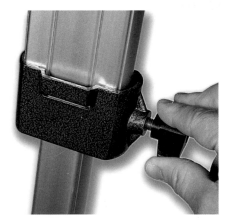

**3.** As an alternative to adjusting the head, simply raise or lower a north- or south-facing tripod leg (loosen the wing nut as shown to adjust the leg) to accomplish the same thing, effectively leveling the scope and aiming the polar axis at the celestial pole.

# TIME TO ALIGN

To track objects properly (or at all!), any equatorial mount must be aligned so that its polar axis points at the celestial pole. The sky rotates around this point, and the mount must too. If you ignore polar alignment, you may as well have bought a cheaper altazimuth mount! For users in the northern hemisphere, the task requires but a few simple steps.

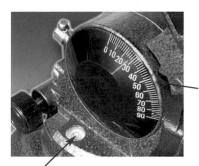

Latitude scale

Bubble level

**1.** Adjust the polar axis so that its angle of tilt equals your latitude. You can look that number up in most atlases and do this during the day.

Altitude or latitude knobs

Azimuth knobs

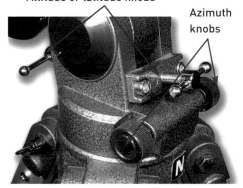

**2.** Place the telescope outside at night so that the polar axis aims due north toward Polaris (that's true north, not magnetic north). Some mounts indicate which side should face north with a large N on the base. Knowing which way is north at your site and how to locate Polaris are essential to finding anything in the sky. See Chapter 11 for tips.

**3.** Use the fine azimuth or altitude adjustments to tweak the aim of the mount, if needed (this is easier than trying to shift the whole tripod to aim the mount due north).

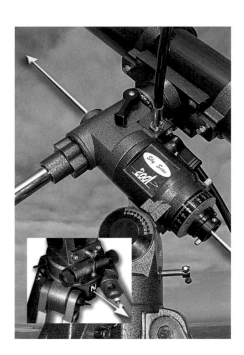

**Variation:** Sighting Polaris through an optional polar scope inserted into the polar axis allows for more precise alignment. But for casual observing, roughly aiming the polar axis due north will suffice; objects will stay in the eyepiece for several minutes at a time.

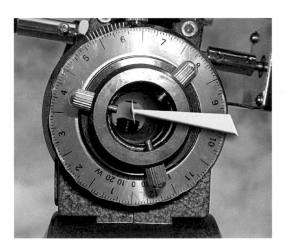

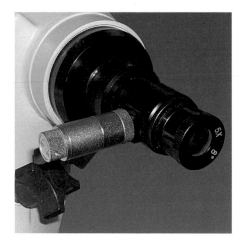

**4.** If your mount has a hollow polar axis (the EQ-3, Celestron CG-5 and Orion SkyView types, for example), you can sight Polaris through the axis to achieve a more precise centering.

On subsequent nights, all you'll likely need to do is simply place the scope at your favorite backyard spot so that the polar axis aims due north. For more advanced alignment methods for astrophotography and setting-circle use, see www.backyardastronomy.com.

## AIMING FOR THE SKY

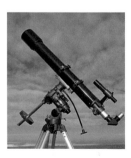

Now you are ready to move the telescope around the sky in search of celestial targets. Once you are lined up on the celestial pole, you do not use the azimuth or altitude adjustments again—they are not for finding objects. Instead, move the telescope around the right ascension and declination axes. Here's the process:

**1.** To swing the tube around to where you want to look in the sky, first loosen the R.A. and Dec. locks. Once near a target (as sighted through the finderscope), lock the axes, then use the slow-motion controls to zero in on the spot.

**2.** Move the telescope around the Dec. axis as shown at left to move it north or south in the sky. Move the telescope in this direction only when you want to change targets.

**4.** If you have a motor on the R.A. axis, you may find the slow-motion control in that axis no longer works (many entry-level scopes have no clutch mechanism to allow the axis to be turned by hand when a motor is attached). If that's the case, leave the R.A. axis just barely unlocked, with a slight tension on it, and nudge the scope by hand until the object is centered. Then lock up the axis, and the motor should engage and start driving.

**3.** Move the telescope around the R.A. axis as shown at right to move it east or west in the sky. The telescope moves from east to west to follow objects across the sky during the night. A motor can attach to this axis for automatic tracking.

**5.** As a final step, if the motor has 8x speed controls, you can fine-tune an object's position by speeding up, stopping or reversing the motor, which makes a variable-speed drive a good choice.

## YOU CAN'T GET THERE FROM HERE

Swinging a German equatorial mount around the sky is where most people get tangled up. There are some places in the sky where such a mount simply won't point easily. For example:

**1.** Here's a German equatorial mount aimed due north at Polaris. No problem.

**2.** Here's the same mount with the scope aimed high in the southeast, where you'll usually be looking (at the Moon and planets, for example). Also fine.

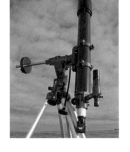

**3.** But here's the mount with the scope aimed high overhead. Oops! The tube hits the tripod. This is inevitable with most scopes on this type of mount and tripod.

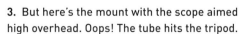

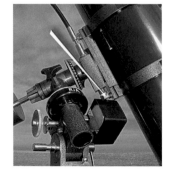

**4.** This EQ-2 mount has an even worse problem. The motor drive is badly positioned, making it impossible to aim the telescope high into the west or northwestern sky (the tube collides with the motor casing). With the better EQ-3 mount, the drive moves with the mount and does not get in the way, no matter where the scope is aimed.

# DOING THE EQUATORIAL TANGO

All German equatorial mounts, even the best, share a limitation: They cannot track an object all the way across the sky in one uninterrupted arc.
Here's the situation:

**1.** The telescope is aimed east and is following an object as it rises into the sky to a position due south. No problem.

**2.** If the telescope stays with the object as it moves into the west, the tube or slow-motion cables will eventually collide with the mount or tripod. What to do?

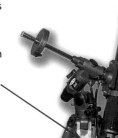

**3.** The solution is to "flip" the tube to the other side of the mount, a little pas de deux all owners of German equatorial mounts must learn to perform.

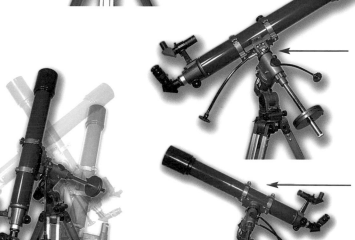

**4.** Now the scope is aimed at the same area of the sky, but the tube is on the other side of the mount and can follow the object into the west. In general, when looking west, the eyepiece should be on the east side of the mount.

**5.** Conversely, when looking at the eastern sky, the eyepiece should be on the west side of the mount. Learning how and when to tango with your telescope will give you unimpeded observing of most of the sky.

# A CHANGE OF LATITUDE

The only time you will need to alter the mount's home position for the polar-axis angle is when you travel far north or south in latitude.

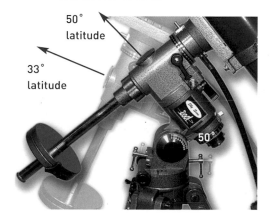

50° latitude

33° latitude

50°

Traveling east or west but staying at the same latitude makes no difference to how you set up your telescope.

In Vancouver, British Columbia, you would set the mount's latitude angle to +50 degrees, but travel south to Phoenix, Arizona, and the angle would have to be adjusted to +33 degrees.

Travel to Sydney, Australia, and the angle would be nearly the same (+34 degrees) as at Phoenix, but the telescope would have to be placed so that its polar axis points due south, at the south celestial pole. The motor drive would need to be switched to the S position so that the scope turns around the polar axis in the other direction.

# First Light

Even after assembling their telescopes and checking them out during the day, new scope owners often have disappointing first-light experiences at night. By heeding a few do's and don'ts, you should enjoy wonderful views your first night out.

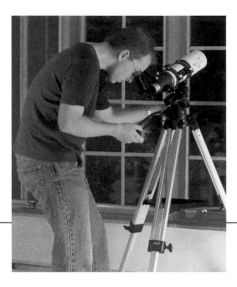

## DON'TS

### DON'T look through a window
Window glass distorts views through telescopes, usually by producing double images. Opening the window doesn't help, because the warm air rushing out of the window blurs the image even more. Telescopes must be placed outside. If the night is particularly cold (freezing or below), allow 15 to 45 minutes for the telescope to cool down, as warm air inside the tube can also blur images.

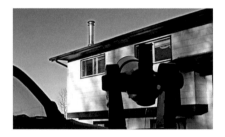

### DON'T look over a heat source
A cooled-down scope will be handicapped if it must peer through warm air rising from a nearby chimney, a heat vent or even a warm car hood. Black asphalt that was searingly hot during the day will also release its warmth at night.

### DON'T take apart lenses
If you disassemble your eyepieces or main refractor lens, chances are that you'll never get the optics back together properly.

### DON'T use your highest-power eyepiece
Higher power isn't better. The 4mm and 6mm eyepieces supplied with many entry-level telescopes yield blurry views, not to mention fields of view so narrow that even finding the Moon becomes a challenge.

### DON'T insert the Barlow lens
The 2x and 3x Barlow lenses often supplied as standard equipment can be just as debilitating to telescopic views. Good-quality Barlows can be a useful accessory; poor-grade Barlows become doorstops.

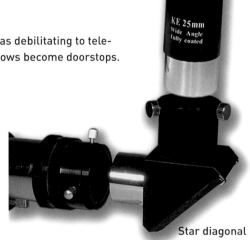

Star diagonal

## DO'S

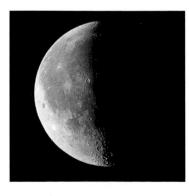

### DO use your lowest-power eyepiece
The 25mm or 20mm eyepiece will produce the brightest, sharpest image for your first views.

### DO use the star diagonal on refractors
Many refractors will not focus without the star diagonal in place.

### DO make the Moon your first target
The Moon presents such a wealth of detail in any telescope, it's hard not to be impressed.

# Why Is the View Upside Down?

Astronomical telescopes, and even their small finderscopes, almost never present right-side-up views. The extra optics needed to flip the image the right way around would just add to the cost and detract from the image by dimming the light and distorting high-power views.

The upside-down views may seem confusing at first, but the only celestial object where the image is obviously flipped compared with the naked-eye view is the Moon. With the stars and planets, an erect image in the telescope has little value, since these objects look like points to the naked eye anyway. Far better to have the sharpest, brightest image at the eyepiece. You'll soon learn to maneuver your telescope so that the image moves the way you want it to without having to give a second thought to which way is up.

Where the inverted image can be confusing at first is in the finderscope. Straight-through finders are easiest to orient; simply turn a star chart upside down to match the view. However, finders with right-angle eyepieces present mirror-image views of the sky. Matching these to a chart requires flipping the chart over and shining a flashlight through the paper or viewing the chart in a mirror. One solution is to print out custom finder charts using computer programs that can flip the sky horizontally. Avoid this nuisance by purchasing a new finderscope with an erecting prism, preferably a straight-through variety for ease of aiming.

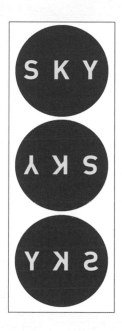

**Erect-Image View as in...**
• Binoculars
• Terrestrial spotting scopes
• Finderscopes with an Amici or erecting prism

**Inverted View as in...**
• Newtonian reflectors
(or any telescope with an even number of reflections (i.e., two mirrors))
• Straight-through finderscopes with no erecting lens

**Mirror-Image View as in...**
• Refractors and catadioptric telescopes with a star diagonal
(or any telescope with an odd number of reflections)
• Finderscopes with a 90-degree right-angle star diagonal

To orient yourself at the eyepiece, remember:

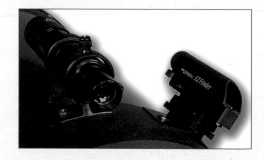

1. With no motor tracking, objects drift across the field from east to west. An object "preceding" another lies to the west and enters or leaves the eyepiece first. An object "following" another lies to the east and follows its companion across the field.

2. Bump the scope toward Polaris, and new sky will appear at the north edge of the field of view.

........................................................................................................................

*To find objects, we prefer to use a straight-through finder, as shown above right, augmented with a window-style finder that projects a red dot or bull's-eye onto a naked-eye view of the sky. The red-dot finder helps you to aim the telescope to the right area of the sky, while a good 6x30 or 7x50 finder allows you to see many targets.*

# Top 10 Newbie Questions

Many of the usual questions that first-time telescope owners ask (How much does it magnify? Can I see the rings of Saturn? and others) are covered throughout this book. But here are some quick answers to a few common questions from those new to the hobby of back-yard astronomy.

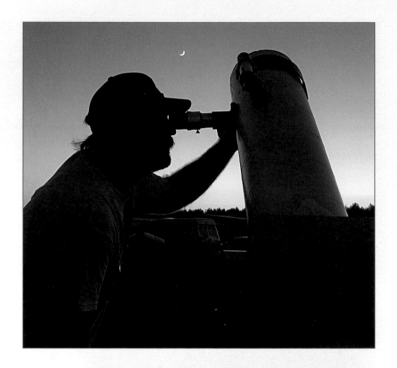

### 1. How far can I see?
Even with the unaided eye, you can see the Andromeda Galaxy, some 2.5 million light-years away. A telescope can reveal fainter galaxies hundreds of millions of light-years away, so distant, in fact, that their light has been traveling to us since long before the Age of Dinosaurs.

### 2. Can I see the flags left on the Moon?
So near and yet so far. Flags, footprints and landing craft on the Moon are simply too small to resolve in any earthbound telescope.

### 3. Is something wrong? Stars look just like points...
When a telescope is in focus, all the stars should look like points. Beyond the Sun, no star is close enough to show itself as a disk. If stars look like large shimmering disks or donuts (as at right), the telescope is out of focus.

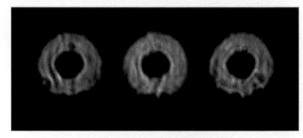

### 4. ...and planets look like tiny dots.
Don't expect the planets to look like the poster-class pictures taken by space probes. Even at 200x, planet disks remain small, but the detail is there. The trick is learning to see that detail, a skill that takes time to acquire but comes with practice.

### 5. In a reflector, doesn't the secondary mirror block some of the light?
Yes, but not enough to dim the image noticeably (the area of the secondary mirror is often no more than 10 percent of the area of the primary). When the image is in focus, the secondary mirror's presence is invisible. Only when you throw a star out of focus do you see the secondary mirror's silhouette as a hole in the center of the donutlike image.

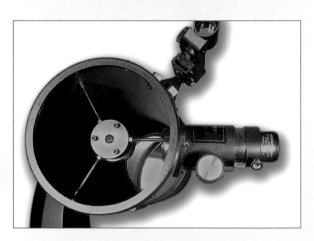

*Almost all reflector telescopes employ a secondary mirror to deflect the light to the side or down to the bottom of the tube. The secondary mirror's presence becomes obvious as a shadow only when the image is thrown far out of focus, as shown top right in a series of images of a defocused star distorted by rippling waves of atmospheric turbulence.*

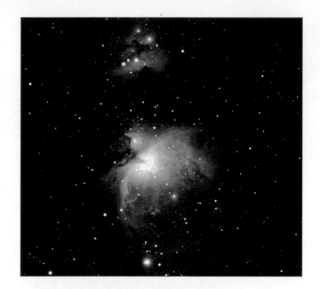

### 6. How do I find objects?

For owners of noncomputerized telescopes, the best method is to star-hop. Using a star map or a guidebook to point the way, first locate a bright naked-eye star near your target, then hop through recognizable star patterns. Place the spot where your target lies at the crosshairs of your finderscope. The object should now be in the field of your low-power eyepiece. See Chapter 11 for more details and recommendations for charts and star-hopping guidebooks.

### 7. Why do objects move so fast across the eyepiece?

This always surprises first-time telescope owners. Telescopes magnify not only the size of an object but also the effect of the sky's east-to-west motion. At low power, objects drift from the center to the edge of the field in two to three minutes. At high power, objects must be recentered every 20 to 30 seconds. That's why slow-motion controls, smooth Dobsonian mounts or motor-driven mounts are so important to the enjoyment of using a telescope.

### 8. Why can't I see colors in nebulas?

Although we cover this elsewhere, the point bears repeating here: Even through large telescopes, nebulas and galaxies are too faint to excite the color receptors in the eye. They appear as monochrome patches of light, as in Adolf Schaller's drawing of the Orion Nebula at left. Only long-exposure photographs can pick up the colors emitted by glowing nebulas.

### 9. Is the view better outside the city?

For faint objects such as nebulas and galaxies, yes. The darker the sky, the better. But the Moon and planets, being bright, can look just as sharp and clear in a light-polluted city sky as they do from the country. The main tip is to avoid looking over heat sources that can produce image-blurring turbulence. See Chapter 8 for advice on picking a site.

### 10. Where can I learn more?

Your local astronomy club, science center or planetarium likely conducts regular stargazing sessions at an observatory, a park or a nature area. These are good opportunities to look through a variety of telescopes and learn what you should see at the eyepiece. At large regional star parties, a smorgasbord of equipment (as at right) provides even experienced amateur astronomers the opportunity to compare various models of telescopes.

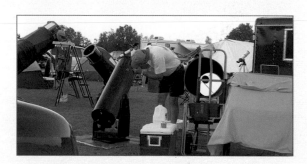

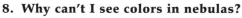

*To make the most of your telescope when viewing deep-sky objects, like the Orion Nebula (top left), travel far from city lights, perhaps to view with a group at a star party (above). While the vivid colors may not be visible, bright deep-sky objects can still reveal a breathtaking amount of detail in dark skies. Nothing beats the live view through a telescope (lower left).*

## PART 2

# The Naked-Eye Sky

The most lasting benefit of
recreational astronomy is not
the lessons in astrophysics you
learn, as valuable as they may
be, but the lifelong awareness
you gain of the sky's wonders.
Whenever you step outside, you
automatically look up. Backyard
astronomers soon learn to look
for, and see, amazing sights that
almost everyone else misses.

The play of light and shadow
in the day sky along with nighttime glows, both subtle and
dramatic, provide a never-
ending sky show unfolding overhead. Enjoying the show doesn't
require elaborate telescopes or
superdark skies. Remarkably,
many of the sky's finest offerings
can be seen from the city with
no more than unaided eyes—
guided by the knowledge of
what to look for under just the
right atmospheric conditions.

..........................

**Billions of six-sided,
ice-crystal prisms
floating in calm, cold
air refract sunlight to
create mock suns,
also known as sun-
dogs, on either side
of the Sun. Photo by
Terence Dickinson.**

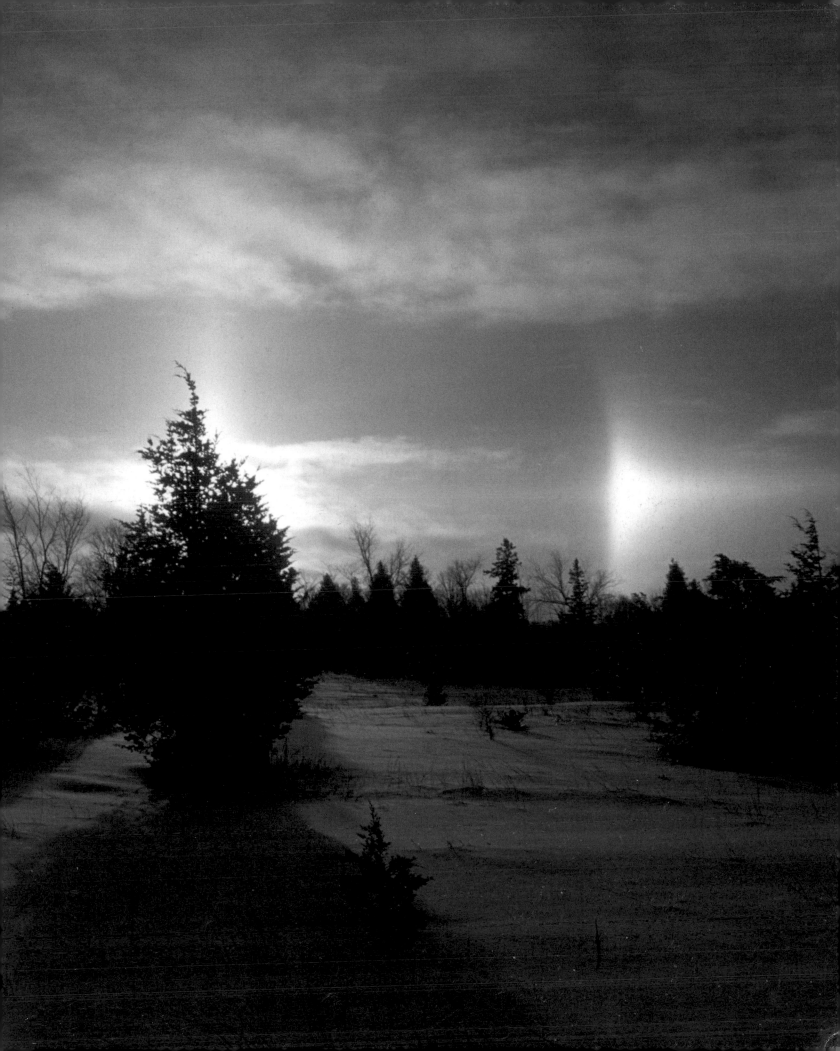

# Phenomena of the Day Sky

We typically think of astronomy as a nocturnal pursuit. But a roster of unusual effects of light and shadow plays across the daytime sky. Almost all are strictly naked-eye phenomena. The most familiar is the one we take for granted: the color of the daytime sky.

## THE BLUE SKY

The sky is blue—not green, yellow or pink —because blue and purple light have the shortest wavelengths of all the colors of the visible spectrum. Short-wavelength light is most easily scattered by the air molecules of our atmosphere. Think of a crowd of bantamweight wrestlers trying to force their way through a line of heavyweight sumo champions. The hapless bantamweights get scattered at random in a wild melee. So it is with blue light waves, in a process called Rayleigh scattering.

How blue the sky appears depends on how clean and dry the air is and on how much air lies above you. Water vapor whitens a sky by scattering all wavelengths equally. Dust and pollution also turn a day sky pale blue, if not brown or gray. Lower-altitude sites are likely to have more contaminants in the air than will mountaintop sites, especially if those summits are in dry desert locations.

On a cloudless early morning or late afternoon, look around the canopy of blue. Where is the sky bluest? You might think it would be the point overhead, where light passes through the least amount of water vapor and contaminants. But look carefully, and you'll see that the bluest sky lies along a band 90 degrees away from the Sun. Scattered light from the sky is naturally polarized (the light waves all vibrate in the same direction), and the degree of polarization is greatest at right angles to the Sun. Even without the aid of a polarizing filter on your camera, this natural polarization darkens the sky, creating a band of deeper blue arcing across a clear day sky.

While it may not be obvious, the sky is also blue on moonlit nights. Moonlight is just reflected sunlight with the same range of colors, simply much fainter and below the eye's threshold of color sensitivity. Take a long exposure, and you'll record a sky painted with scattered blue sunlight that has reflected off the Moon first.

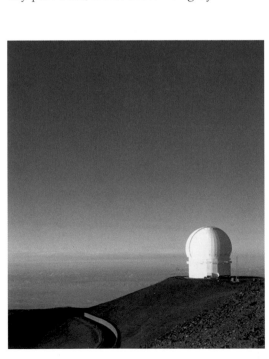

**Reaching for the Sky ▶** The Earth's atmosphere is an ocean of life-sustaining air that cloaks our planet. As serene as the atmosphere appears on a clear day, astronomers know that the more air they look through, the more layers of natural turbulence they encounter. That means researchers favor high-altitude sites, such as this 14,000-foot summit in Hawaii. Recreational astronomers need not go to such extremes.

# RAINBOWS

If a rainstorm has just passed by and sunlight is breaking through, look around for a rainbow. A sunbeam shining through a raindrop is usually reflected once and heads back in nearly the same direction as it entered. In the process, the beam of light is split into its component colors by the prism-like qualities of the raindrop. Multiply this effect by millions of raindrops, and the result is a curving swath of color arching around the point in the sky directly opposite the Sun. To be precise, rainbows always lie at a radius of 42 degrees from the antisolar point. To find the antisolar point, stand with your back to the Sun and imagine a line extended from the Sun through your head toward the ground in front of you.

From ground level, the antisolar point always lies below the horizon. So we never see a rainbow as a complete circle. Instead, we see the top arc of the full circle. The closer the Sun is to the horizon, the more of the rainbow we see. However, from an aircraft or a mountain peak, it is possible to see a rainbow as a full circle. On the other hand, if the Sun is high overhead, you can never see a rainbow. For a rainbow to be visible against the sky, the Sun cannot be more than 42 degrees above the horizon. For that reason, rainbows are a late-afternoon or an early-morning phenomenon.

Double rainbows occur when the sunlight is particularly strong and the sky saturated with raindrops. A second rainbow appears as a result of light bouncing through two reflections inside the raindrops. The fainter second bow shows up outside the primary bow at a radius of 51 degrees from the antisolar point, with its colors in reverse order—red on the inside of the arc rather than the outside, where it appears in the main rainbow.

Other rainbow-related effects to watch for are the brightening of the sky inside the main bow and the darkening of the sky between the primary and secondary rainbows, an effect called Alexander's dark band, named for the Greek scientist who first described it in 200 A.D. Also watch for the appearance of supernumerary arcs—purple and green bands on the inside edge of the main bow. Created by interference effects among the various beams of light

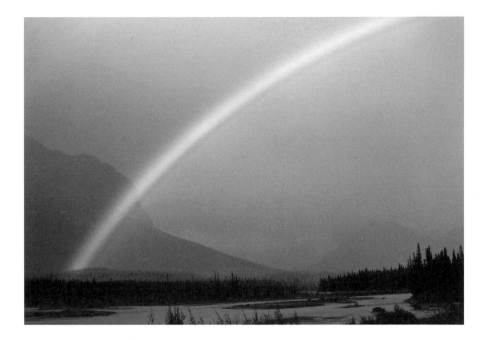

exiting the raindrops at slightly different angles, supernumerary arcs appear only when the main rainbow is especially intense.

A rare type of rainbow can be created by the full Moon at night. Moonbows are faint and colorless to the naked eye. A long-exposure photograph, however, reveals all the colors of a moonbow as if it were a daytime rainbow.

Another unusual but more obvious form of rainbow is the fogbow. When the air is filled with fine droplets of mist or fog, look for a white bow smaller than a conventional rainbow opposite a brilliant light source, perhaps the Sun breaking through the mist or, at night, a bright artificial light.

▲ **After the Storm**
The refractive properties of water combine with the reflectivity of the inner surface of a water droplet to produce one of the sky's premier optical gifts. Photo by Alan Dyer.

▼ **Moonbow**
A time exposure in moonlight reveals a moonbow, seldom seen because moonbows are so dim. Photo by Matt BenDaniel.

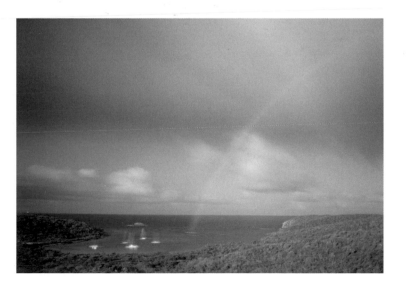

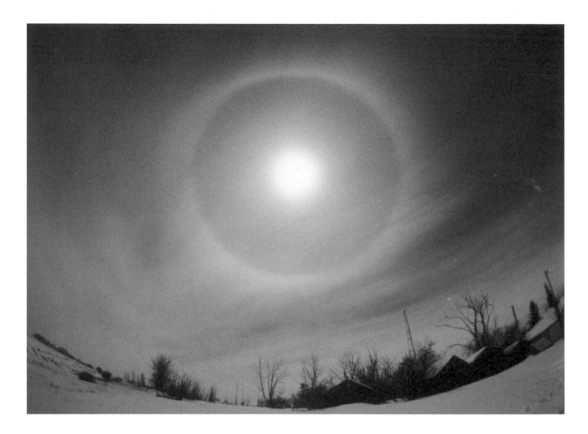

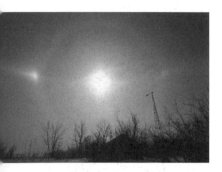

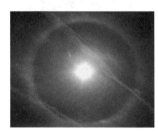

## HALOS AND SUNDOGS

Planetariums and observatories often receive calls from people reporting unusual rings of light around the Sun or the Moon. Solar and lunar halos are more common than rainbows but are not nearly as well known. They are caused by light refracting through hexagonal ice crystals. Unlike rainbows, most halo phenomena are centered around, not opposite, the Sun or Moon. Halos are usually a cold-weather phenomenon but can occur anytime the sky is covered with high-altitude cirrus clouds or icy haze.

The most common halo is a ring of light 22 degrees from the Sun or Moon. A much larger and fainter circle can sometimes be seen 46 degrees from the Sun.

Sundogs (the formal term is parhelia) appear as bright spots, sometimes colored, on either side of the Sun. (At night, look for rare moondogs.) When the Sun is low, as it is in winter, sundogs lie 22 degrees from the Sun and show up as intense areas on the inner halo. When the Sun is higher in the sky, sundogs appear just outside the inner halo.

The next most common halo phenomenon is the circumzenithal arc, a rainbow-like arc high in the sky curving away from the Sun. It is part of a circle centered on the zenith. It often appears tangent to the large 46-degree halo.

When the Sun is high in the sky, a horizontal arc can sometimes be seen passing through the Sun, running parallel to the horizon around the sky. Bright spots can appear on this horizontal arc at 90 degrees or 120 degrees to the Sun or even directly opposite the Sun. Other arcs can sometimes be seen tangent to the sides, bottoms or tops of the inner or outer halos.

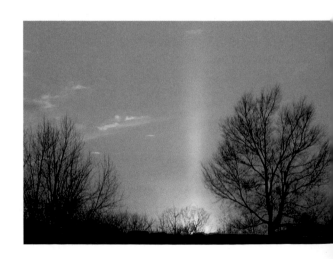

It all depends on the way the light refracts through the combinations of facets on the six-sided ice crystals. If you see any sort of halo, be sure to scan the sky—there may be other rare and subtle effects of refraction shimmering nearby.

Light can also reflect off flat ice crystals, creating a pillar of light that rises up from the low Sun. On cold, calm nights, light pillars can form above bright streetlights, creating the effect of a sky filled with searchlights.

## GLORIES AND CORONAS

Glories and coronas are colored rings that form when light is diffracted by water droplets, ice crystals or even pollen grains in the air.

The corona is a small, circular glow surrounding the Sun or Moon, usually with a diameter of no more than 10 degrees. It is often plain white but, at times, can appear as a series of colored rings—diffraction rings—much like those found around star images in a telescope. In order for a corona

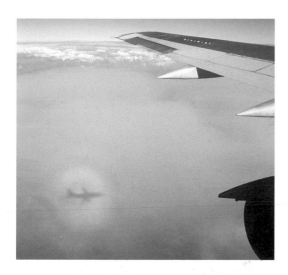

to form, the Sun or Moon must be embedded in a light haze. When distinct clouds are nearby, they may be fringed with iridescent colors (these are part of the corona). Dark sunglasses are essential for picking out coronas and iridescent clouds in the bright sky around the glaring Sun.

A similar effect, the glory, occurs around the point opposite the Sun. Your best chance of seeing a glory is from an aircraft. Sit on the side of the plane away from the Sun. As you break through the nearby clouds, look for the plane's shadow on the clouds below. The shadow may be surrounded by colored rings. A form of glory known by its German name *heiligenschein* can sometimes be seen as a glow of light around the shadow of your own head when it is projected onto a dewy lawn or low-lying fog in the morning.

## DAYTIME SIGHTINGS

A popular misconception is that the Moon "only comes out at night." But on a late afternoon around first-quarter phase, look 90 degrees away from the Sun to the east, and there you'll see the rising Moon, plain as day in the blue sky. At last-quarter Moon in the early morning, look 90 degrees from the Sun to the west for the setting Moon. Both quarter moons make fine daytime telescope targets. Use a red filter or, better yet, a polarizing filter to darken the sky and increase contrast.

When Venus is near one of its greatest elongations, roughly 45 degrees east or west of the Sun, the brilliant planet becomes another daytime target. The challenge is finding it. Two high-tech methods involve the use of either a computerized telescope to point the way or a conventional telescope's setting circles to offset from the Sun by the correct number of degrees (a planetarium computer program can provide that angle).

A low-tech method is to wait for a day when the crescent Moon is nearby. Find the Moon first, and let it lead you to Venus. Binoculars may be necessary to glimpse Venus initially, but once you have located it, look with your unaided eyes. The trick is getting your eyes to focus on the pointlike Venus surrounded by nothing but blank blue sky. Once your eyes "autofocus" properly to infinity, Venus suddenly becomes obvious.

Far more challenging is Jupiter. When it is near the point in the sky 90 degrees away from the Sun (at quadrature), the giant planet lies in the sky's natural polarization band, where the daytime sky is darkest. The air must be clean and clear, but it is possible to sight Jupiter under these conditions with binoculars and, with perseverance, no more than your naked eyes.

▼ **Surprise Moon**
The Moon can be seen in the daytime sky 10 or more days a month, yet many urban residents who tune out their natural surroundings have never seen a daytime Moon and are amazed that such a sighting is possible. Photo by Alan Dyer.

◀ **View From a Height**
This pale, circular rainbow, known as the glory, is caused by the diffraction and reflection of sunlight by water droplets in the clouds. The geometry of this phenomenon favors sightings from airplanes. Photo by Alan Dyer.

If you have a computerized telescope that can be left outside aligned from the previous night, try waking it in the daytime and slewing to some bright stars. Focus the telescope first on something far away on land. Then, if it's late summer, punch in Sirius. Zip! There it is in your eyepiece, a sparkling gem set in the blue sky of day. The stars you see in the daytime in summer are the same stars you see at night in winter. Similarly, in January, your computerized scope can direct you to daytime views of the stars of the Summer Triangle.

# Phenomena of the Sunset Sky

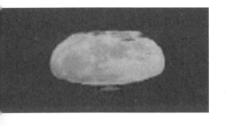

**Green Flash** ▶
The rarely seen lunar red flash (above) has origins similar to the more famous solar green flash (right) rimming the top limb of the Sun's disk when it is on the horizon. Photos by Leo Henzl.

**Squashed Sun** ▼
Atmospheric refraction can generate a layer-cake effect when the Sun is within a degree of the horizon. Photo by Alan Dyer.

You have driven to a hilltop site, anxious for a night under the stars. The Sun is going down, and you are busy setting up your telescope gear. But wait. Take a moment to watch the sunset. Close inspection will reveal some beautiful effects beyond the familiar red undersides of clouds that everyone notices.

## GREEN FLASH

As the Sun sets, its disk can dim and redden enough that it can be observed safely with binoculars or a telescope—usually. Exercise caution. If you must squint when looking at the Sun, then it is too bright.

On most occasions as the Sun sinks

below the horizon, its disk becomes red and flattened, often with a boiling upper edge. Watch this edge—you will likely see it rimmed with blue or green. In the last moments before the Sun disappears, a vivid green blob of light sometimes appears at the top of the disk and breaks off. It lasts only a second or two. This is the green flash.

The flash is caused by a prismlike dispersion of colors created by our atmosphere. The bottom of the setting Sun's disk becomes red, while the top becomes blue. Well, not quite. The short-wavelength blue light that would normally rim the top of the Sun becomes so scattered, it disappears, leaving the top edge of the Sun rimmed with green. If the atmosphere is layered into regions of varying temperatures, as in a mirage, the normally thin green rim can stretch and detach into a short-lived green

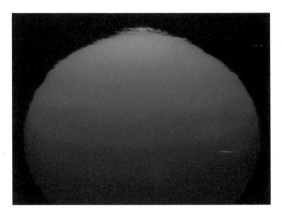

blob at the top of the Sun's disk. Under such conditions, a separate patch of red light can appear at the bottom of the Sun, creating an extremely rare red flash. The same effects can be seen at sunrise but are easy to miss, as the rapidly rising Sun takes observers by surprise.

To see the green flash, you need a clear view of the true flat horizon, over either land or water. Lucky observers have even witnessed rare green and red flashes on the rising and setting Moon and on Venus.

## CREPUSCULAR RAYS

When the Sun shines from behind clouds or distant hills, another effect can appear: crepuscular rays. The rays are usually seen as shafts of sunlight beaming down through

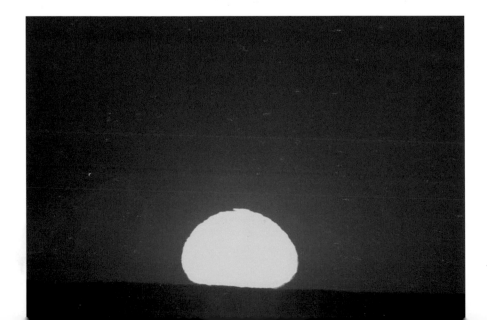

holes in a cloud deck. Crepuscular rays are especially evident with the setting or rising Sun. The rays then appear as shafts of sunlight and shadow spreading out from the sunset or sunrise point and arcing across the sky. They sometimes converge opposite the Sun, where they are called anticrepuscular rays. The diverging and converging effect is due to perspective, as the rays and shadows are actually parallel beams.

## TWILIGHT AND THE EARTH'S SHADOW

Once the Sun has set, watch the changing colors of the sky. If the atmosphere is clean and clear, you will see the western sky painted the entire spectrum, from reds and yellows near the horizon through

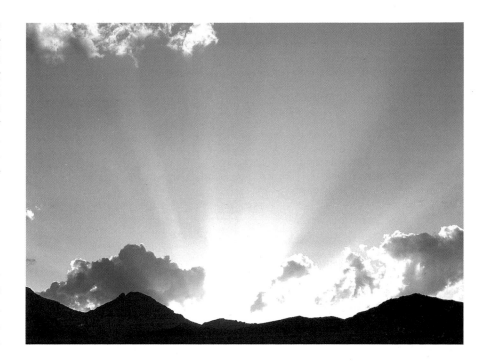

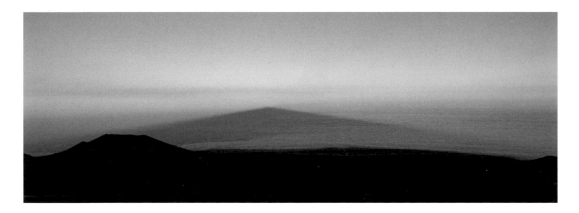

**Shafts of Sunlight**
Crepuscular rays can lend an awesome appearance to an afternoon studded with cumulus clouds.

**Summit Shadows**
Mauna Kea, the 14,000-foot dormant volcano that marks Hawaii's highest peak, can cast its shadow for 60 miles or more at sunrise and sunset. Such mountain shadows are always triangular, regardless of the shape of the mountain. In this scene, the horizontal blue band is the shadow cast on the atmosphere by Earth itself. Both photos by Alan Dyer.

green-blues a few degrees up to deep blue-purples at 20 degrees altitude or more. Under crystal-clear skies, this twilight purple can linger for 30 minutes or more in the west. If the atmosphere is filled with high-altitude smoke from forest fires or aerosols from distant volcanic eruptions, the postsunset sky will look redder than usual. Volcanic sunsets can appear around the world for months following a major eruption.

Now turn and face east. Look for a dark blue arc rising along the horizon. This is the Earth's shadow cast out across our atmosphere and into space, the same shadow that intersects the Moon's orbit and creates a lunar eclipse. In a clear sky, look for the so-called belt of Venus, a pinkish glow rimming that shadow, caused by the last

red rays of the Sun lighting up the high atmosphere to the east.

As the Sun sets farther below the horizon, the Earth's shadow climbs higher in the east. It is easiest to see when the Sun is about five degrees below the horizon. As the sky darkens, the boundary of the Earth's shadow becomes invisible, but it is still there, evidenced by its effect on orbiting satellites, which will be discussed later.

## HARVEST MOON

On full-Moon nights, you'll see the Moon rising embedded in the blue shadow of Earth. Because it lies opposite the Sun, the full Moon comes up as the Sun goes down, rising at a point on the horizon directly opposite the Sun. Like the low

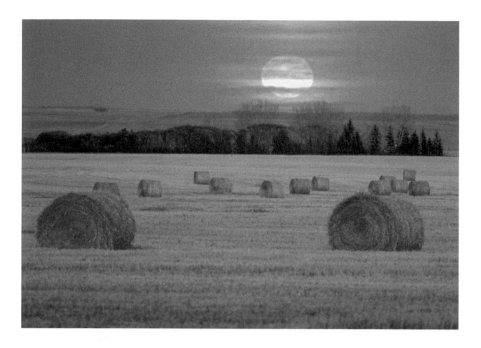

the pastel pink belt of Venus above to create one of the sky's most colorful scenes.

The best time to see this panorama is at Harvest Moon—the full Moon closest to the autumnal equinox. In autumn, the Moon rises only 20 minutes or so later each night (compared with as much as an hour later at other times of the year). For two or three nights in a row, we see a golden Moon rising nearly due east in the early evening, a combination of factors that makes the Harvest Moon an obvious sight to even the most casual skywatcher. Smoke or dust in the atmosphere at harvesttime also contributes to the Moon's golden hue and dramatic appearance.

Adding to the scene is the Moon's apparently large size when it is near the horizon. And the Moon isn't the only object to provide this illusion. A constellation rising in the east also looks much larger than it does several hours later, when at its highest altitude. It's as if we perceive the sky above us not as a semicircular dome but as a flattened arc—close to us overhead but far away at the horizon.

Sun, the rising Moon looks flattened by atmospheric refraction and reddened by atmospheric absorption of the shorter blue wavelengths. The orange tint of the Moon contrasts with the deep blue shadow and

# Why Does the Moon Look So Big?

One of the greatest puzzles of astronomy isn't the origin and fate of the universe or the existence of alien life—it's "Why does the Moon look so big on the horizon?" It is an optical illusion, something you can prove by taking photos of the Moon when it is both low and high in the sky. The Moon's disk is the same size (about 0.5 degree) no matter where it is in the sky. Yet the majority of the population will report that a rising full Moon looks about 1.5 to 2 times larger than a normal high Moon.

The Moon is no closer to us when it is rising or setting than when it is high up. In fact, it is slightly farther away by an amount equal to the Earth's radius. So what causes the effect?

Every few years, a new theory is offered, each claiming to be the final word. While no one theory has ever been fully accepted, the most common explanation suggests that when the Moon appears near the horizon next to familiar earthly objects, our brain interprets the Moon as being nearer to us. Therefore, we see the lunar disk as subtending a larger angular size than when we see the Moon high, isolated and apparently farther away in the night sky.

The illusion can be made to go away by crossing your eyes or by looking at the Moon with your head upside down. Stand with your back to the rising Moon, then bend down and put your head between your legs. Don't let your neighbors see you—they won't believe you are conducting an experiment in naked-eye astronomy!

*The Moon's extraordinary size when seen near the horizon is one of the most powerful visual effects in nature. Yet it is pure illusion. Hold an aspirin tablet at arm's length when the full Moon is near the horizon and again when it is well up. When you compare sizes, you'll see that there is no size difference at all between a low Moon and a high Moon. Photo by Terence Dickinson.*

# Phenomena of the Darkening Sky

Cinematographers refer to the brief time of twilight as the magic hour. So it is for astronomers. While you are waiting for dark to settle, look for unique phenomena that appear only in the deepening blue of a twilight sky.

## EARTHSHINE AND THIN MOONS

You have seen the effect many times but may not know what it is. A crescent Moon hangs in the west. Only a thin sliver of the disk is illuminated by the Sun, yet you can see the entire disk of the Moon. The "dark side" of the Moon appears faintly visible, an effect colloquially known as the old Moon in the new Moon's arms.

The dark portion of the lunar disk is nighttime on the Moon. But in the lunar night sky hangs a large and brilliant Earth displaying a nearly full phase. Sunlight reflecting off the oceans, clouds and polar icecaps of Earth lights up the nightside of the Moon with a blue-white light. Some of this Earthshine is reflected back to Earth, enabling us to see the dark side of the Moon.

Don't confuse the "dark side" of the Moon with the far side of the Moon—there is a side of the Moon we never see from Earth, but the lunar far side experiences the same cycle of day and night as does the side that perpetually faces Earth. The only places on the Moon "where the Sun never shines" are a few deep craters at the lunar north and south poles.

Spring is the best season for seeing Earthshine. In springtime, the waxing crescent Moon stands at its highest above the horizon, putting it into clearer air and allowing it to hang in the sky well after the sky is fully dark.

Spring is also the best season for seeing the youngest Moon possible. A young Moon appears low in the sky as an ultrathin crescent embedded in the bright twilight, making it a challenge to see. According to the U.S. Naval Observatory, the record for a naked-eye sighting is of a Moon only 15.5 hours old (after the official moment of new Moon). For a telescopic sighting, the record is a 12.1-hour-old Moon. Most of us would consider any sighting of a Moon less than 24 hours old an accomplishment to add to our "life list" of astronomical phenomena. Use binoculars to find the Moon, then try it with your unaided eyes.

## SIGHTING MERCURY

Another twilight challenge is locating the innermost planet, Mercury. It has the reputation of being difficult to sight, but once you see Mercury under good conditions, you'll be amazed how easy it is. The key factor is "good conditions."

Mercury never appears more than 28 degrees from the Sun, and then only for a little more than a week at a time. It darts back and forth from the evening to the morning sky, putting in six or seven appearances at dusk and dawn each year. Some apparitions of Mercury are better than others. It appears highest in the evening sky in spring and highest in the morning sky in autumn. Even at such favorable apparitions, however, Mercury shines just 10 to 15 degrees above the horizon, making cloudless skies a must.

Binoculars are also essential, allowing the planet to be spotted in a darkening sky not long after sunset and a good 15 to 20 minutes before it becomes visible to the unaided eye. There is usually little con-

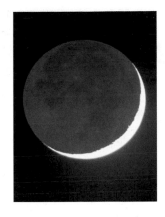

▲ **Dark Side of the Moon?** The subtle illumination of the nightside of the Moon is impressive and often puzzling when observed for the first time. Known as Earthshine, the ghostly light is sunlight reflected off Earth and toward the Moon. Binoculars show it well. Photo by Alan Dyer.

▼ **Sighting Mercury** The crescent Moon is a favorite guidepost to the inner planets Venus and elusive Mercury, as seen here. In this case, a binocular sighting of the Moon led to Mercury less than three degrees to the right. Photo by Murray Paulson.

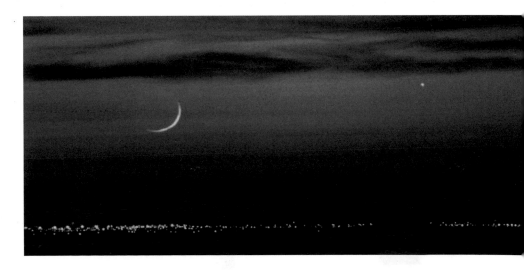

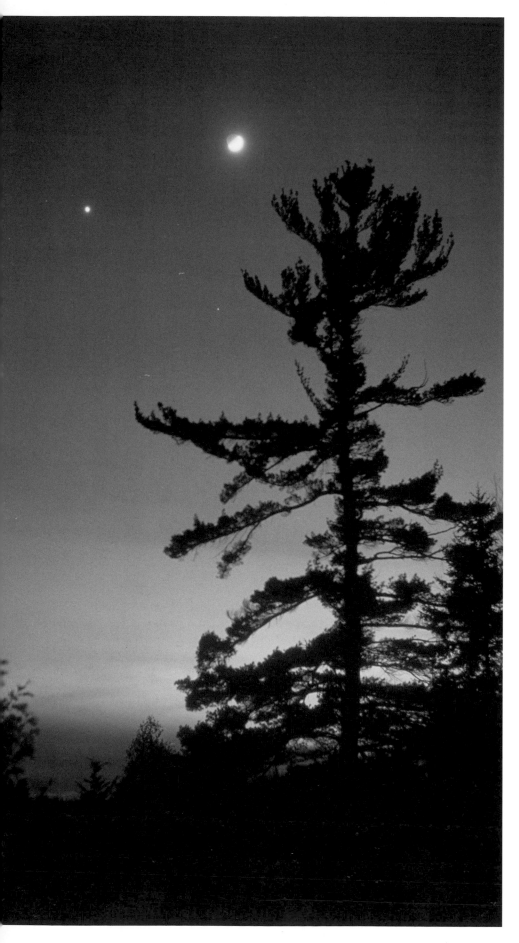

fusion—Mercury outshines all but the brightest stars found near the ecliptic, such as Aldebaran, Betelgeuse, Procyon, Regulus, Spica and Antares, and can be a full magnitude brighter. Once you see Mercury, you'll wonder how you missed it before. If there is a star nearby for comparison, use it to note the change in Mercury's brightness over a few days.

## MOON-PLANET CONJUNCTIONS

As the planets go about their wandering ways among the stars, they occasionally meet up with one another. In common

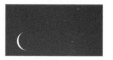

| **DATE** | |
|---|---|
| April 14 – May 14 | 2002 |
| December 1 | 2002 |
| July 17 | 2003 |
| November 4 – 5 | 2004 |
| December 7 | 2004 |
| June 25 | 2005 |
| June 17 | 2006 |
| May 19 | 2007 |
| June 30 | 2007 |
| February 1 | 2008 |
| December 1 | 2008 |
| February 27 | 2009 |
| October 13 | 2009 |
| August 5 | 2010 |
| May 11 | 2011 |
| March 13 | 2012 |
| November 27 | 2012 |
| May 28 | 2013 |
| August 18 | 2014 |
| February 22 | 2015 |
| July 1 | 2015 |
| October 26 | 2015 |
| December 7 | 2015 |

usage, the term conjunction often refers to any close gathering of planets or of planets and the Moon. But, to be technically correct, two planets are not in conjunction unless they share the same right ascension (i.e., form a line north and south of each other).

The naked-eye outer planets—Mars, Jupiter and Saturn—can meet in conjunction high in a dark sky. Meetings between slow-moving Jupiter and Saturn are rare, only once every 20 years. The last was in May 2000. Although Mars can meet Jupiter or Saturn in a dark sky, all Mars-Jupiter and Mars-Saturn pairings between now and 2015 are morning or evening twilight events.

So it is with any conjunction that in-volves Venus or Mercury. They can pair up with any of the three naked-eye outer planets, but twilight conjunctions of Venus and Jupiter, the two brightest planets, are the most spectacular. Tightly spaced three-planet gatherings are noteworthy enough to make the nightly news reports. Just as popular are events when four or even all five of the naked-eye planets gather in the same region of the morning or evening sky.

Though not a conjunction, the planets under these circumstances create a line of worlds strung along the ecliptic in the twilight. Add the Moon to the lineup, and you have a memorable night, indeed.

A close conjunction of the crescent

◀ **One-Night Stand**
This distinctive triangle of the Moon, Venus and Saturn (faintest) lasted only one evening, because the Moon moves about 12 degrees eastward every 24 hours. Photo by Terence Dickinson.

# Superb Conjunctions 2002 – 2015

*A selected list of the best planetary conjunctions of the early 21st century*

**PLANETS INVOLVED**

All five naked-eye planets lined up in evening sky; the tightest visible five-planet gathering until 2040
Crescent Moon, Venus and Mars all within 2° in morning sky
Gibbous Moon 0.5° below Mars at its brightest
Venus and Jupiter 0.6° apart in dawn sky
Crescent Moon passes in front of Jupiter (as seen from eastern North America)
Mercury, Venus and Saturn within 1.5°, followed by Mercury and Venus just 0.1° apart on June 27
Mars and Saturn 0.5° apart and 1° from Beehive star cluster low in evening sky
Crescent Moon and Venus 1° apart in evening sky
Venus and Saturn 0.8° apart low in evening sky
Venus and Jupiter 0.6° apart in dawn sky, with crescent Moon nearby on February 4
Crescent Moon, Venus and Jupiter form 3°-wide triangle in evening sky
Crescent Moon and Venus 2° apart in evening sky
Venus and Saturn 0.5° apart in dawn sky with Mercury nearby
Venus, Mars and Saturn form 5°-wide triangle low in evening sky
Mercury, Venus and Jupiter form a 2°-long vertical line low in dawn sky with Mars nearby
Venus and Jupiter 3° apart in evening sky
Venus and Saturn less than 1° apart in dawn sky
Venus and Jupiter 1° apart in evening sky with Mercury nearby
Venus and Jupiter a mere 0.25° apart and 1° from Beehive star cluster low in dawn sky
Venus and Mars 0.5° apart in evening sky
Venus and Jupiter 0.5° apart in evening sky
Venus and Jupiter 1° apart high in dawn sky; they form a triangle with Mars on October 28
Crescent Moon and Venus 2° apart in dawn sky

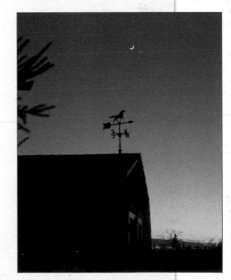

 *Readily visible conjunctions of the night sky's two brightest objects, the Moon and Venus, can be real head turners if they are less than two degrees apart. They occur infrequently enough to prompt an extra effort to see. Photos by Terence Dickinson.*

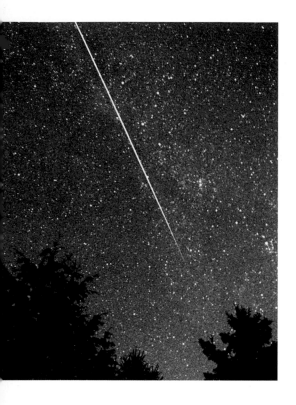

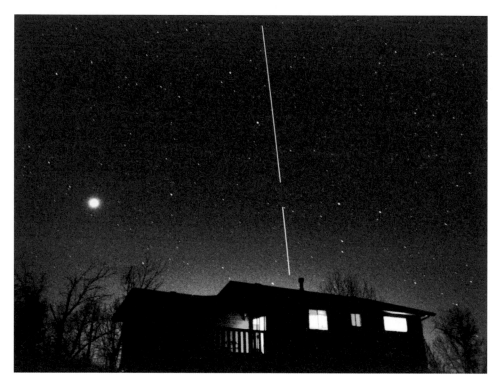

**Celestial Streaker** ▲
A time exposure turns the space station into a streak that fades away as it enters the Earth's shadow. Photo by Terence Dickinson.

**Rendezvous in Space**
Above right: The space station (just above house) follows the shuttle as the two prepared to dock in early 2001. Photo by Alan Dyer.

**End of Mir** ▼
A familiar sight for 15 years, the Mir space station plunged to a fiery end over the Pacific Ocean in March 2001. Reuters Photo.

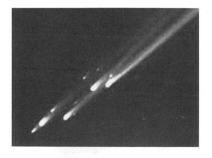

Moon and Venus, with both set in the deep blue of twilight, is another top-ranked event in the list of naked-eye spectacles. During each nine-month-long morning or evening apparition of Venus, the Moon passes it once a month. One or two of these conjunctions will be a close and special event worth seeing and photographing.

## SATELLITES AND SPACE STATIONS

Between 1957, when Sputnik 1 was launched, and the end of 2000, humans have sent about 4,000 payloads into space, with a launch rate that puts 100 more objects into orbit each year. That translates into a population of thousands of satellites, spent boosters and chunks of space debris circling Earth. By the turn of the 21st century, the number of orbiting objects larger than the size of a grapefruit was 8,700, though only a few hundred were working satellites. Spend any time looking through a telescope, and you will see some of these objects zipping across the field of view. Satellites resemble stars, even through a telescope.

Three or four satellites will be easily visible above your horizon every hour. But you won't see them at any time of night—satellites are best observed during the 90 min-

utes after sunset or before sunrise. The Sun is then below the horizon for earthbound observers but is still shining at the satellite's altitude. That's how, and only how, we see satellites—they reflect sunlight.

Satellites in low Earth orbit (160 to 1,600 kilometers up) can appear as bright as first- or second-magnitude stars. The space shuttle can become as bright as magnitude –1 or –2, as did the Russian Mir space station during its 15 years in orbit. The brightest object in Earth orbit is the International Space Station (ISS), which, at an early stage of its construction, was peaking at a magnitude of –2.5, as bright as Jupiter but with a yellow tint from its gold-colored solar panels. As the ISS grows in size, its reflecting area, particularly from the large winglike solar panels, will also increase. A brightness equal to that of Venus or better (about magnitude –5) is likely once this complex, which will be the size of a football field, is completed in 2006. With an orbital inclination of 51.6 degrees and an altitude of 400 kilometers, the ISS can be seen from anywhere on Earth between 60 degrees north and 60 degrees south latitude.

How quickly a satellite crosses the sky depends on its altitude: The higher the altitude, the slower it moves. Satellites in low Earth orbit take 90 to 200 minutes to complete one orbit around the planet. Such a satellite can take two to five minutes to traverse your local sky. Most objects travel from west to east (never east to west), but polar-orbiting objects, such as Earth-observing and spy satellites, glide across the heavens from north to south or south to north.

Satellites provide some unusual visual phenomena. In the evening sky, objects brighten as they head east and become more fully illuminated, just like a full Moon opposite the Sun. Sometimes, they flash briefly as sunlight flares off a reflective surface. Tumbling objects often pulse in brightness. Now and then, two or more objects can be seen traveling together—the shuttle arriving at or departing from the ISS provides the best opportunity to sight double satellites. If you see a triangle of satellites flying in formation, don't panic. It isn't an alien invasion. You're witnessing a pass of one of the U.S. Navy's Naval Ocean Surveillance System (NOSS) triplets. Orbiting 1,000 kilometers up, the NOSS satellites travel in threes to triangulate the positions of ships at sea.

In the evening, a satellite might fade out halfway down the eastern sky. This occurs as the object enters the Earth's shadow. It has orbited into our planet's nightside and has just experienced a sunset. As the night goes on, our shadow rises higher to engulf the entire sky, so no satellite can be in sunlight. Only people living at northern latitudes (above 45°N) can see satellites crisscrossing the sky all night long, and then just in summer, when the Sun can light high-altitude objects even at local midnight.

## NOCTILUCENT CLOUDS

As the name suggests, noctilucent clouds are seen at night. They look like silvery blue cirrus clouds across the northern horizon, glowing with an opalescent quality unlike any other clouds. Though they have been sighted from as far south as Colorado, noctilucent clouds are usually confined to latitudes between 45 and 60 degrees north. The strange apparitions appear only around summer solstice, when from northern lati-

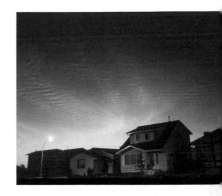

▲ **Clouds of Summer**
Noctilucent clouds shimmer over Edmonton, Alberta, one of the best places in the world to see this high-latitude phenomenon. Photo by Alan Dyer.

# Iridium Flares

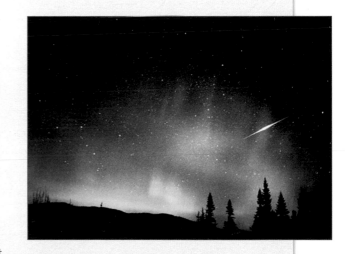

The most famous flashers in the night are the Iridium satellites. A fleet of more than 60 of these communications satellites was launched in the late 1990s to provide satellite cellphone service. The costly service never caught on, and the private consortium that operated the Iridiums went bankrupt. The satellites were to have been de-orbited and intentionally burned up in the atmosphere but were saved at the eleventh hour by the U.S. Department of Defense.

Orbiting 780 kilometers up, the Iridium satellites have a set of three highly reflective antennas, each the size of a door. The antennas act as flat mirrors, creating brief but intense flashes of sunlight. For a few seconds, an Iridium can rise from its normal magnitude of +6 (barely visible to the naked eye) to peak as high as magnitude –8, which is 25 times brighter than Venus. A brilliant star literally appears out of nowhere. Iridium flares can be seen almost every night but are highly localized. A friend a few dozen kilometers away won't see the same flare you do. Websites provide customized predictions. See www.backyardastronomy.com for links.

 *While shooting a display of northern lights from a wildlife preserve in Quebec, photographer Dominic Cantin also captured the brief flare of an Iridium satellite.*

# Decoding Orbital Elements

Desktop-computer programs allow you to enter all the terms (called the orbital elements) necessary to describe a satellite's orbit. Computerized telescopes, such as Meade's ETX Autostar units and Vixen's SkySensor 2000, also allow you to enter a satellite's orbital elements into the hand controller. While the data can be downloaded through an Internet connection, it is often easier, when updating or adding just one or two satellites, to type in the numbers. (Orbital elements are always changing and must be updated regularly to be accurate.) Once entered, the computer program can then plot the path of that satellite across your sky. Or a computerized telescope can slew to where the satellite will appear, then begin tracking it across the sky as it rises out of the west.

The latest orbital elements for satellites are available through websites (go to www.backyardastronomy.com for links) but are often displayed in a cryptic format called Two-Line Elements (TLE), used by NORAD and NASA. Here's how to translate the relevant TLE data into the terms your computer program or telescope requires. For an example, let's look at the TLE for the International Space Station (ISS) on February 4, 2001:

```
ISS
1 25544U   98067A   01035.20358796   .00037579   00000-0   30612-3 0   5935
2 25544      51.5755 40.3812 0010147 69.7279  313.1640 15.70196435126290
```

Line 1: The first two sets of numbers refer to the official satellite designation. The 98067A term indicates this was the 67th payload launched in 1998.

Line 1: The third set of numbers provides the year and date to which these data apply; in this case, decoded as follows:

Epoch Year = 01 (for 2001; the year is the first two digits)
Epoch Day = 035.20358796 (the 35th day of the year, plus decimal day)

Line 2: The first set of numbers repeats the satellite designation. The rest of the line provides all the terms you need to enter:

| | |
|---|---|
| 51.5755 | = Inclination (in degrees) |
| 40.3812 | = Right Ascension of Ascending Node (in degrees) |
| 0010147 | = Eccentricity (the decimal point before the first 0 is assumed) |
| 69.7279 | = Argument of Perigee (in degrees) |
| 313.1640 | = Mean Anomaly (in degrees) |
| 15.70196435 | = Mean Motion (revolutions per day) |

The last six digits (underlined) indicate the number of orbits the satellite had completed on the specified date and provide a checksum figure (the last digit). These are not normally required by your computer. Remember, this is just an example. The actual numbers you encounter, even for the space station, will be quite different, though their order and format will be the same.

---

 *The International Space Station, shown here as it appeared in 2001, shines as the brightest artificial star in the sky. NASA's website at www.spaceflight.nasa.gov provides sighting predictions for cities around the world. Photo courtesy NASA.*

tudes, the Sun is a mere 6 to 16 degrees below the horizon, even at midnight.

Noctilucent clouds occur at an altitude of 80 kilometers, five times higher than 99 percent of our planet's weather systems. This amazing height puts them well above the stratosphere, at the very fringes of the Earth's atmosphere, where the clouds remain sunlit all through a summer night.

These are not normal clouds. They may be made of ice crystals precipitated around dust from incoming meteors or man-made pollutants wafted into the mesosphere and trapped in cold polar regions. If you are at a high northern latitude near the end of June or early July, be sure to look north between midnight and 3 a.m. for the eerie forms of noctilucent clouds.

Although these clouds appear in the middle of the night, we still think of them as a twilight phenomenon, because in early summer at high latitudes, twilight never ends. A glow can be seen along the northern horizon all night long. And it is within this perpetual twilight of a short summer night that noctilucent clouds are seen.

# Phenomena of the Dark Sky

Once the sky is completely dark, a new set of glows and lights in the heavens is on display for the watchful observer.

## METEORS

About 1,000 tons of dust and rock enter the Earth's atmosphere every day. A particle the size of a sand grain produces a typical meteor (called a falling, or shooting, star by nonastronomers) as it penetrates the atmosphere and incinerates. A brilliant, shadow-casting meteor slashing across the starry dome might be created by an object the size of a basketball. Even these are so short-lived, you will be lucky to be outside looking in the right direction to see one more than once or twice in a lifetime.

During a prolonged watch on any given night, you will inevitably see a handful of

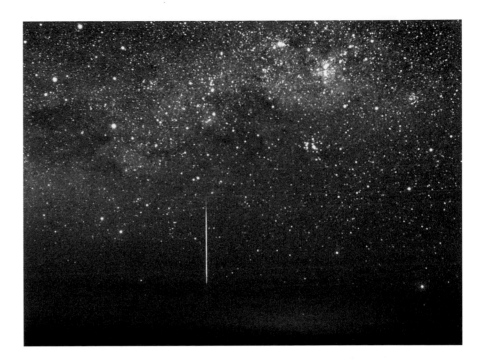

meteors appear randomly across the sky. Astronomers call these sporadics. The typical meteor, at magnitude 1 to 4, produces a streak of light lasting only a second or two, the classic "falling star." A bright meteor, at about magnitude –1, might travel more slowly over a longer path and be visible for two to three seconds. Such a meteor often leaves an ionized trail that glows long after the meteor itself has burned up and faded.

Most meteoric material comes from old comets spreading a trail of dusty debris throughout the solar system. When a comet approaches the Sun to within the orbit of Mars, its icy surface begins to vaporize from solar radiation. Dust and debris encased in

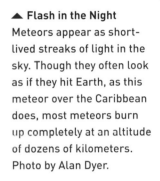

▲ **Flash in the Night**
Meteors appear as short-lived streaks of light in the sky. Though they often look as if they hit Earth, as this meteor over the Caribbean does, most meteors burn up completely at an altitude of dozens of kilometers. Photo by Alan Dyer.

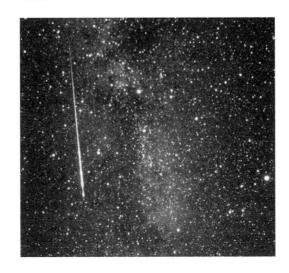

◀ **Falling Stars of Summer**
A brilliant Perseid meteor streaks down the summer Milky Way. Seen on warm summer nights each August 11 or 12, the Perseids are the most popular of the year's eight or so major meteor showers. Photo by Alan Dyer.

# Main Meteor Showers

**Pre-Christmas Meteors ▶**
A Geminid meteor zipping through the constellation Taurus was captured by accident in this shot by Terence Dickinson, which was originally intended to show the planet Mars beside the Pleiades star cluster. Meteors from specific showers can be identified by tracing their trails backwards to a common origin—in this case, the constellation Gemini.

the ice since the solar system formed 4.6 billion years ago are released to drift into space, and some of those particles plunge into the Earth's atmosphere as meteors. A few large meteors originate in the asteroid belt between Mars and Jupiter. These are literally chips off the large rocky objects that orbit in this zone by the thousands.

## METEOR SHOWERS

Next to eclipses and bright comets, the celestial events that receive the most publicity are meteor showers. These are predictable annual events during which, for one or two nights, the normal sparse count of meteors jumps to around 20 to 80 meteors per hour. There are about eight major meteor showers each year.

During a meteor shower, Earth crosses the orbit of a comet, passing through the dust left in the wake of the comet's previous trips around the Sun. The Perseids are thought to be the flotsam left behind by Comet Swift-Tuttle, last seen in 1992. The Geminids are now identified as debris strewn along the orbit of the asteroid Phaethon, which is more likely a tailless, fizzled-out comet.

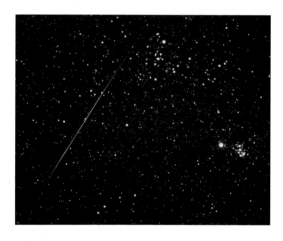

Meteor showers tend to be a disappointment for many first-time viewers, especially if the news media publicize one that occurs when the Moon is visible. Even the best showers, the Perseids (August 11-12) and the Geminids (December 13-14), produce only one meteor per minute on average. Of course, meteors never appear on a precise one-per-minute schedule. At the

peak of a shower, several minutes may go by without any meteor; then there will be a flurry of six or seven within a minute or two and nothing again for 5 to 10 minutes. Some showers can try an observer's patience. In the late 1980s and early 1990s, the Perseids put on wonderful shows, but their intensity has declined after the parent comet, Swift-Tuttle, headed back to the outer reaches of the solar system. Currently, the Geminids provide the best annual show, with the Perseids a close second.

For any shower to be seen at its best, the site must be dark, with no Moon in the sky. The observing equipment is simple: a lawn chair, blanket or sleeping bag, a hot beverage and some favorite music. Just sit back and watch the skies. If you are with long-time amateurs, you will hear "Time!" yelled out with every meteor spotted. It becomes an automatic reaction, a useful habit for sessions when someone is recording the time of each meteor seen by a team of observers.

The first thing you will notice about shower meteors is that their trails point back to the same spot in the sky. For the Perseids, this radiant point is in the constellation Perseus, hence the name. For the Geminids, it is in Gemini. The Quadrantids of early January are named for the obsolete constellation Quadrans the mural, a pattern once charted in the Draco-Boötes area.

The apparent convergence of meteor trails is due to perspective. Earth is actually passing through a parallel stream of meteors, but their paths in the sky can be more than 160 kilometers long. Because a meteor's end point is closer to the Earth's surface—and you—than is its beginning point, meteor paths undergo the same perspective effect as do railroad tracks or any parallel lines stretching off into the distance.

Shower meteors can appear anywhere in the sky. Meteors seen near the radiant are short and slow; meteors far away from the radiant are faster and leave lengthier trails. A meteor entering head-on at the radiant appears as a brief starlike flash.

A common practice of veteran meteor watchers is to wait until after midnight (or 1 a.m. when daylight time is in effect). More meteors, both shower and sporadic, grace the skies of the postmidnight hours. At that time, the side of Earth we are on is turned in the direction of our planet's orbital motion around the Sun. Therefore, we face "into the wind." Any meteoric debris we run into hits the atmosphere with greater speed and produces a brighter, hotter trail.

## FIREBALLS AND METEORITES

A night spent watching a meteor shower inevitably prompts the question, Do shower meteors ever hit Earth? The answer is a qualified no. No meteor-shower debris has ever been known to strike the Earth's surface. Since shower meteors are made of fine, crumbly comet dust, they all burn up at altitudes of 60 to 120 kilometers. Of course, everyone has heard of meteorites, the correct name for objects that do hit Earth. These rocky chunks have a different origin than do most meteors: They are fragments of asteroids that have collided somewhere between the orbits of Mars and Jupiter.

Any meteor brighter than –4, the brightness of Venus, is called a fireball. Few produce meteorites. A meteor that appears to explode in a fireworkslike display is called a bolide. If a bolide produces several glowing pieces that carry on past the main explosion, it is possible that some of those small

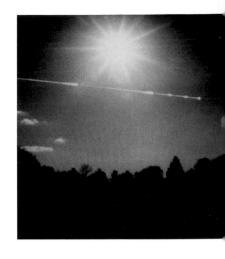

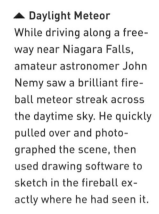

**▲ Daylight Meteor**
While driving along a freeway near Niagara Falls, amateur astronomer John Nemy saw a brilliant fireball meteor streak across the daytime sky. He quickly pulled over and photographed the scene, then used drawing software to sketch in the fireball exactly where he had seen it.

# The Leonid Storms

One annual shower, the Leonids, enjoyed a rise in activity, and publicity, in the late 1990s. This shower typically produces no more than 30 to 40 meteors per hour. However, every 33 years, the Leonids' parent comet, Tempel-Tuttle, returns to the vicinity of Earth as it passes through the inner solar system. In 1833, 1866 and 1966, the Leonids produced once-in-a-lifetime meteor storms of tens of thousands of meteors per hour. Observers reported an effect that must have resembled a science fiction starship warping into hyperspace.

In 1998, the Leonids produced a fireball shower, a display of unusually bright meteors, though their numbers were far below storm proportions. In 1999, the predicted storm year, observers in the Middle East and Europe did see a brief burst of up to 3,000 meteors per hour, but the storm was so short-lived, the show was over by the time darkness fell in North America. Remarkably, astronomers were able to predict this brief peak with unprecedented accuracy.

For 2000, the models of the Leonid debris streams suggested an off year, with only a few hundred meteors per hour at best, which is just what happened. For 2001, the same models predicted a good show for North America—another successful forecast. Observers throughout the United States and Canada reported up to 20 meteors per minute, by far the best display seen in decades by backyard astronomers on this continent.

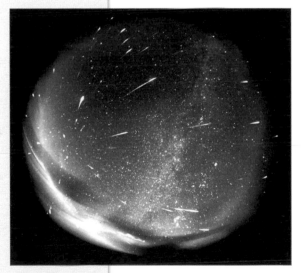

*This 4-hour exposure on November 17, 1998 (above right; courtesy of Slovak Academy of Sciences), shows the streaks of dozens of brilliant meteors emerging from the constellation Leo. The Leonid showers of 1998 to 2001 were the closest we have come in recent times to duplicating the colossal meteor storms of 1966, 1833 and 1799 (woodcut above left).*

fragments will survive to the surface. A very slow but brilliant (magnitude −10 or brighter) fireball with little sign of a final burst is another candidate for a meteorite fall. In either case, report such a sighting to your local planetarium, observatory or college astronomy department. Note the meteor's direction of travel, height above the horizon in degrees (or hand widths at arm's length) and cardinal directions of both the start and end points. Record the time and your location (many sightings are made from moving cars). Make note of any unusual sounds—rumbling, hissing, whistling or short sonic booms—and any time delay between sight and sound.

Keep in mind that while the bolide may appear to have fallen just a few hundred meters away, bolides explode at altitudes of 12 to 25 kilometers, far into the stratosphere and well above the cruising altitude of commercial jets. What looks as if it were just over the next hill could be in the next state.

How do you distinguish a natural bolide from a reentering artificial satellite? Experienced observers have found that satellites burn up more slowly; they last longer (at least 30 seconds) and traverse a greater angle (100 degrees or more) than do natural bolides. Even very bright bolides exhibit a quick terminal burst and burnout.

## ZODIACAL LIGHT AND GEGENSCHEIN

A far more subtle effect of interplanetary dust can be seen in the evening skies of spring and the morning skies of fall (true for both northern and southern hemispheres). On a moonless night in spring, for example, wait until the bright glow of twilight has left the western sky. If your site is dark, with little horizon glow from distant cities, look for a faint pyramid-shaped glow stretching 20 to 30 degrees above the horizon. It is fainter than the brightest parts of the Milky Way and is often taken for the last vestiges of atmospheric twilight. The glow is actually sunlight reflecting off comet-strewn dust in orbit around the Sun in the inner solar system. It is known as the zodiacal light, because it appears along the ecliptic, the region of the zodiac. The closer you live to the equator, the better your chance of seeing light pyramids, both morning and evening. On clear nights, however, sharp-eyed observers as far north as 60 degrees latitude can pick them out.

Much tougher to see are the other zodiacal light effects: the gegenschein and the zodiacal band. The gegenschein appears as a large (about 10 degrees wide), subtle brightening of the sky at the point directly opposite the Sun (*gegenschein* is German for counterglow). It is produced by sunlight scattering off meteoric dust beyond the Earth's orbit. Late March to early April and early October are the best times for detecting the gegenschein, since it is then projected onto star-poor regions. Seeing this elusive glow is a naked-eye observing challenge. Even more difficult to observe is the zodiacal band, a stream of light connecting the eastern and western zodiacal pyramids to the gegenschein.

## AURORAS

Northern observers are often treated to (some say plagued by) displays of northern lights, or aurora borealis. Auroras can be the most entertaining of all the naked-eye celestial phenomena, providing light shows that ripple and swirl like waving curtains or billowing plumes of colored smoke.

An aurora is a phenomenon of the upper atmosphere. Auroral curtains extend from

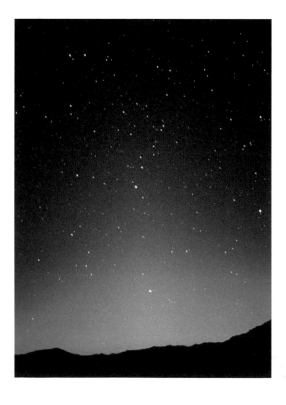

**False Dawn** ▶
Seen here from a mountaintop site in central Chile, the zodiacal light appears as a subtle vertical band of light emerging from the pinkish glow of morning twilight. From temperate latitudes, the zodiacal light appears tilted over at an angle as it follows the path of the planets, the ecliptic. Photo by Alan Dyer.

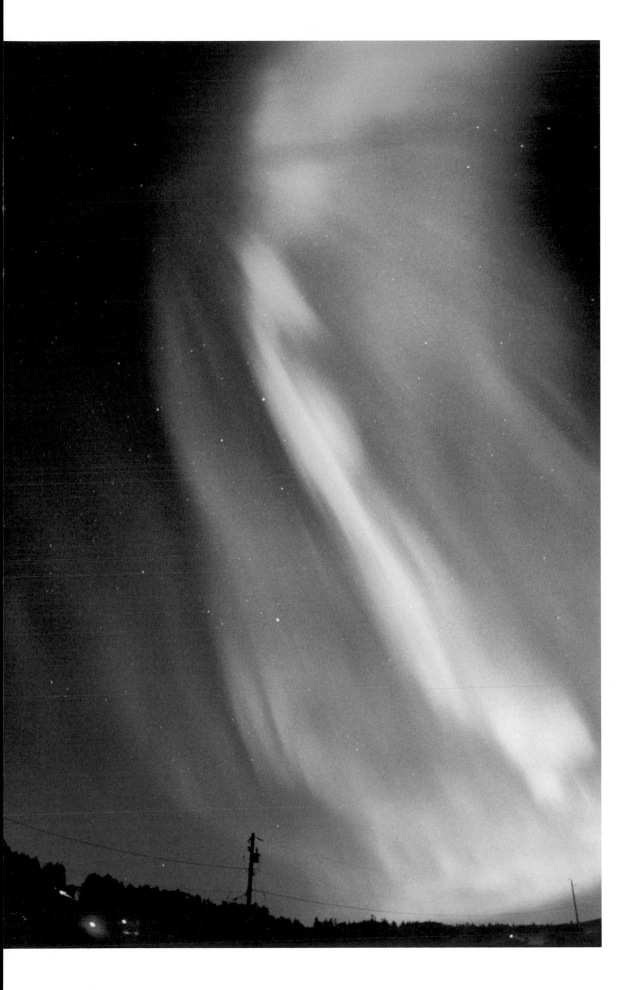

**◀ Nature's Light Show**
Among the most awesome
of celestial displays,
auroras can fill the sky
with waving curtains of
light. Green light from
glowing oxygen molecules
predominates, but pinks,
reds and blues can appear
during intense displays.
Photo by David Lee, taken
from near Victoria, British
Columbia, August 12, 2000.

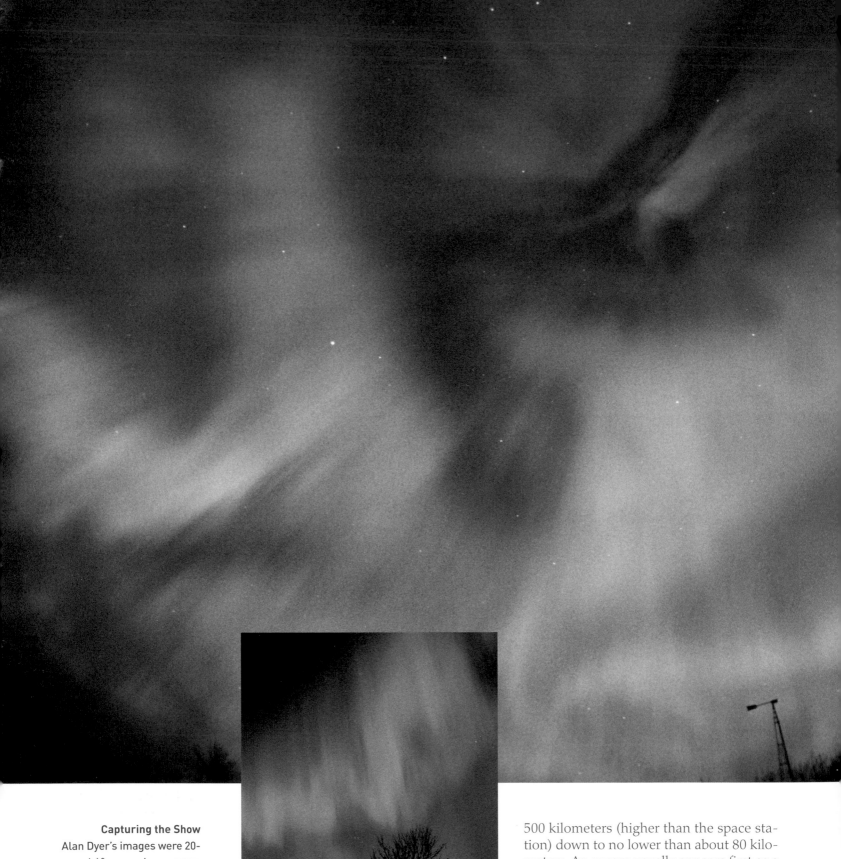

**Capturing the Show**
Alan Dyer's images were 20- and 40-second exposures on Ektachrome E200 using 16mm (above) and 28mm lenses. Terence Dickinson used Konica Centuria 800 for 8 seconds at f2. Dominic Cantin used Fuji 800 at f2.8.

500 kilometers (higher than the space station) down to no lower than about 80 kilometers. An aurora usually appears first as a greenish band of light low along the northern horizon. But during a full-blown display, luminous shafts climb high into the sky, eventually filling the heavens with curtains and streamers. An aurora can reach up to the zenith, forming a coronal burst that

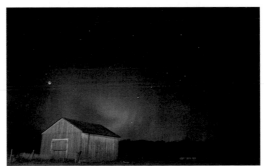

looks like the tunnel effect in the classic movie *2001: A Space Odyssey*. After an outburst, called a substorm, a typical display often turns into patches of light pulsing on and off all over the sky.

The predominant green color, at a wavelength of 557.7 nanometers, comes from glowing oxygen atoms. Very energetic displays exhibit blood-red tints from a much fainter emission line of atomic oxygen, at 630.0 nanometers. A combination of blue-green and red light from ionized nitrogen can add pink fringes to the bottoms of auroral curtains.

Displays of aurora borealis are a common sight in Alaska, in Canada's northern territories and Prairie Provinces and in northern Ontario and Quebec, where up to

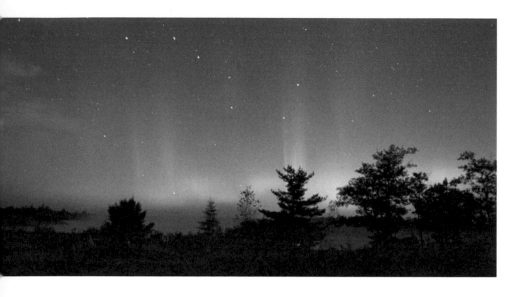

200 displays a year are visible. In Europe, extreme northern Norway and Sweden record similar numbers. In the northern United States and southern Canada, the skies shimmer with auroras a few dozen times a year. The east and west coasts of North America and the southern United States have auroras 5 to 10 times a year. A rare superaurora can extend as far south as Mexico and the Caribbean, but such a display occurs only once a decade on average.

Despite the relatively high geographic latitude of the populated parts of Europe, auroras are rarely seen there, compared with the equivalent latitudes in North America. Canada and the northern United

**Painting the Sky With Light**
Although there are obvious similarities, no two auroras are identical. Indeed, the beauty of the phenomenon is in the subtle variations from one display to another, as demonstrated by the photos on these pages, taken from southern Ontario (above) by Terence Dickinson, southern Alberta (right) by Alan Dyer and eastern Maryland (far right) by Paul Gray.

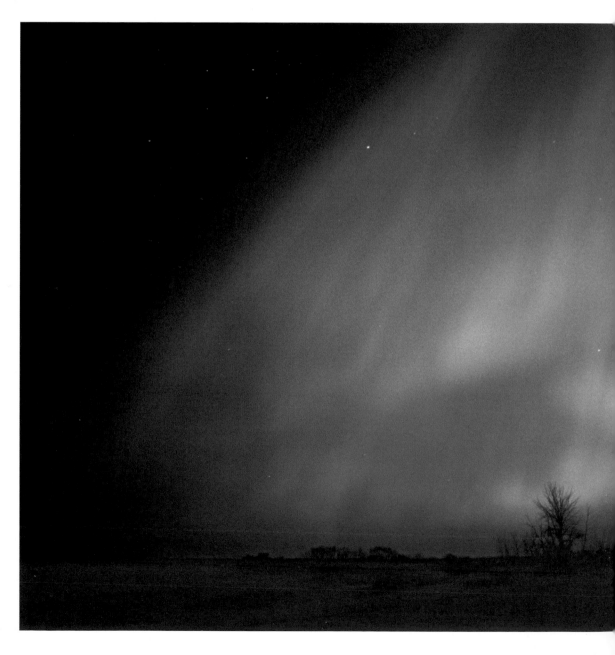

States are much closer to the magnetic north pole, located in the Canadian high-Arctic islands. Auroras form in an oval-shaped zone with a radius of roughly 2,400 kilometers centered on the magnetic north pole. Oddly enough, the Earth's true geographic North Pole gets no more auroras than do the northern Prairie Provinces.

The same situation exists in the southern hemisphere, where the aurora australis forms around the south magnetic pole, in Antarctica. Because there are few populated landmasses underneath the southern auroral zone, most displays of southern lights go unappreciated except by research scientists and penguins.

A common misconception about auroras is that they occur most often in winter. In fact, March, April, September and October usually host the best auroral displays. Brilliant auroras can also appear in summer skies. Their cause has nothing to do with the weather on Earth. The trigger is a bombardment of the upper reaches of our atmosphere by electrons and protons that originate at the Sun. Occasionally, the outer atmosphere of the Sun, its corona, lets loose with a coronal mass ejection, and part of the Sun's atmosphere is literally blown into space, perhaps set off by a powerful explosion on the surface. Some mass ejections shoot streams of charged particles toward Earth, where they saturate the radiation belts surrounding our planet. The exact process is only now becoming understood, but it seems that the radiation belts act as particle accelerators, beaming intense currents of energy onto the Earth's rarefied upper atmosphere. Like a planet-sized television screen, our atmosphere glows when hit by the electron beams.

A typical aurora requires an energy input of about 1,000 billion watts, hundreds of times greater than the output of the largest hydroelectric power plants. During an intense display, up to one million amperes of current flow along an aurora, enough to

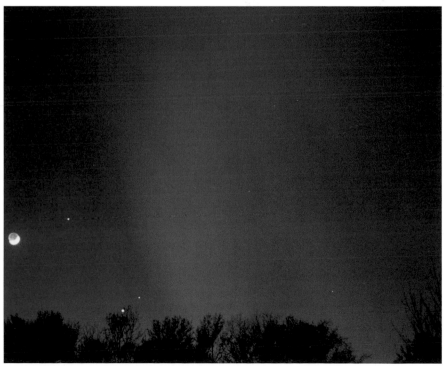

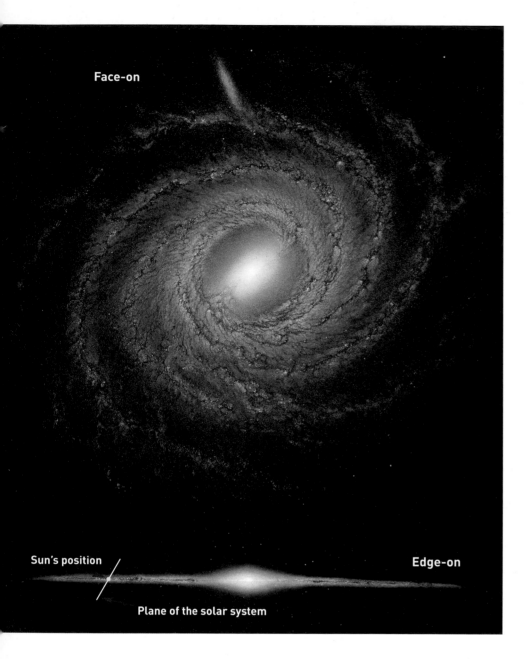

**Face-on**

**Sun's position**

**Plane of the solar system**

**Edge-on**

**Our Galactic Home ▲**
The Earth's orbit around the Sun is tilted at nearly right angles to the galactic equator and the disk of the Milky Way. For this reason, each season presents a unique view of the Milky Way: toward the galactic downtown core in northern summer; toward the galactic city limits in northern winter; out the top and bottom of the Milky Way's flattened disk in spring and fall. Art by Adolph Schaller.

create fluctuating magnetic fields on Earth. This in turn produces damaging currents that flow along extended electrical conductors, such as Arctic pipelines and power-grid networks. Auroral storms have been known to short-circuit satellites in orbit.

Auroras follow the rise and fall of solar activity, peaking every 11 years. The solar maximum of 1989 produced auroral displays so energetic, one shut down the Quebec power grid, plunging the Eastern Seaboard into darkness. The maximum of 2000/2001 was less intense but was the first to be observed by a flotilla of dedicated satellites. The Solar and Heliospheric Observatory (SOHO), the Advanced Composition Ex-

plorer (ACE), Cluster and the Imager for Magnetopause-to-Aurora Global Exploration (IMAGE) satellites watch the Sun, the solar weather and the Earth's magnetic domain minute by minute. They provide warnings of impending solar storms and geomagnetic upsets. By logging onto their websites (NASA's www.spaceweather.com is a good place to start), backyard astronomers can retrieve advance warnings of possible displays of northern lights.

# The Best Dark-Sky Sight: The Milky Way

Only one other sky-spanning phenomenon can rate with an aurora for spectacle. Unlike displays of northern lights, this wonderful sight appears almost every night of the year yet, sadly, goes unnoticed from urban backyards. But from a site away from light pollution, there's little that can surpass the naked-eye view of the Milky Way.

## OUR HOME IN THE GALAXY

The Milky Way appears as a delicate, misty band of light punctuated with bright glowing clouds of stars and split by obscuring lanes of dark interstellar dust. Aim any optical aid at the Milky Way, as Galileo discovered in 1609, and it resolves into what it really is: thousands of stars too faint and far away to be seen as individual stars with unaided eyes.

When you look at the Milky Way, you are seeing a giant galaxy—ours—edge-on. All the stars you see at night with the unaided eye belong to the Milky Way Galaxy and lie relatively close to us, within a few thousand light-years of the Sun. But our entire spiral-shaped galaxy stretches 90,000 light-years from side to side. It is the light of the more distant stars lining our galaxy's spiral arms that blends together to form the band we call the Milky Way.

Our solar system lies about 27,000 light-years from the center of the Milky Way Galaxy, a little more than halfway from the galaxy's hub to the visible edge. When you

in a 360-degree panorama. Overhead, we look straight up toward a galactic pole—the north galactic pole in northern spring and the south galactic pole in southern spring. It is around these galactic poles, far from the Milky Way's obscuring dust clouds, that we can peer farthest into space. Here, we find the richest collection of galaxies beyond ours. For telescope owners, spring is galaxy-hunting season, whether you live in Sydney, Nova Scotia, or Sydney, Australia.

## WHERE TO SEE THE MILKY WAY

To see the Milky Way at its best, head as far from city lights as you can. Under the darkest August skies, for example, our galaxy's subtle glow can be seen extending far from the main star clouds into constellations such as Delphinus and Lyra. The Great Rift of dark interstellar dust that splits the Milky Way through Cygnus and southward becomes obvious. Diagonal dark lanes in southern Ophiuchus and Sagittarius form what looks like a prancing horse. Under superb conditions, the dark lane that delineates the horse's front legs can be traced as far as Antares, in Scorpius.

The Milky Way is most spectacular through Sagittarius and Scorpius, home of

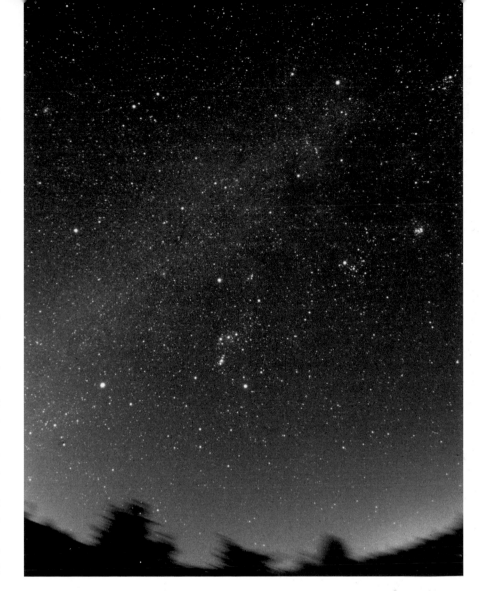

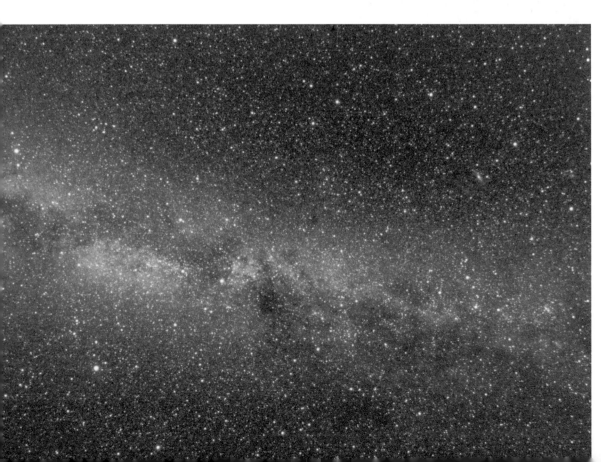

▲ **Winter Milky Way**
In January, from those same midnorthern latitudes (Canada, the United States and Europe), a fainter Milky Way, created by the outer spiral arm of our galaxy, spans the sky.

◀ **Southern Milky Way**
In May, from Australian latitudes (autumn in the southern hemisphere), the Milky Way spans the sky from east to west. The same is true in October from the northern hemisphere. Photos on these two pages by Terence Dickinson.

153

the center of the galaxy. From Canadian and northern European latitudes, these constellations crawl along the southern horizon, denying northerners decent views of the best part of our galaxy. To see this area of the sky well, go south. From the southern states or Caribbean latitudes, the glowing star clouds of Sagittarius climb high enough that they shine as the dominant naked-eye feature of a dark summer sky.

Even a trip to the Caribbean, however, gets you less than halfway to a full view of the southern sky. The magic latitude is 30 degrees south: central Chile, Australia and southern Africa. From there, under dry desert skies, the center of the galaxy passes overhead on winter evenings (June to August), glowing so brightly, it can cast a shadow. Lie back, gaze up at the galactic core and enjoy a three-dimensional experience unlike any other in astronomy. Our location on the outskirts of a spiral galaxy suddenly becomes obvious. You sense your place in the universe. This is naked-eye astronomy at its finest.

Our galaxy's wheel-shaped structure is

# Recording Your Observations

*By Russell Sampson*

On June 3, 1989, as I was walking in a park near my home, I happened to glance skyward and saw an elaborate and peculiar solar halo. The sky was full of colorful circles and arcs. I made a quick sketch in a small notebook I carry and later produced a finished drawing. One of the arcs was an extremely rare and mysterious eight-degree-radius halo I had never seen before. Trying to describe this event accurately from memory would have been difficult at best.

Amateur astronomers record sky phenomena for many reasons, but the main one is simply as a personal reminder of what was seen and when. Drawings, data tabulations and written notes have the added benefit of sharpening an observer's skills. Whether I am sketching the planet Jupiter or jotting down a series of variable-star estimates, I keep my records in a small coil-bound artist's sketchpad. The thick paper withstands the effects of dewing better than ordinary notepaper. A small paper clip prevents the pages from flapping in the wind. On cold winter nights, a pencil inserted through a piece of one-inch wooden doweling provides a better grip with gloves on.

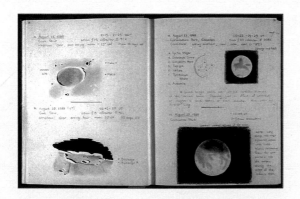

Preparation before going outside may be necessary. If you plan to observe a planet, draw its outline. For deep-sky observing, draw a circle to mark your eyepiece's field of view. Dividing this circle or planetary outline into quadrants helps when positioning objects or features. Once outside, record the date, the time, the observing conditions and the instrument used. Write things down as you see them; try not to rely too much on your memory. Instead of finishing your sketch on the spot, draw outlines of features and use a numerical scale for brightness. For planetary detail, try a scale of one to five, where five is the darkest. The same method can be used for deep-sky drawings.

When drawing large-sky phenomena, such as a solar halo or an aurora, use an extended fist to estimate angular size or separation. A bare fist, viewed at arm's length, is between 8 and 11 degrees from little finger to thumb. For some observers, these quick field notes are enough. I recopy and finish my drawings onto the pages of a bound artist's sketchbook as soon after the observation as possible. For finished planetary sketches, use a soft pencil, a white drafting eraser (the pencil-shaped erasers are the best) and a blending stump. As its name suggests, a blending stump is used to smear or blend graphite onto paper. It costs less than a dollar and is sold in art-supply stores.

One of the most challenging aspects of planetary sketching is making a realistic outline of the planet. Saturn's complex system of rings, Jupiter's equatorial bulge and the phases of the inner planets are difficult to render in a so obvious when you see the core overhead, it doesn't take much of a leap to imagine that the Greek philosophers would have figured it out more than 2,000 years ago had they been positioned at the appropriate latitude. As it happened, it was not until early in the 20th century that astronomers were able to settle the question of the Milky Way Galaxy's shape, burdened as they were with northern-hemisphere viewpoints.

After a few minutes under the southern Milky Way, the first order of business for the northern astronomer is coping with the dis-orientation in the sky caused by viewing our galaxy from the opposite end of the planet. What is upright in our sky is upside down in theirs. Thus Orion, Canis Major and Leo—to mention only a few—are seen turned on their heads.

Overshadowing the novelty of the flipped constellations is a huge sector of new sky: dozens of unfamiliar constellations and a great swath of Milky Way littered with the sky's best star clusters, globular clusters, nebulas and galaxies. (For more on this, see Chapter 12.)

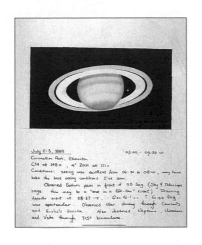

lifelike manner. A technique used by modern graphic designers can be adapted to make a planet outline. First, find an image of the planet in its proper phase or orientation, such as the line drawings of planet disks that appear in every issue of the *Astronomical Calendar*. Then photocopy selected images, and cover the back side of the photocopy with a thick layer of pencil graphite. Carefully tape the photocopy onto your sketchbook, and trace over the image with a pen or pencil. The graphite is transferred onto the page in the form of an outline. The photocopies can be used over and over again.

Dark planetary or lunar features are added with the pencil and blending stump. Black areas, such as the background sky or shadows on the Moon, can be applied using an opaque watercolor called gouache. You will need a very fine brush to outline the planet and a wider brush to fill in the background.

Color drawings can be done with pencil crayons; they are inexpensive and easy to use. The best and most widely available are Prismacolor Crayons by Berol. The choice of paper is important. Smooth papers, like loose-leaf, are not abrasive enough to take the pigment off the crayon.

The shape of the crayon tip is also critical to keep the colors diffuse and uniform. With a sharp knife, sculpt the tip of the crayon into a broad, slightly rounded stump. For large color fields, like the background sky, keep the broad face of the crayon flat on the paper and use a gentle circular motion. If you apply only light pressure, the crayon will produce a soft airbrushlike quality. To color small markings or features with sharp edges, angle the crayon tip off its broad face to its edge. For deep-sky objects, try a white pencil crayon on black construction paper.

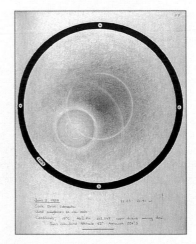

Keep your ambitions and plans in perspective. Attempting to draw the entire face of the Moon as seen through a telescope is unrealistic. Try sketching one interesting lunar feature at a time. The satisfaction of having a "hard copy" of your observation will be a reward in itself.

 *Russ Sampson is an astronomy educator and longtime amateur astronomer originally from Edmonton, Alberta, and now teaching in Connecticut.*

# Observing Conditions: Your

When our grandparents were children, the splendor of a dark night sky thronged with stars and wrapped in the silky ribbon of the Milky Way was as close as the back door. From almost any backyard anywhere, city or country, the majesty of the starry night sky was visible.

Not anymore. Giant domes of yellow light cover every city in North America at night. Major metropolises, such as Los Angeles, Chicago, Dallas and Toronto, are visible 100 miles away as glows on the horizon. From 40 miles, they wreck most of the sky. Any closer, and the night is no longer dark.

We are not suggesting that night lighting isn't needed in modern society. The problem is waste lighting. It's everywhere: Empty parking lots are floodlit all night; security lights pour into neighbors' windows rather than being confined to the target area; and inefficient streetlights spill as much as 30 percent of their output horizontally, reach-

ing only the eyes of distant drivers and the air above our heads.

At the request of astronomers, environmentalists and citizens who note the energy wastage as well as the loss of nature's night sky, some local, county and state governments have introduced sensible lighting legislation in Arizona, New Mexico, California, New Jersey and New York State. Other regions are considering the idea.

The longest-standing lighting ordinances, dating from the 1970s and early 1980s, are in Tucson, Arizona, and San Diego, California. All light fixtures are designed to aim down to bathe the street, sidewalk, parking stalls or other targets and not the sky or neighboring property. Older fixtures have been retrofitted with shields. Highway signs and billboards are illuminated from the top down, rather than the reverse. Drivers see the road instead of interfering glare from road lighting, as is often the case in other cities.

The Moon and Venus pose above the Calgary skyline for an impressive picture. As this chapter reveals, however, the rest of the story of outdoor night lighting is not pretty. Photo by Alan Dyer.

te and Light Pollution

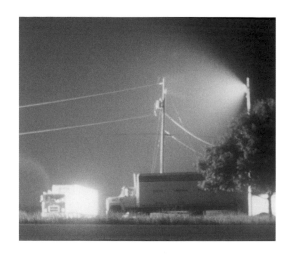

# The Eroding Sky

If stargazing were the exclusive casualty of the growth in inefficient night lighting, a call for action might be considered trivial. But astronomy is just part of it. In the United States, poorly designed or badly installed outdoor lighting wastes more than $1 billion worth of electricity annually by producing light that streams into the sky, illuminating nothing but airborne dust and water vapor.

Municipal authorities, like most bureaucrats, tend to resist change. Unshielded streetlamps are mass-produced and inexpensive, and they do the job. Why replace them? People are accustomed to glaring streetlights because that is all they have ever seen. "Nobody's complaining to me about too much light," one township engineer told us. "People want more light, not less." This attitude will change only when people are presented with alternatives and a reason for change. Shielded lights are simply more efficient: By eliminating the waste component, each fixture can have lower-wattage lamps, yet the target area still receives the same amount of light. Shielded lights are the environmentally friendly alternative.

"Yes, all that may be true," said the township engineer, "but they cost more, and they look dimmer. People don't like it." However, the reality is that residents of San Diego and Tucson readily accepted the new lights once they understood the reasons for the change —thanks to an advertising campaign conducted by both the city and the environmental groups.

Fortunately, there are signs of a slow but definite shift in attitudes toward lighting. Sales of full-cutoff fixtures that redirect horizontal or higher beams to road level

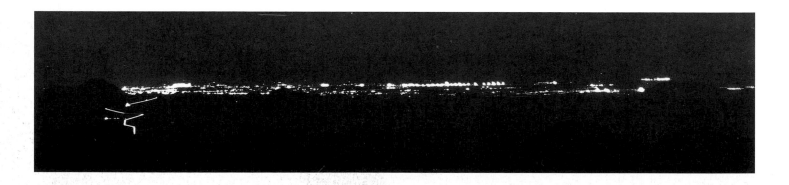

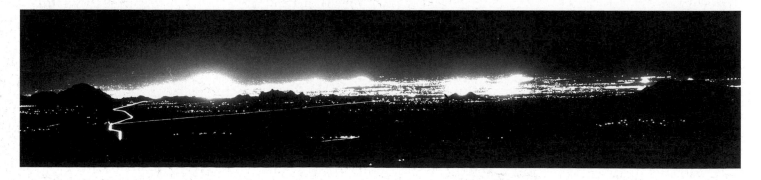

are growing every month. While researching this chapter, we spoke with several major outdoor-lighting suppliers. In many cases, the price difference between fully shielded lights and the former standard cobra-head fixtures is now close to zero. Most manufacturers are convinced that shielded equipment will eventually take precedence in roadway lighting and assured us that there is a slowly growing commitment to eliminating the glare produced by the old fixtures.

Professional astronomers urge municipalities near their observatories to use low-pressure sodium lamps, because the narrow spectrum of the illuminated sodium vapor can be filtered out at the telescope more easily than other light sources. For what backyard astronomers enjoy doing, however, these filters are of limited advantage. Containing the overall brightness of the sky and eliminating the direct interference from specific lights are the main issues. In that regard, shielding is much more important than the type of light source.

The inspiration of a dazzling starry night is unknown to most children today and is a dim memory to seniors who saw the spectacle from the front porch in their youth. We cannot go back to the good old days, but as with any other aspect of our planet's natural heritage, we should preserve at least some of the night sky for future generations.

## LIGHT VS. NIGHT

There is no way to reveal again the stars over the city as our grandparents once saw them. But now the situation is deteriorating deep in the country too. A single dusk-to-dawn, pole-mounted, mercury-vapor lamp dims the starscape for hundreds of yards in every direction. One of these lamps several miles away appears brighter than Sirius, the brightest star in the night sky.

## DARK-SKY PRESERVES

Inefficient and wasteful outdoor lighting—light pollution—is an environmental issue whose time has come. Something must be done soon, primarily for energy efficiency but also to restrict light trespass and to preserve what little dark sky remains in areas within a reasonable driving distance of populated centers.

One of the most interesting approaches we have seen is the establishment of Dark-Sky Preserves that encompass an already-existing state, provincial or national park. Preserving the natural night sky meshes well with the overall goals of the park. Of-

▼ **This Is What It's All About**
Stargazers enjoy the view from a park about a 90-minute drive from the nearest city (whose glow is seen near the horizon). Photo by Terence Dickinson.

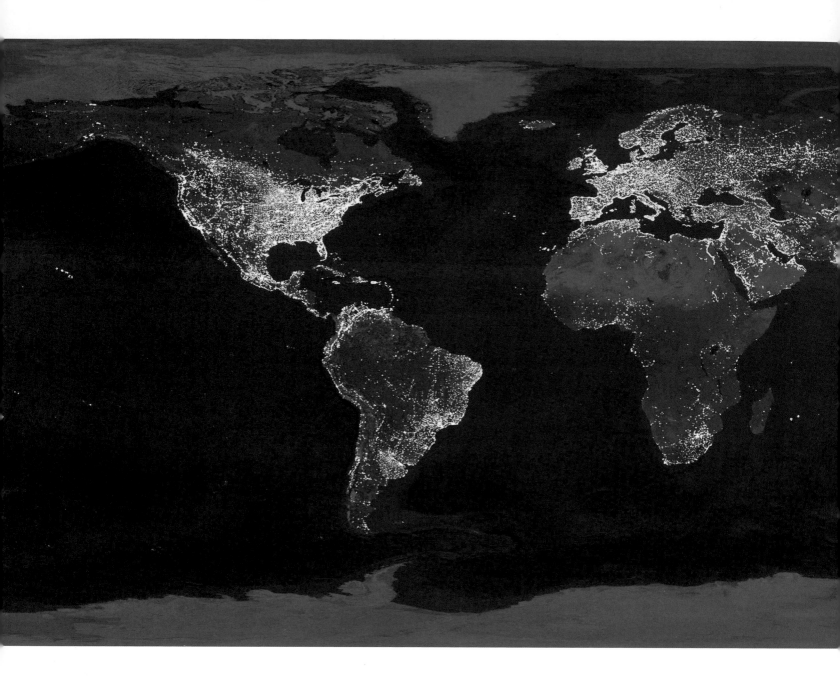

**Global Night Lighting** ▲
Satellite imagery of our planet shows that night lighting closely matches regions of dense population or industrialization—usually both. Courtesy NOAA.

ten, the park is already dark, and all that needs to be done is to shield a few existing lights and have the Dark-Sky Preserve added to the park's mandate by legislation. This requires time and legwork to arrange because of the necessary government wheels that have to turn, but it can be a worthwhile project for an astronomy club or a less formal group.

What can an individual do? Take a good look at the outdoor lighting at your home or business. If it involves dusk-to-dawn security lights, calculate their yearly operating cost. Electricity is not inexpensive anymore. Can the job be done with fewer lights? Are the fixtures shielded? For secu-

rity, consider floodlights with an infrared motion-detector switch. Infrared systems use negligible electricity and, compared with all-night lighting, pay for themselves in a year or two. If the lighting is primarily decorative rather than functional, smaller-wattage lamps will usually do the job. Does your outdoor light spill into your neighbors' yards or windows? They may not appreciate it. Does a streetlight or other powerful light reduce your quality of life or prevent you from sleeping? Politely but firmly complain to the municipality or to the light's owner. All bad lights can be shielded. Unwanted light is in the same nuisance category as a blaring stereo in the neigh-

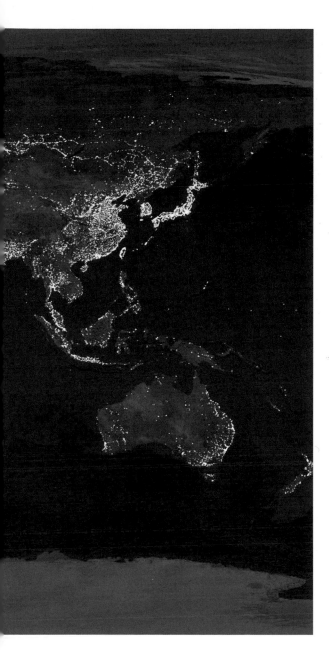

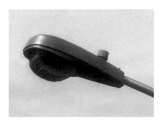

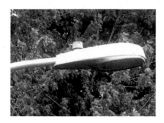

ation, a nonprofit organization established to advance awareness of the problem (visit www.darksky.org). In 2001, The George Wright Society devoted a full issue of its journal to light pollution, particularly in relation to national parks and other dark sites in the United States. This entire issue of *The George Wright Forum* (Vol. 18, Issue #4) is available at the following website: www.georgewright.org/pubslist.html

## YOUR OBSERVING SITE

Astronomy can be conducted from just about anywhere. A view of the sky, however restricted or veiled by lights and haze, still shows something. Occasionally, astonishing results emerge under even the most adverse conditions. English comet and nova hunter George Alcock proved this in 1983 when he took a break during an outdoor observing session to have a cup of tea in his kitchen. Sitting at the table, he picked up his binoculars and began scanning, through a closed window, a familiar field of stars in the constellation Draco. Resting his elbows on the back of a chair to steady his 15x80 binoculars, he spied a fuzzy patch he knew did not belong. It was a comet, his fifth discovery.

Alcock's comet-hunting prowess suggests that ideal weather is not a prerequisite either. The British Isles are notorious for cloudy conditions. Perseverance lies behind Alcock's success.

Ideally, of course, all backyard astronomers would like to live on a mountain where the sky is clear more than 200 nights a year. Realistically, however, even if the sky were perfectly clear that often, few of us would be able to make full use of it. The frustration arises when nature's schedule and the observer's do not harmonize. Take a cue from Alcock: Accept the local weather, and make the best of it. Think of how the professional research astronomer must feel when, after booking time a year in advance on one of the world's largest telescopes and traveling thousands of miles to use it, the site is clouded out. It happens to everybody.

In any case, the chief drawback is usually not the number of clear nights but the local observing conditions. Most of us live in or near urban areas where light intensity

▲ **Street Smart**
The standard cobra-head streetlamp (top), affixed to millions of poles throughout North America, hasn't changed much since the basic design was developed four decades ago. Then, as now, illumination from the fixture's central lamp not only lights up the street below but also shines horizontally out the sides of the hemispheric lens. This side glare shooting off into the distance is a complete waste of illumination. A more efficient design (above) is the full-cutoff streetlamp, which has a flat lens that eliminates light escaping horizontally. This design directs the light more efficiently where it is needed and reduces glare in the eyes of approaching drivers and nearby pedestrians. Next time you are in an aircraft at night, notice how you can see the illumination from the streetlamps below shining directly into your eyes.

borhood. In some cases, when such approaches have failed, several annoyed citizens have taken the dispute to small-claims court and won. But this type of extreme action remains very rare.

Sharing your telescope and the wonders of the universe with your neighbors can yield hidden benefits in this regard. In an outdoor situation at night, gently point out the reality of light pollution. Nearly everyone is open to learning more about environmental issues. Do not preach; simply inform. Most people are receptive. They may even turn off their lights for you.

For more information on light pollution, contact the International Dark Sky Associ-

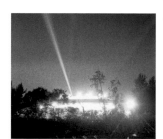

grows worse every year. In the center of a large city, the light pollution can be so intense that only the Moon, Venus, Jupiter and a few first-magnitude stars poke through. Yet even heavy light interference can be circumvented to a degree.

## OBSERVING FROM THE CITY

Amateur astronomer Ted Molczan lives in a 33-story high-rise apartment a few blocks from the heart of Toronto. From the roof of the building, with the city lights blazing below, Molczan has seen, with the naked eye, fifth-magnitude stars in the overhead region and even a vague hint of the Milky Way in Cygnus at the zenith. Using 11x80 binoculars, he has no trouble seeing ninth-magnitude stars. Most people in a similar situation would have given up without even trying. By making the best of what is at hand, Molczan has conducted a program of artificial-satellite observations that has led to the recovery of several lost satellites and to the refinement of the orbits of others.

Some apartment and condominium dwellers may even be limited to observing through a window. While this is better than nothing, the window glass always introduces some distortion and/or multiple imaging to binocular and telescopic viewing.

Observers in suburban situations can position their telescopes in a part of their yard that is shielded by fences or bushes from surrounding porch and street illumination. After a few minutes, when the observer's eyes have adapted to the semi-darkness, there is lots to see. Typically, fourth-magnitude stars are visible from the suburbs of a large city, and fifth-magnitude stars can be seen from the outer reaches of a smaller metropolis. Of course, it depends on local conditions, but once a spot is found that is protected from direct glare (or action is taken to block such light by erecting a fence or planting a row of dense evergreen trees), the result can be surprising.

Suppose, for example, that the local observing site shows 4.5-magnitude stars at the zenith, third magnitude at 40 degrees altitude and nothing much below 25 degrees. What does this offer? Plenty. The Moon, Venus, Mars, Jupiter and Saturn are bright and largely unaffected by light pollu-

tion. An atmospheric inversion layer induced by big-city heat and pollution sometimes steadies the air so that seeing is occasionally better than in the country. Metropolitan telescopic views of Jupiter or Mars often show as much detail as those at a dark location and make wonderful showpiece objects for visitors to your telescope.

Other kinds of observing, such as looking at lunar occultations and bright variable stars, examining the brightest star clusters and tracking the paths of asteroids, are affected by urban conditions to some extent but are generally possible from moderately light-polluted environments. The toll exacted by urban glow is a brightening of the sky background; the telescope's resolution is not affected. An 8-inch telescope under suburban fourth-magnitude skies, for example, will be limited to roughly the same deep-sky targets as a 3-inch telescope under black sixth-magnitude skies. But the resolution of the 8-inch instrument is unchanged, so what *is* seen can be studied in greater detail. However, such comparisons should not be carried too far. Much depends on the specific object being observed. The use of light-pollution filters changes the equation too, but only by a limited amount. Light-pollution filters make nebulas easier to see but cannot transform urban skies into rural dark-site skies.

## EVALUATING THE OBSERVING SITE

When you are considering a telescope purchase, it's a good idea to have a specific observing site in mind. Unless there is absolutely no alternative, do not rely exclusively on a remote, albeit ideal, site. Carefully evaluate the types of observing you can do from a site near home as well as the convenience of the site. Consider the following:

• Is the local site limited to binoculars, or is there room for a telescope to be used in reasonable privacy? (If you will be easily visible to neighbors or passersby "doing something" with a telescope, don't assume that the first thing people will think of will be astronomy!)

• If a telescope is useful at the local site, how far will it have to be carried?

• How many pieces must the instrument

# Rating Your Observing Site

How can you evaluate an observing site? Experience is the best guide, but here is a checklist that will help you rate your site. If you use more than one site, each should be assessed separately.

• Convenience: If you can observe in comfort from your own yard, take 5 points; 3 points for a short walk; 2 for a short drive; 0 for an hour or longer drive.

• Ground Level: If your site is outside a ground-level entrance where your equipment is stored, take 5 points. Take 3 points if car loading is required, but score 0 if any full flights of stairs are involved.

• Privacy: No possibility of surprise interruptions by people, animals or unwanted car lights is worth another 5 points. Score 0 if you sometimes feel nervous at the site.

• General Light Pollution: Score 10 if, on a good night, you can see magnitude 6.5 overhead, the Milky Way is obvious and the dome of light from the nearest city bulges less than 10 degrees above the horizon. Score 0 if you cannot see the Milky Way at all or any star fainter than magnitude 5.0. Estimate values for intermediate conditions.

• Local Light Pollution: A crucial factor. Zero points if you cannot avoid a local light as bright as the Moon. Score 4 points for a site that requires moving around to remain protected from light while observing. For a full 10 points, the brightest unobscured light should be fainter than Venus.

• Horizon: A clear, flat horizon to the south earns 5 points. South obstructions higher than 30 degrees rate just 2 points. Similar obstructions in all directions score 0.

• Insects: Mosquitoes are the enemy. Subtract up to 5 points if they are predictably annoying for more than one month each year.

• Snow: Snow cover has no redeeming value in astronomy. Apart from the cold weather that accompanies it, snow reflects light and increases overall light pollution. Subtract 1 point for each month of likely snow cover at the site.

   Maximum possible score (for a site next to your home, at a perfectly dark location, with little chance of mosquitoes or snow) is 40 points. Any score over 20 should be considered perfectly acceptable for regular use.

 *The full Moon looks down on the myriad lights of the Los Angeles basin. Photo by Leo Henzl.*

**Where Are Dark Skies?** ▶

The answer depends on where you live and what you mean by "dark." While true dark skies could be many hours away, a moderately dark environment might be relatively nearby. Experienced observers regard a limiting naked-eye magnitude of seven overhead as superbly dark. Those skies are visible from the black areas on this map. The gray regions are also excellent, with stars of magnitude 6.5 regularly visible to the naked eye. The mauve sections yield at least sixth magnitude—very good by today's standards of rampant light pollution. The green areas are afflicted with noticeable light pollution on the horizon, but overhead is still good, with close to sixth magnitude at the zenith. Yellow indicates that even overhead, the sky has deteriorated so that stars of magnitude 5.5 are usually the faintest visible. Orange is distant suburbs, where the worst source of light pollution wipes out the sky in the direction of the largest city and the limit overhead is fifth magnitude. Red is near-suburban environment that seldom allows fourth-magnitude stars to shine through. White is severe urban light pollution—typically, the faintest stars visible are magnitude 2.5. (The above are averages only, and intrinsic visual-magnitude limits vary from person to person.) Courtesy P. Cinzano et al, copyright Royal Astronomical Society.

be broken into for setup at the local observing site versus the remote site?

• How many trips will have to be made back and forth to the car or house during a typical telescope setup?

• Can the equipment safely be left unattended while the telescope is being assembled and disassembled? If not, can everything be carried at once?

• Is power available for the motor drive (if applicable)?

Considering the above, the key question becomes: Is one telescope suited to both the local and the remote site? A telescope that can be carried outside without being disassembled, even into just a few pieces, will be used more frequently than one requiring a several-step assembly and disassembly. At first, this might seem like a minor point, but the process of setting up and breaking down the instrument looms as a much bigger factor once the initial euphoria of the new telescope has worn off. The "paraphernalia effect" makes the pieces seem to grow larger and more awkward each time the telescope is transported to the observing site —local or remote. Frequently, the solution is two telescopes: one suited for the less-than-ideal local site and another for use at a remote dark site.

## REMOTE OBSERVING SITE

Few people live where sixth-magnitude stars are visible from their backyards. Most amateur astronomers must hunt for such a site, and reaching it can be an expedition if you live in or near a city of more than a million people. A 90-minute drive simply to get a reasonable view of the Milky Way is, unfortunately, a commonplace experience.

Well outside the city, there are still obstacles. More and more homeowners have installed dusk-to-dawn security lights that pump light horizontally into the corner of a sky observer's eye perhaps two miles away. But beyond the aggravation of rural farm and home lighting is the question of where to observe from once you are in the country.

Stopping on an infrequently traveled country road is fine for binocular gazing, but it's less than ideal for setting up equipment that cannot be retrieved and put in the car in a matter of seconds. There is the slight

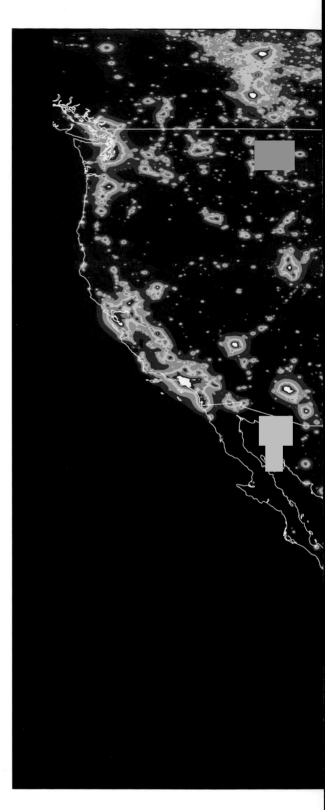

but very real possibility of being mistaken for a trespasser or some other form of law-breaker. Amateur astronomers tell horror stories about being routed by suspicious landowners (who can blame them?) or, worse, by a carload of troublemakers.

Heading out on your own and driving

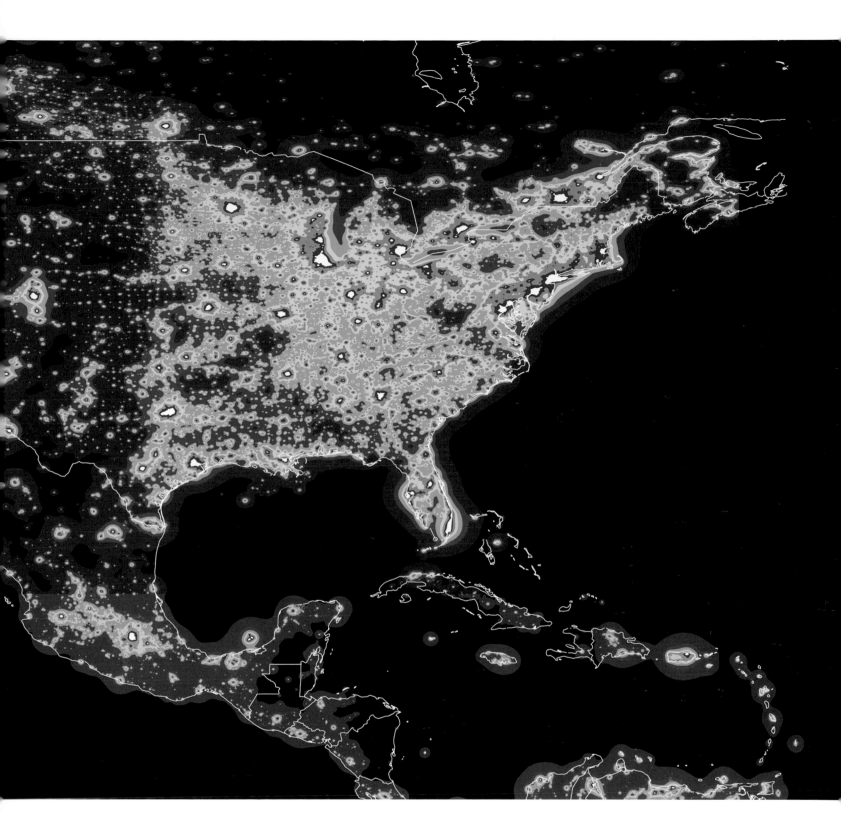

country roads in search of a good dark site from which to observe an aurora, a meteor shower or a bright comet is always a gamble. But sometimes, it is the only way—especially in the case of a comet that is situated near the horizon or some other specialized quarry that needs specific viewing geometry. In general, though, a predetermined safe and dark site, free of intruders, should be a long-term goal. Begin with inquiries at the local astronomy club, or if there is no club, ask other amateur astronomers in the area. Find out where they go for dark skies. They may have a private

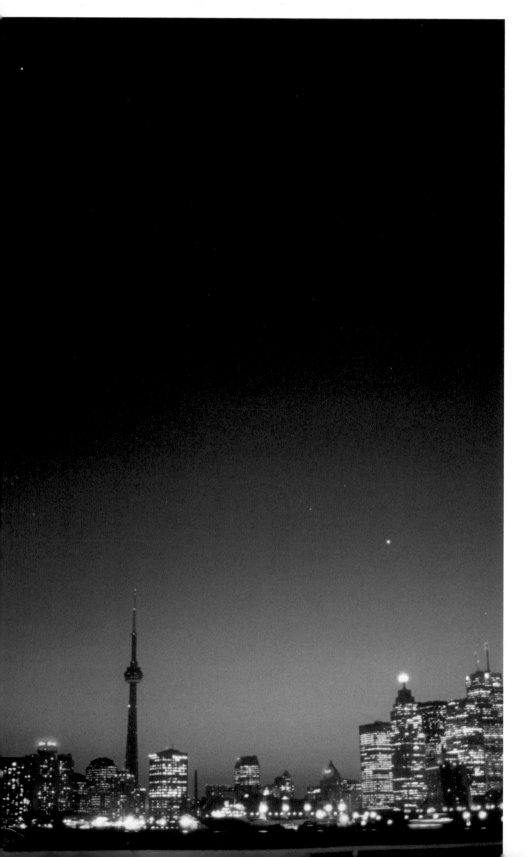

observatory in an ideal location, or perhaps they have found a state park or a campsite that has an area perfect for stargazing.

What constitutes an ideal observing site? The Milky Way should be distinctly visible. Under the very best conditions, the Milky Way has a textured appearance to the naked eye, with many levels of intensity and obvious rifts from dark nebulas. Third-magnitude stars should be visible less than five degrees from the horizon, and binoculars should reveal stars right down to the true horizon. It is almost impossible not to have at least one dome of light somewhere on the horizon from a nearby town or a more distant city, but if the largest such dome is in the northern sector of the sky, it will be least annoying. Objects in that direction are visible near overhead two seasons later.

The specific terrain of the observing site can influence sky conditions as well. Snow cover is the worst situation, not just because it is cold but because it reflects light. Even if there is little light pollution, the starry sky itself illuminates the ground. When that light is reflected back toward its source, it illuminates the dust and moisture particles in the atmosphere from below. Observing sites under a blanket of snow rate at least half a magnitude worse because of this reflected-skylight effect. Cold, crisp nights may look good on first inspection, but the overall brilliance of starry winter nights is partly an illusion caused by the brightness of Orion and its star-rich neighbor constellations.

The ideal dark site is an isolated, slightly elevated clearing in an area with fairly heavy vegetation: dense grass, scrub shrubbery or trees or a combination of these. Coniferous vegetation is preferable to leafy trees because the former releases less moisture (i.e., dew) into the atmosphere. A further advantage is the skylight-absorbing qualities of dark vegetation. A thick ground cover also acts as an insulating blanket, slowly releasing the ground's heat at night and protecting it during the day from soaking up as much heat as would bare ground.

Deserts may seem to be an ideal place from which to observe, and in one important respect, they are: lots of dry, dew-free, clear nights. But pale desert soil has undesirable skylight-reflecting qualities, and the large day-to-night temperature fluctuations in arid areas can work against a stable atmosphere. The sky may be clear, but layers of convection turbulence can affect the stability of high-resolution planetary images and the sharpness of stars. Also, a wind-prone desert site and its associated dust can put an end to an otherwise clear observing session. All that being said, we have had

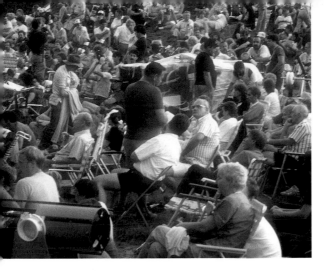

many outstanding arid-climate observing experiences in Arizona, California, Utah and New Mexico. The chief attraction is the percentage of clear nights, which is easily double what we enjoy at home.

## CONVENTIONS AT DARK-SKY SITES

One significant sign that recreational astronomy has come of age is the explosive growth in observing conventions at good dark-sky sites. The agenda is to have fun observing and interacting with fellow amateur astronomers. As recently as the 1970s, there was only one main convention, Stellafane, in southern Vermont; but now there are at least eight major conventions and another two dozen that each attract about 100 enthusiasts. One of them should be within driving distance of your home.

Stellafane was the first and is still the best-known meeting of amateur astronomers in North America. Each summer, several thousand people gather for a weekend atop a granite knoll named Stellafane, shrine to the stars. The convention has grown from a tiny gathering of 20 enthusiasts at its first meeting in 1926 to crowds of up to 3,000 who swarm over the rocks, bulge out of the lecture tent and devour thousands of hamburgers and hot dogs while examining the display telescopes with gem inspectors' eyes and, undoubtedly, a certain amount of envy. During the convention, which runs from Friday evening until early Sunday morning, telescopes are set up to be judged for optical and mechanical performance. Other telescopes are assembled for viewing purposes alone. At one time, Stellafane was a magnificent

dark site, but as is happening almost everywhere, encroaching urbanization is beginning to take its toll on the once pristine skies. However, the skies still rate a "B."

Unlike Stellafane, most amateur-astronomy conventions at favorable observing sites are of recent vintage. Some trace their roots to the tireless efforts of a single enthusiastic individual, such as the late Cliff Holmes, the driving force behind the Riverside convention held each May near Big Bear Lake, northeast of Los Angeles. Officially called the Riverside Telescope Makers Conference, the meeting focuses on telescope making, but not exclusively. Prominent amateur astronomers give talks on observing techniques, astrophotography and the use of equipment, as well as nuts-and-bolts telescope making. An important drawing card for the Riverside meeting is the commercial sales area in which all major telescope manufacturers are represented, many with discount offerings. (Commercial exhibits are still prohibited at Stellafane.) The night skies are hindered by the glow from the L.A. basin, but the altitude makes up for part of it, providing "B+" conditions overall. (The convention is always held on the Memorial Day weekend, so the Moon interferes some years.)

Another big annual astronomy meeting is the Texas Star Party, held since the late 1970s in May at the Prude Ranch in southwest Texas, near Fort Davis. It lasts for one week and is a 2½-hour drive from Midland or El Paso, the nearest good air services, but it is the meeting with the greatest potential rewards for dark-sky-hungry backyard astronomers. The southern latitude (31 degrees north) and the site's extreme isolation from major urban areas provide some of the best skies in North America. It merits a dark-sky rating of "A+."

◀ **Popular Astronomy Conventions**
Every year for the past four decades, more than 1,000 amateur astronomers have gathered at Breezy Hill, Vermont, for the Stellafane astronomy convention. This photo, taken in 1980, shows participants jockeying for good seats at the traditional Saturday-night twilight talk. More than a dozen similar large conventions are now held annually across North America.

▼ **Equipment New and Used**
Although Stellafane has never permitted commercial exhibits, all other amateur astronomers' conventions feature them prominently. Both used and new equipment are offered for sale, sometimes at bargain prices.

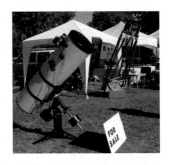

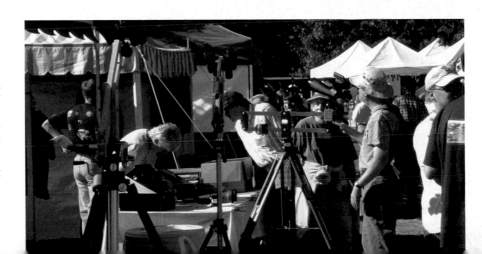

Astrofest, which is held each September in central Illinois, is another major convention at a campsite with good observing conditions (rating "B"). A weekend meeting

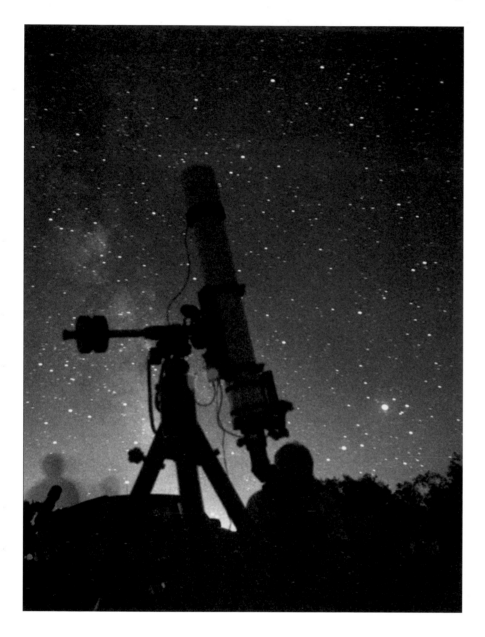

**Star-Party Observers** ▲
Star parties offer unique opportunities for dark-sky personal observing, chatting with fellow enthusiasts and comparing the views through many telescopes. Photo by Terence Dickinson.

that has become the largest amateur-astronomy convention in the Midwest, Astrofest always has an excellent range of astronomical equipment displayed by both amateurs and commercial exhibitors.

Other meetings at fine dark sites that are beginning to draw large crowds are the Winter Star Party in the Florida Keys in February (skies "B," with excellent seeing); Starfest, near Mount Forest, Ontario, in August (skies "B") ; and the Mount Kobau Star Party in south-central British Columbia

in August (skies "A"). Dates, locations and addresses for further information on these and other meetings can be found three to six months in advance in *Sky & Telescope*, *Astronomy* and *SkyNews* magazines. There is no better place than a star party for sharing enthusiasm for astronomy.

## LIMITING-MAGNITUDE FACTORS

How faint are the dimmest stars visible through a telescope? It depends on much more than the instrument's aperture. Factors include seeing, the transparency of the atmosphere at the observing site, the quality of the telescope's optics, their cleanliness, the type of telescope, its magnification, the observer's experience and use of averted vision and the type of celestial object being viewed.

Many amateur-astronomy guidebooks deal with limiting magnitude in one or two short paragraphs and a table. The table lists a telescope aperture and a corresponding limiting magnitude, not taking into account many of the factors mentioned above or clarifying which factors are considered. In our table on page 57, the magnitudes listed are based on good-quality optics, transparent dark skies and a reasonably experienced observer looking at a stellar object at high magnification, 20x to 30x per inch of aperture. If any of the conditions are not met, expect to see less for the reasons mentioned below.

An experienced observer can generally see a magnitude fainter than can a novice. Hundreds of hours of observing through a telescope train the eye to detect threshold detail, whether it is definition of features on a planet or the subtle wisps of a nebula. Of all factors, experience is the most important. Visual acuity varies from person to person, but the difference seldom amounts to more than half a magnitude.

Young people's eyes generally have slightly more sensitivity to objects at the threshold of vision, but veteran observers who are in their fifties or older can usually come within two-tenths of a magnitude of eyes 30 years younger. The ability of a youthful observer's eyes to dilate to 7mm or 8mm, compared with 6mm or less for more senior eyes, has no bearing on the equation—higher magnification, which reveals fainter objects by darkening the sky background, is achieved by smaller exit pupils.

One of the many long-standing assumptions of backyard astronomers is that faint deep-sky objects are best seen at low magnification operating at maximum exit pupil. While this does apply to some large, diffuse nebulous objects, such as the Helix Nebula, it is completely untrue when viewing faint stars.

Even on the darkest nights, the sky background is not black but gray. At low power, more sky is included in the view, so the overall brightness of the background sky actually increases as magnification decreases. Conversely, the sky background can be darkened by increased magnification —up to a point, of course. Magnification beyond 50x per inch of aperture seldom produces any further advantage.

Simply stated, the advantage of high magnification is that the sky background is darkened while the apparent size of the star image stays the same or is only marginally increased. A point source on a blacker background is easier to detect. Visual acuity is enhanced at higher magnification as well. Above 25x per inch of aperture, the exit pupil is down to 1mm or less, so the light cone, if centered on the eye's pupil, is passing through the most optically perfect part of the human vision system.

Everyone who has looked through a telescope is familiar with the effects of poor seeing, manifested as ripples on the Moon or undulations distorting the face of a planet. Stars and deep-sky objects are affected by seeing as well. Tiny stellar point sources are fuzzy and distorted at high power, and their light is spread out rather than concentrated into as small a point as possible, as it is with good seeing. Faint stars at the threshold of vision are significantly easier to detect in perfectly steady air than under turbulent, poor-seeing conditions. Bad seeing can remove a full magnitude from the penetration limit of a steady-air night.

Do different types of telescopes have different magnitude-penetration limits? Yes, but there is not much variation. It depends more on quality of optics than on the type of telescope. High-quality optics yield pinpoint star images instead of the tiny puff-

balls that never quite come into focus in mediocre telescopes. The more a star's light is concentrated into a point, the easier the point is to see; the star's per-unit surface-area brightness is higher than when its light is spread out by poor-quality optical systems or improperly collimated optics.

Finally, what about the unaided eye? The standard naked-eye limit for most people is sixth magnitude. In rare instances, under superb skies, people with abnormally good vision can see to 7.0 and even 7.4. Typically, the limit is 6.5, but it depends on

▼ **Visual-Magnitude Limits** Ursa Minor, the Little Dipper, is an ideal eye chart for determining your naked-eye magnitude limit. Use the clearest, darkest night at your favorite site to establish your limit. Star magnitudes are shown with the decimal point omitted to avoid confusion with star images. Inset chart can be used for binoculars.

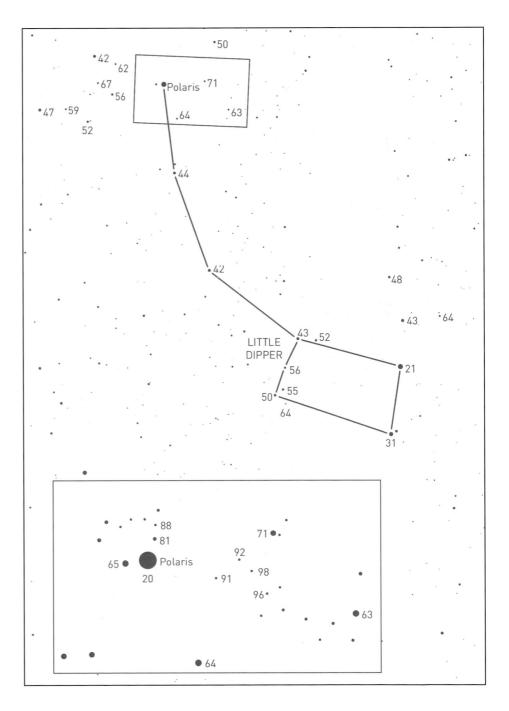

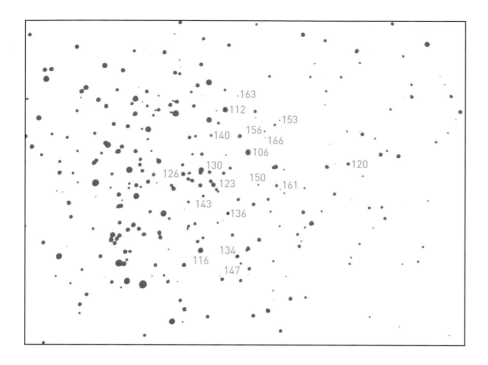

## THE DARKEST SITES IN THE WORLD

By this, of course, we mean the darkest *accessible* sites in the world. If you can't get there with your telescope, the main point is lost. We also impose another restriction: altitude. For most people, higher altitude causes some dizziness and breathing difficulty. Just as important, though, reduced oxygen above 9,000 feet begins to have a negative impact on dark adaptation—you slowly lose the ability to detect faint objects at night. For instance, on the 14,000-foot summit of Hawaii's Mauna Kea, which is dotted with giant observatories, the sky to the naked eye looks less starry than at the 9,000-foot level.

Thus our definition of a truly great dark-sky site is one between 3,000 and 9,000 feet altitude, with arid low-humidity climate, well away from sources of light pollution and accessible by paved road. (Lengthy treks on dirt roads raise intolerable amounts of dust that inevitably coat your optics and work into your mount and drives.) Well-known sites that fit this description are sectors of Australia and the American West

the specific sky conditions. Binoculars are more limited than telescopes of the same aperture because of their fixed low power. It is an achievement to reach magnitude 9.8 with standard 50mm binoculars and 10.8 with 80mm glasses.

# Averted Vision

Averted vision allows the observer to pick up fainter objects than can be seen by looking directly at them. The technique is simple. Look slightly away from the object under study while continuing to concentrate on it. Averted vision is most effective if the observer looks at a point halfway from the center to the edge of the field of view (the object in question is presumably at the center). The technique works especially well for diffuse objects such as comets, nebulas and galaxies, but it helps reveal fainter stars too.

It is a good practice to use the averted-vision technique from a variety of angles, because the highly effective dim-light sensors in the peripheral areas of the eye have different sensitivities. The overall gain achieved with averted vision can amount to more than half a magnitude. However, most backyard astronomers do not consider a sighting to be definite unless it is seen with direct vision. A notebook may read: "Glimpsed with averted vision but uncertain with direct vision." Such an observation is usually regarded as "probable." Definite sightings of faint objects need to be at least "apparent with averted vision and glimpsed directly."

The chief advantage of averted vision is gaining initial awareness of the existence of a threshold object. Then, once the target is detected, vision can be concentrated on it to attempt a direct confirmation.

and Southwest, parts of Hawaii, British Columbia and Alberta. These locations are home to some of the largest and best-equipped research observatories on this planet. We have observed from all of them.

We are trying to be realistic here, limiting our discussion to North America. Even the most well-heeled enthusiast can't go flying off to a world-class site for every observing session. To place comparisons of observing sites in perspective, we first need to examine the factors affecting all sites.

• Natural sky glow. No matter where you are on Earth, the night sky has a natural brightness, a gray luminance that becomes obvious to fully dark-adapted eyes. There are three sources that cause natural sky glow: (a) zodiacal light (sunlight reflected off dust particles that orbit the Sun in the plane of the solar system); (b) the atmosphere's permanent low-level auroral glow, which varies with solar activity; and (c) illumination of the Earth's atmosphere by starlight. Natural sky glow is a given; it's always there.

• Atmospheric extinction. Particles in the air scatter and absorb light coming to our eyes from celestial objects. Chief among these are dust and water vapor (humidity). The illumination of these particles by sources of light pollution accounts for most of the sky glow seen from the average observing site. Even assuming no light-pollution interference, the absorption factor amounts to at least 0.3 magnitude at a site near sea level and 0.15 magnitude at 7,000 feet. This assumes an arid site. A typical site in eastern North America loses another 0.2 magnitude to extinction from higher average humidity (added water vapor in the air compared with arid sites).

At superb sites, such as the Atacama Desert in Chile and the best locations in the American Southwest, all these factors add up to a loss of one-third to one-half magnitude compared with observing from, say, the space shuttle. This has been confirmed by astronomer-astronauts who have made careful visual comparisons to the finest Earth-based sites. However, there are many locations across North America where experienced observers, whose feet remain firmly on the ground, have confirmed that the legendary sites don't have a monopoly on the best skies on this planet. For exam-

# The Magnitude Scale

The brightness of a star—its magnitude—is rated on a scale that runs backward to what might be expected. The brighter the star, the lower its magnitude number. A difference of five magnitudes in brightness is equal to a brightness difference of 100 times.

| Magnitude | Celestial Object |
| --- | --- |
| –27 | Sun |
| –13 | Moon |
| –4.2 | Venus at its brightest |
| –2.9 | Jupiter at its brightest |
| –1.4 | Sirius (brightest star) |
| 0 to +1 | The 15 brightest stars |
| +1 to +6 | The 8,500 naked-eye stars |
| +6 to +8 | Deep-sky objects for binoculars |
| +6 to +11 | Bright deep-sky objects for amateur telescopes |
| +12 to +14 | Faint deep-sky objects for amateur telescopes |
| +15 to +17 | Objects visible in large amateur telescopes |
| +18 to +22 | Objects visible in large professional telescopes |
| +24 to +26 | Faintest objects imaged by largest ground-based telescopes |

ple, observers and astrophotographers in California don't all swarm around Palomar Mountain on a clear night. They know that there are many accessible places in the state which compare favorably to Palomar. Some are even better. The key is light pollution— or, more precisely, avoiding it.

On the North America light-pollution map on pages 164-165, all the black and dark gray regions are excellent hunting grounds for skies of pristine quality. Admittedly, there are no black or dark gray zones within an easy drive of the light-saturated Boston-Washington megalopolis or the Rust Belt states (Illinois, Indiana and Ohio). Conversely, note that parts of Arizona, Texas and New Mexico are darker than Kitt Peak Observatory. By using this map, you can be prepared when you travel to areas of lower population density.

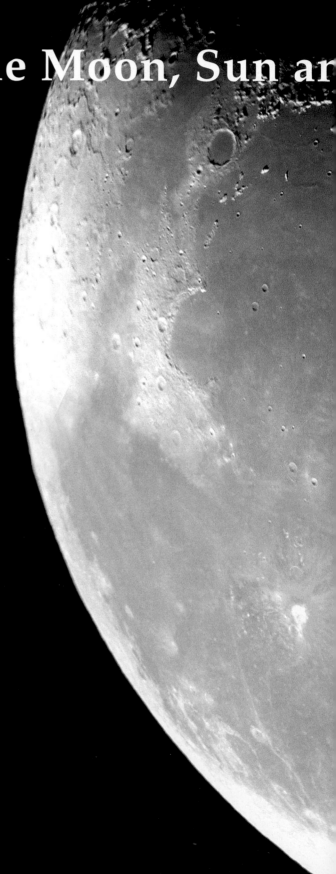

# Observing the Moon, Sun an

The Moon and the Sun were probably the first celestial objects Galileo looked at with his telescope nearly four centuries ago. The excitement that he must have felt as he gazed at the Moon's rumpled face for the first time is part of the legacy of the telescope as an instrument of exploration. Even in the crude 32x instrument, crippled by nearly every aberration known to optics, Galileo still saw features never before observed: craters on the Moon and spots on the Sun.

So it is today. Sunspots or lunar craters are usually the first details brought to the focus of a new telescope. Many newcomers to recreational astronomy are surprised to learn that all the lunar details visible in the photo at right can be seen in the eyepiece of a modern entry-level telescope.

.........................

Using a digital camera, Gordon Bulger took five images of sections of the last-quarter Moon, then "stitched" them together into this seamless single high-resolution image.

# Comets

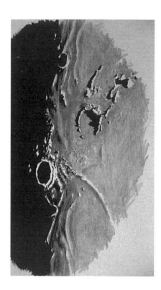

# Lunar Observing

Everybody's first look at the Moon through even the simplest of optical instruments is instantly rewarded by a wonderfully detailed image of our nearest cosmic neighbor. The eye and the mind are overwhelmed by detail—wrinkled plains, rugged fields of craters jumbled together, mountain ranges, valleys—all in stark relief, undistorted by even a wisp of haze, fog or mist (on the Moon, that is). The satellite is so close, its features so easily visible and the detail so abundant that regardless of the effects of poor seeing, there is always something of interest to examine.

Today, telescopic observation of the Moon is limited almost exclusively to introductory observing and to showing off the wonders of the universe to people who rarely have an opportunity to look through a telescope. Along with Saturn, the Moon is the number-one showpiece object—near, yet clearly alien. Without an atmosphere to protect it, the surface of the Moon has

been bombarded for billions of years by meteorites, comets and asteroids—debris left over from the formation of the solar system. Its cratered face is testament to this. On a smaller scale, the powdery material kicked up by the Apollo astronauts is the result of micrometeorites that grind down the surface into fragments as fine as dust.

As recently as the early 1960s, the lunar surface still had many secrets to divulge to backyard astronomers. By the late 1930s, much of the Earth-facing side of the Moon had been photographed to a resolution of two kilometers, and a few exceptional pictures showed features near the shadow line, or terminator, to a resolution of a few hundred meters. But unlike time exposures of deep-sky objects, which reveal far more detail than the eye can see through the same telescope, Earth-based lunar photographs always show less. Exposures of 1 to 2 seconds, typical of high-resolution lunar photography, are somewhat degraded by the Earth's atmospheric turbulence. (By high resolution, we mean close-up images of small areas of the Moon, rather than full-disk images.)

Even today, lunar photography from Earth can never quite equal what the eye

**Sketching the Moon** ▲
A rendering of the region around the 40-kilometer-wide lunar crater Aristarchus captures abundant subtle detail on the Moon's sunrise terminator. Amateur astronomer Matthew Sinacola used 175x on his 6-inch Criterion Newtonian reflector, like the one pictured on page 31.

**Moon in Detail** ▶
Even the smallest telescope will show wonderful detail on the Moon. But if the seeing is good, as it was here, the scene in a high-quality, large-aperture telescope can be nothing short of awesome. Alan Dyer was using Fuji Velvia 50 slide film on a research-grade 24-inch Cassegrain.

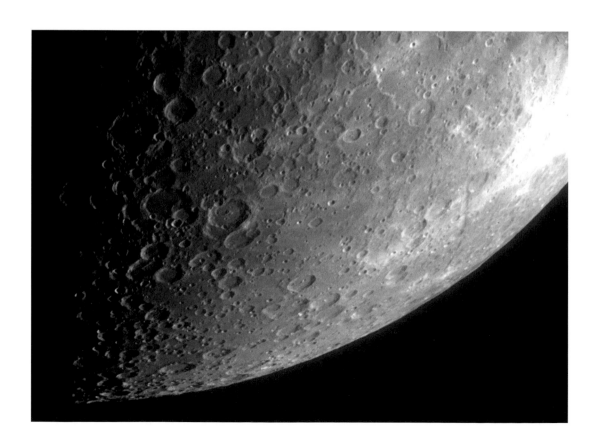

can discern through the same telescope. Only under rare instances of perfect seeing do photography and CCD imaging begin to come close. The experienced observer can watch for the momentary flashes of perfect seeing that occur during typical observing situations. At such times, resolution of better than two kilometers is commonplace in 6-inch telescopes, and occasionally, amazing amounts of fine detail almost magically emerge. But you have to wait for it.

## IS THERE ANYTHING LEFT TO DISCOVER?

Today, the vast majority of backyard astronomers regard the Moon as little more than a source of natural light pollution—a celestial nuisance whose light spoils views of dimmer objects. Other than as a showpiece for relatives and neighbors, the Moon is seldom deemed a worthy telescopic subject. The reality is, too many amateur astronomers have been duped by the notion that if there is nothing new to discover, there is no point looking. The Moon is a wonderland of alien landscapes; to see them, the observer needs to know *how* to look more than what to look for.

A high-resolution photograph of the Moon gives only an inkling of the truly impressive views that are possible with even moderate-aperture telescopes. Indeed, lunar observing is probably the one case in which aperture works almost in reverse —less is sometimes better than more. The Moon's image is so bright that even small apertures used at very high powers provide enough light to show the displayed features clearly; when using apertures greater than 6 inches, what can be seen is limited not by the telescope but by the steadiness of the atmosphere. Of course, at exorbitantly high powers—more than 60x per inch of aperture—the image often becomes fuzzy because of the limits of resolution of the optical system. As a general rule, however, the Moon can be viewed with higher magnifications than any other celestial object.

For instance, we have used a 5-inch apochromatic refractor up to 450x for lunar observation. However, the most common lunar-observation magnification used with that telescope is 220x, and the images are sharp and detailed—the equivalent of looking out the porthole of a spacecraft orbiting the Moon at an altitude of 1,600 kilometers. With such a telescope on a solid, smoothly driven equatorial mount, our satellite becomes a fascinating world to explore. The crawling terminator alters the appearance of the features in deep shadow within minutes, and in fine seeing, it is possible to distinguish every feature shown in the best pictures taken from Earth.

What will keep drawing you back to the Moon, though, is the exquisite beauty and the barren alien nature of the lunar surface, which constantly changes with the continuous sweep of the terminator.

## EQUIPMENT FOR LUNAR OBSERVING

At first glance, the telescopic image of the Moon is often overwhelmingly bright. But the glare is easy to control with a lunar neutral-density filter (about $15), which screws into the base of a 1.25-inch eyepiece in the same way that deep-sky and color filters do. The neutral-density filter blocks 90 percent of the light passing through it yet affects the image in no other way. This low-tech accessory reduces lunar glare to a comfortable level on most scopes.

Another way to reduce brightness is to use a twin polarized filter (about $40), sold as two separate 1.25-inch filters or, better, as two filters in a housing that fits on top of the eyepiece. An adjusting lever varies the amount of polarization between the dual filters, regulating the quantity of light entering the eyepiece from 50 percent to less than 1 percent. It works well for casual scanning, but the critical observer will notice a slight loss in resolution that does not occur with a single neutral-density filter. However, for casual lunar observing, the variable polarizer is an easy way—sometimes the only way besides extreme magnification—to reduce the Moon's brilliance in a large telescope to a satisfactory level.

A more serious glare-reduction device, advantageous with 10-inch and larger Newtonians and Schmidt-Cassegrains, is an off-axis diaphragm. Placed over the front of the instrument, it's a simple cardboard mask that has a one-third-aperture hole —precisely circular and smooth-edged— located toward the edge, where it will not

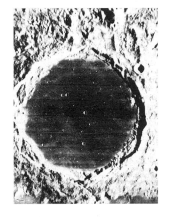

▲ **Plato's Craterlets**
A traditional test of good optics and steady seeing is to examine the floor of the crater Plato for craterlets when the Moon is one or two days past first quarter. Several craterlets are clearly shown in this 1965 Lunar Orbiter image. The two largest, about three kilometers wide, have been detected with a 4-inch apo refractor. NASA photo.

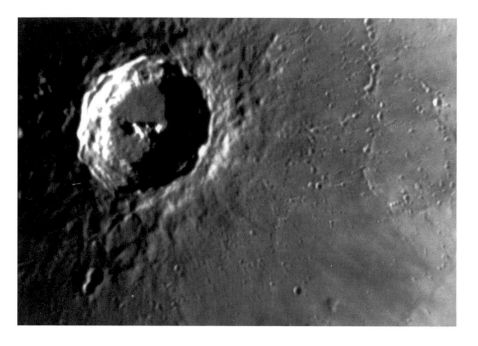

be obstructed by the secondary mirror or its support vanes.

Off-axis diaphragms reduce the amount of light going through a telescope to about 13 percent of the full-aperture value, similar to the reduction obtained when using a standard neutral-density filter. Because the aperture is now unobstructed like that of a refractor, contrast is enhanced, a definite bonus for lunar viewing.

Some lunar observers who use achromatic refractors prefer color filters. The favorites are green, deep yellow and orange. These inexpensive color filters have the advantage of dimming the Moon while greatly reducing chromatic aberration and thereby sharpening the view. The overall color cast is soon easy to ignore. Color filters have no advantage over a neutral-density filter in other types of telescopes.

### PROBING LUNAR VISTAS

Anyone who has offered views of the Moon to the general public at stargazing events or to friends has heard the question, Can we see where the astronauts landed? The answer is, yes and no. The location can be pinned down by using a lunar map, but what the person really wants to know is whether it's possible to see the footprints, lunar lander and other paraphernalia left on the surface. A good way to answer is to point out that the smallest crater visible

under ideal conditions in a large telescope is several times larger than the biggest sports stadiums in the world, whereas the largest piece of hardware left by the Apollo astronauts is smaller than a two-car garage.

The likelihood of seeing any changes on the Moon—for instance, a meteorite hitting the surface or possible volcanic activity—is infinitesimally small. A meteorite large enough to produce a cloud of ejecta visible from Earth would strike the Moon no more than once or twice per century—at least on the daytime areas. The lunar nightside is another matter. Low-light video cameras viewing through moderate-aperture telescopes recorded impact flashes on the lunar nightside during the heavy Leonid meteor showers in 1999 and again in 2001. These were caused by meteors perhaps the size of tennis balls. It would take a much larger object to create a visible flash on the lunar dayside. In any case, none has ever been observed.

A few decades ago, a small band of enthusiasts maintained that searches for so-called transient lunar phenomena (clouds of dust which could arise with the release of gases from the Moon's interior) were viable observing programs for backyard astronomers. However, there's no unequivocal proof that this type of activity has ever been observed.

# Solar Observing

Examining the intensely brilliant surface of the Sun is possible only with proper filtration. Never look at the Sun through any optical device unless you are sure that it is safely filtered and that you know what a safe filter is. A very dense filter is necessary. Not only must it reduce all visible wavelengths to a safe level, but it must also block infrared and ultraviolet light. This is serious stuff. These invisible wavelengths can damage the retina of the eye and, in extreme cases, cause partial or complete blindness. *Do not take chances.* Astronomy is a benign hobby with few opportunities for personal injury. But this is one of them. Read the following carefully.

Over the years, especially around solar-eclipse events, many materials have been recommended as solar filters. Most are unsafe. Do *not* use smoked glass, sunglasses, layers of color or black-and-white film (no matter how dense), photographic neutral-density filters or polarizing filters. People have even been known to look through the bottoms of beer bottles. This seems laughable, yet many novice amateur astronomers use a type of solar filter that is far more dangerous than anything listed above.

The filter in question, which comes with many small department-store telescopes

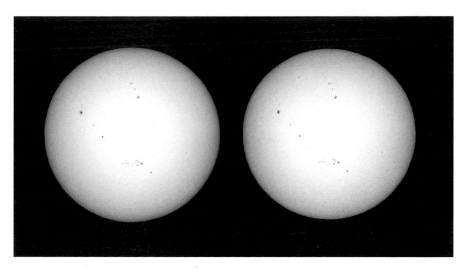

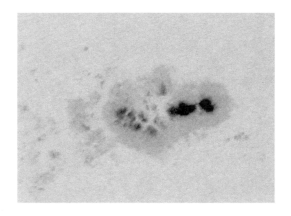

especially older models picked up at garage sales—is a piece of dark green glass that screws into the bottom of an eyepiece. Such "sun filters" are unsafe because they sit at the focus of the telescope, where all the light and heat are concentrated. When the instrument is aimed at the Sun, the temperature near the filter can reach hundreds of degrees, cracking it and letting through a blinding wash of sunlight. Eyepiece solar filters are rarer than they once were but are still included as an accessory with some small telescopes given to children for Christmas.

The simplest, safest and most readily available filters for gazing at the Sun with unaided eyes are No. 14 welders' filters. They are sold in two-by-four-inch rectangles for a few dollars at well-stocked welding-supply outlets. Because the filters are made to exact specifications for the welding trade, they are reliable and have the right density to be completely safe. Only the No. 14 grade is appropriate; the more common No. 12 welders' filter is too light.

To observe the Sun with a No. 14 welders' filter, place the filter in front of your eyes before you gaze up at the Sun. Individuals with 20/20 vision or better will see sunspots that are Earth-sized or larger. People with exceptional vision, 20/12 or better, can see spots almost daily with no optical aid other than the welders' filter.

A pair of seldom-used binoculars can become a permanent sunspot device if two welders' filters are securely taped over the front of the main lenses. Spots the size of Asia can be seen. However, for detail on the spots, a telescope is needed. A 60mm-to-80mm refractor is ideal and, if properly filtered or used for projection, reveals a wealth of fine structures in and around sunspots.

**Sunspot Observations**
For full-disk solar viewing—either projection or filtered—a small, inexpensive refractor will do the job. To prove it, Bob Botts took these identical photos (above) with two 70mm refractors, one worth $150, the other 10 times as much. The image of a sunspot at left was taken with a 70mm apo refractor. The superb close-up (below), taken with front-line research equipment, shows impressive detail. Courtesy SSVT.

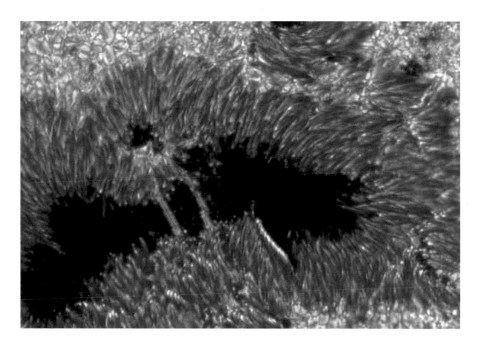

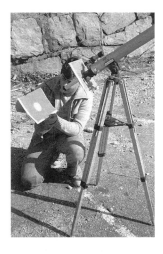

# SOLAR VIEWING BY PROJECTION

Solar projection is the filterless way of observing the Sun. Aim the telescope at the Sun by watching the instrument's shadow, not by gazing up the tube at the Sun and especially not by peeking into the finderscope. (Always cover the front of the finderscope during any solar observing.) When the telescope's shadow becomes circular, the telescope is aimed approximately sunward. Hold a white card a foot behind the eyepiece to catch the Sun's image. Using an eyepiece yielding about 30x, focus the projected image. This method is ideal for group observing and is particularly effective if the white card is mounted on an easel or a tripod and shaded from direct sunlight to increase contrast. However, the detail visible in direct filtered viewing cannot be equaled.

Even so, projection is preferred for making full-disk drawings of sunspot positions and their relative sizes. For orientation con-

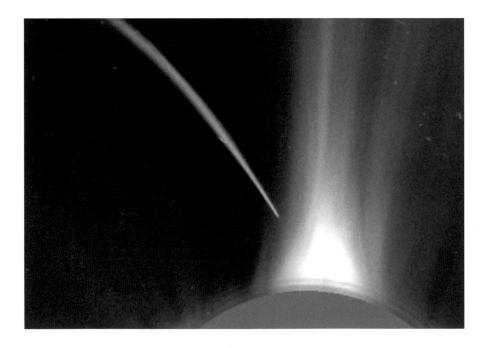

sistency, determine the celestial east-west axis for each drawing by watching the solar disk drift into or out of the undriven field of view. The centerline of the drift motion is celestial east-west. A few months' worth of drawings reveal fluctuations in the numbers of spots. Small spots may last several days or grow into large spot groups and stay

on the solar face for weeks. Because of the 3½-week solar-rotation period, the scene changes a little every day and a great deal in a week.

For projection, the telescope's aperture should be limited to 60mm or less to avoid damage to the eyepiece. Small refractors are ideal; larger telescopes must be covered with a diaphragm. A hole in a piece of cardboard covering the aperture is fine for refractors or Newtonians, but never practice solar projection with a catadioptric (Schmidt-Cassegrain or Maksutov-Cassegrain). Heat will quickly build up inside these instruments before you are aware of a problem, causing unwanted seeing effects or even damaging the telescope.

Sunspots wax and wane in an 11-year cycle. The last maximum was in 2001. Solar activity continues throughout the cycle, and there are usually a few spots no matter when you look. A surprising amount of detail is visible around a large sunspot and is best seen by direct viewing with the appropriate filter.

# SOLAR FILTERS FOR TELESCOPES

A proper solar filter is made to fit snugly over the front of your telescope, where it safely reduces the intensity of sunlight before it enters the tube. Known as full-aperture filters, or prefilters, the most durable are made with optical plane-parallel glass coated with a nickel-chromium alloy. Thousand Oaks Optical is one of the major suppliers of these filters for any size telescope. Prices range from $50 for a 60mm refractor to $150 for a 12-inch Schmidt-Cassegrain.

Metal-coated thin Mylar film, a sturdy plastic material about the thickness of shrink-wrap, is an alternative to the glass filter. The Mylar must be in a cell that is press-fit over the front of the telescope. Several telescope manufacturers make these filters, or you can make your own from a piece of the filter material, available as a stock item from large telescope dealers. We have found that the Mylar filters equal or exceed the optical quality of glass filters. This is especially true for the Baader solar-filter material, which revealed the sharpest sunspot detail in our tests with 4-inch apo refractors. As you would expect, you will pay a premium

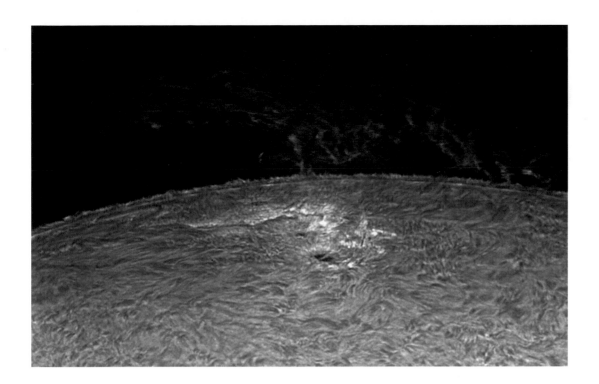

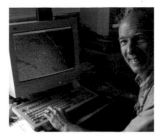

▲ **Jack Newton at Work**
Using a Coronado 90mm
Hydrogen-alpha filter
attached to a 5-inch refractor, Jack Newton captures
amazing solar portraits
with a Meade Pictor CCD
camera. Newton takes two
pictures—a short exposure
of the solar surface and
a longer exposure of the
fainter prominences—
then digitally stitches them
together for the final
image (left). For visual observation (below), the tiny
40mm-diameter Coronado
H-alpha filter, seen at front
of this Meade ETX 70mm
refractor, provides surprisingly detailed views.

for a filter made from the Baader material.

Some Mylar filters impart an unnatural blue cast, though this characteristic is correctable with a No. 23A eyepiece filter, commonly used for Mars observations, which absorbs the blue and gives the solar disk a yellow cast. Most metal on glass types also present a more natural-looking yellow Sun.

Another word of caution: Not just any Mylar will do. Do not run down to your local hardware or automotive-supply store to buy sheets of Mylar for homemade filters. Most metallized Mylar sold for use on van and camper windows and incorporated into materials like "space blankets" is unsafe for solar viewing. It is not dense enough and does not necessarily block harmful infrared and ultraviolet light. Use only filters sold specifically for astronomical applications by telescope manufacturers. With them, you and your friends and family can watch sunspots for hours in complete safety.

All the solar filters discussed so far show the Sun in white light; that is, they reduce the amount of light across the entire spectrum. A very specialized type of solar filter works a little differently. It eliminates all light from the Sun except the single wavelength emitted by hydrogen atoms—656 nanometers. When the Sun is viewed in the light of hydrogen atoms, previously invisible features such as solar prominences, fila-

ments and flares appear. Prominences are normally seen only during a total solar eclipse. With a Hydrogen-alpha filter, they are visible on any cloudless day.

H-alpha filters add a new dimension to the hobby of astronomy, but they are expensive, costing between $700 and $10,000, depending on the aperture and bandwidth. Coronado and Daystar are the principal manufacturers. The most common sizes have apertures of 40mm to 90mm and work best on refractors in the same range. Because seeing in the daytime is poorer than seeing at night (owing to solar heating of both the air and the ground, a double whammy), filter apertures over 90mm have a diminishing advantage. Also, H-alpha filters are more fragile than a typical astronomy accessory. Always transport them with care and lots of foam protection.

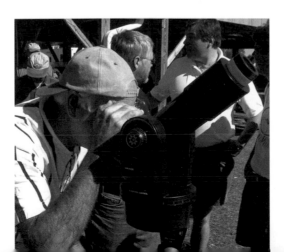

# COMETS

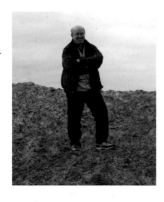

**Alan Rides a Comet ▲**
Satisfied that he has conquered the final frontier, author Dyer stands atop a cometlike "dirty snowball" —the melting remnants of a winter's worth of parking-lot snow piled at the edge of one of North America's largest shopping malls.

**The Great Hale-Bopp ▼**
Members of a 1997 college astronomy class watch with binoculars as the most readily visible comet since Halley's in 1910 graces the northeastern skies. Photo by Terence Dickinson.

Comets are agglomerations of ice impregnated with dust. Not glacier-type ice—more like a pile of dirty snow about the size of a small city. They orbit the Sun as the planets do, but in more exaggerated, elongated paths. Comets that venture closer to the Sun, within the orbit of Mars, are the quarry of the backyard astronomer. At this distance, sunlight vaporizes the comet's icy surface, releasing gas and dust that are swept into a tail by the solar wind and the pressure of sunlight in the vacuum of space.

Apart from a few historical exceptions, comets are named for their discoverers. Comet Hale-Bopp, for example, was spotted on the same night in July 1995 by New Mexico comet hunter Alan Hale and amateur astronomer Tom Bopp of Arizona.

Comets that boast tails easily visible to the unaided eye are relatively rare, averaging two per decade. Of course, that is just an average. Two such comets were seen six months apart in 1957 and two others just 12 months apart in 1996 and 1997.

Halley's Comet, by far the most famous comet of all time, made its most memorable appearance in April 1910, when it reached first magnitude, sported a 30-degree tail and was watched by millions. The comet's fame stems from its 76-year orbit, which closely matches the average human life span, making the feathery cosmic visitor a true once-in-a-lifetime event. Although astronomers correctly predicted its feeble return in 1986 (magnitude 3.5 at best in the northern hemisphere), huge crowds turned out whenever a science center or astronomy club announced a public telescope viewing. The throngs came because of the name, and many went away disappointed with the reality. Even though correct brightness predictions were widely disseminated through the news media, we live in a celebrity culture, and Mr. Halley's comet is as big a celebrity as astronomy has.

## BRIGHT COMETS: 1950s to 2002

In 1957, two bright, naked-eye comets made nearly back-to-back appearances. The first, Comet Arend-Roland, became a brilliant first-magnitude object with a 15-degree tail during the last week of April. It rapidly faded to sixth magnitude by the middle of May as it receded from the Sun and Earth. The comet was notable for its luminous antitail, a dust tail that, because of our perspective from Earth, appeared extended from the comet's nucleus toward the Sun.

The second comet was Comet Mrkos, a first-magnitude object discovered with the unaided eye in twilight by a pilot on July 29. During the initial week of August, the comet was well placed in the evening sky and had an overall magnitude near 1 and a five-degree tail. By the middle of August, the comet was third magnitude in darker skies and had a straight 10-degree tail.

Comet Seki-Lines of 1962 was a respectable entry that briefly reached third magnitude in the April evening sky and had a tail more than 10 degrees long when viewed through binoculars.

The brightest comet of the 20th century was Comet Ikeya-Seki of 1965, a member of a rare class of sungrazers that swoop to within one or two solar diameters of our star's surface. The explosive vaporization of the comet's ices during its rapid hairpin turn around the Sun generated a dense coma and a huge tail that made Ikeya-Seki visible to the unaided eye in broad daylight if the Sun was blocked from view. When

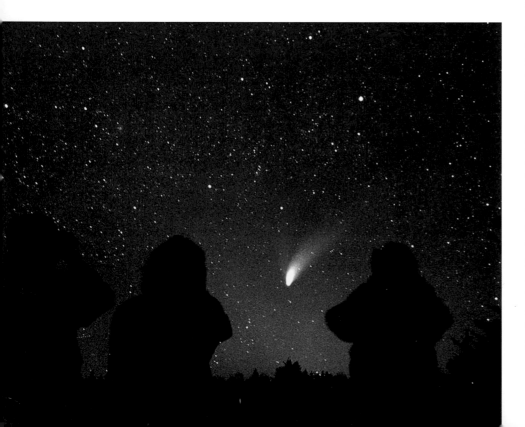

only two degrees from the Sun, the comet was estimated at an astounding magnitude –10 and had a two-degree tail.

As Ikeya-Seki swung into the morning sky during the last few days of October, an amazing 45-degree tail extended from a small but brilliant nucleus. Although the comet was stupendous in the southern hemisphere, it was poorly seen north of 40 degrees north latitude, because the tail was angled low toward the horizon. It remained a naked-eye object for only a week, rapidly fading as it pulled away from the Sun. Nevertheless, it holds the title of the brightest comet of the 20th century.

Comet Bennett was discovered in December 1969 by John C. Bennett of South Africa during a deliberate comet search. Throughout April 1970, Comet Bennett was a conspicuous object in the evening sky in the northern hemisphere, shining at second magnitude overall with a 20-degree-long tail. It was one of the most widely observed comets of the 20th century and the first to be photographed extensively in color by backyard astronomers.

Three years later, Comet Kohoutek arrived. Although it was not a great comet, it is so well known in the history of comet observation that it deserves to be dealt with in more detail.

## THE SAGA OF COMET KOHOUTEK

On March 18, 1973, Lubos Kohoutek of Germany's Hamburg Observatory discovered a tiny 16th-magnitude smudge—a comet—on a photograph he had taken two weeks earlier during a routine asteroid search. Orbit calculations soon showed that the comet was nearly five times the Earth's distance from the Sun and would not reach perihelion, its closest point to the Sun, until December 28, 1973, when it would pass within the orbit of Mercury. In early January 1974, Comet Kohoutek would emerge into the evening sky near Venus and Jupiter.

How bright would it be? In September 1973, NASA published a guide to Comet Kohoutek that was widely circulated to teachers and the news media. It predicted that the comet would be magnitude –4 in early January—equal to Venus—which would make it the brightest comet in four

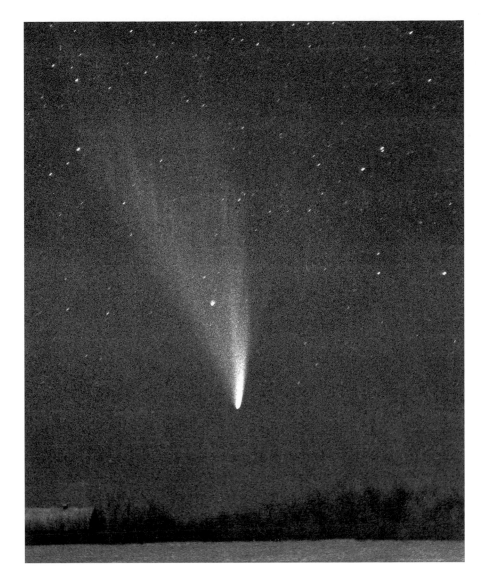

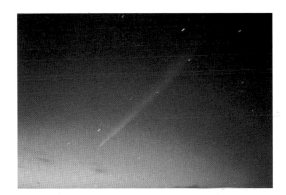

centuries. Astronomy enthusiasts were bursting with anticipation. Even the most conservative estimates made in the summer and early fall of 1973 suggested that the comet would be at least zero magnitude.

By December, though, something was clearly wrong. Kohoutek, at its best, was

**Catching Comets**
William Krosney captured more than Comet West (above) in this fine celestial portrait taken on March 5, 1976, from Manitoba. You can feel the stillness and cold of the night and share in the exquisite beauty of this rare phenomenon. Comet Ikeya-Seki (left) had the longest tail of any comet in the 20th century. Richard Keen photographed it on November 1, 1965, using a 55mm lens.

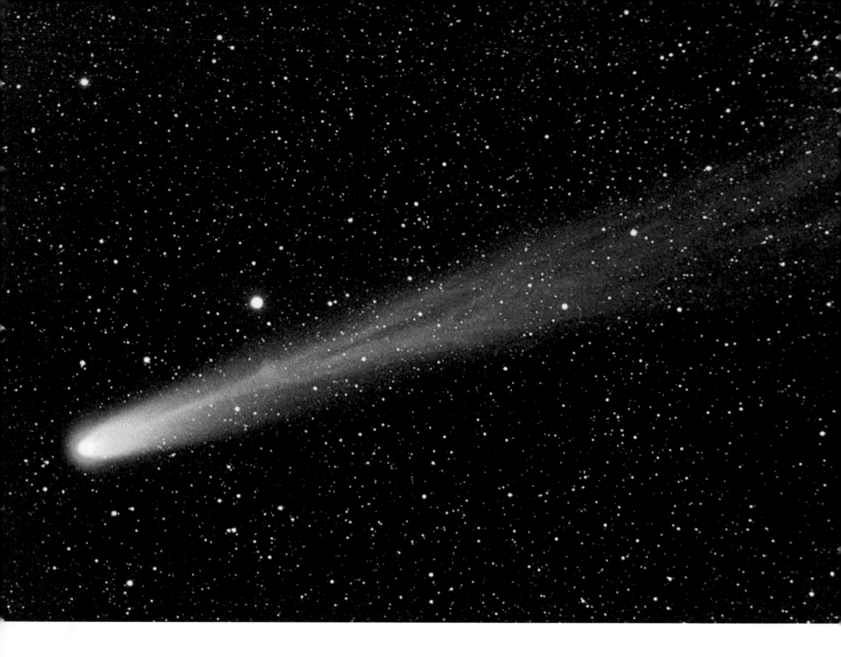

**Comet Hyakutake** ▲

After a comet "drought" lasting a generation, Comet Hyakutake made a dramatic appearance in northern-hemisphere skies in the spring of 1996. Sporting a slender 55-degree-long tail, it reached first magnitude in late March and was visible all night when it was nearest Earth. This 5-minute photo of the comet was taken through a 180mm f2.5 telephoto lens on Fuji 400 print film by Terence Dickinson.

fourth magnitude, in the same league as Halley on its swing by the Sun in 1985-86, and was a huge disappointment to the thousands of backyard astronomers out in force to greet it in early January. The public, primed for a "blazing spectacle" or a "cosmic searchlight," saw nothing.

A major contributor to the inflated predictions was the fact that Kohoutek was a comet on its initial visit to the inner solar system from the Oort cloud, the comet reservoir beyond Neptune. It was a pristine chunk of cosmic flotsam that had never been exposed to solar heating before. Readily vaporized materials, such as methane, hydrogen cyanide and methyl cyanide, formed a cloud around the frozen comet nucleus at huge distances from the Sun, giving the false impression that the

activity would continue unabated closer to our star. Once these materials had dissipated, though, the rate of brightening declined sharply.

Comet Kohoutek still ranks among the top 25 comets of the 20th century, but it will always be remembered as the comet that fizzled. In doing so, however, it provided valuable information that allowed astronomers to avoid making the same predictive goof ever again.

## COMET WEST

The story of Comet West is the Kohoutek saga turned on its head. It was discovered in November 1975 as a small puff on survey plates taken at the European Southern Observatory in Chile. As it approached the

Sun, it was much dimmer than Kohoutek was at similar distances beyond Mars, so it attracted comparatively little attention. But as fate would have it, in January 1976, on its way to perihelion inside the orbit of Mercury, Comet West brightened much faster than predicted. Ten hours past perihelion, the comet was visible to the naked eye 10 minutes before sunset at an estimated magnitude of –3, making it an extremely rare daylight comet. But because this was both unexpected and unpredictable, few astronomers—professional or amateur—witnessed it.

In early March, the cosmic visitor moved into the morning sky and dimmed only slightly to magnitude –1. On March 7, one of the most beautiful comets of all time adorned the heavens in every part of the northern hemisphere that had reasonably dark, cloudless skies. Comet West had a yellow main dust tail 30 degrees long, which hung like a ghostly feather above the horizon just before dawn.

By March 13, the show was over, unobserved by the general population. With the Kohoutek fiasco still fresh in their minds, news editors had ignored Comet West, so few people beyond the amateur-astronomy community even knew about it. Others (including the authors) missed Comet West because of bad weather.

For the comet to become 50 times brighter than initial predictions (by about the same factor that Kohoutek had been

dimmer than predictions), something unusual must have occurred. Around the time of perihelion, three huge chunks broke away from Comet West's nucleus, releasing far greater amounts of gas and dust than would be expected of an intact comet.

The key is the dust. Tiny particles like those which typically float in the air (illuminated in a room by shafts of sunlight) are wonderful reflectors of light. Comet dust is pushed away from the nucleus into the sweeping tail by the pressure of sunlight. Lots of dust results in a bright tail.

## RECENT COMETS

Four years after Halley's return, the sixth visual discovery of North American comet sleuth David Levy reached magnitude 3.5 in the northern hemisphere in the August 1990 evening sky. Because it was 1.3 astronomical units (1 AU = Earth-Sun distance) from the Sun at the time, its tail was never more than a faint fan, but it was observed and enjoyed by amateur astronomers throughout the northern hemisphere.

By 1995, there was talk of a comet drought, as no bright, long-tailed comets had been seen since West, a generation earlier. In early 1996, the drought ended. Using giant 25x150 comet-hunting binoculars, Japanese comet hunter Yuji Hyakutake picked up a comparatively small comet, later determined to have a two-kilometer-diameter nucleus, compared with Halley's

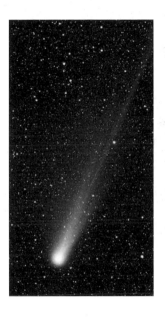

▼ **21st-Century Comet**
Third magnitude at its brightest, Comet Ikeya-Zhang was visible in spring 2002. It was a beautiful binocular comet, with a bright seven-degree tail. Photo by Alan Dyer using a Nikon 180mm f2.8 lens.

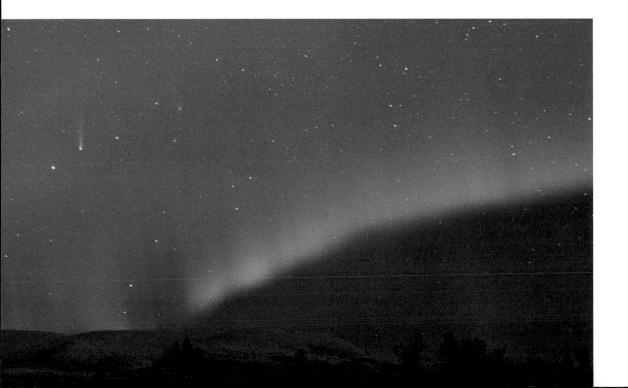

◀ **Northern Dancers**
In this shot taken from just outside Whitehorse, Yukon, Phil Hoffman recorded Comet Ikeya-Zhang floating behind a graceful auroral curtain. He used Fuji Provia 400F slide film and a 50mm f2 lens for the 30-second exposure.

183

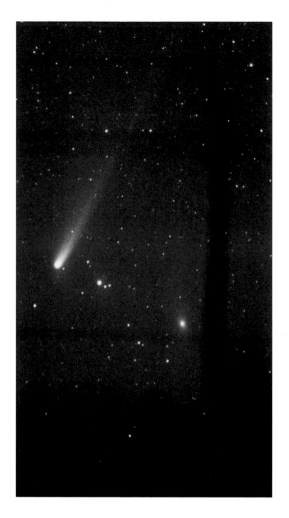

11-kilometer-wide icy core. But Comet Hyakutake came within 14 times the distance to the Moon. On a cosmic scale, that's close. Even a small comet looks big at that distance. (For comparison, Comet Halley remained more than 400 times farther away than the Moon during its 1986 visit.)

In late March and early April, Comet Hyakutake put on a great show, growing a 60-degree tail—more than twice the length of the Big Dipper—and reaching first magnitude as it hurtled through the constellations. Some veteran observers still consider Hyakutake the most beautiful comet they have ever seen.

Then came the undisputed king of the comets: Hale-Bopp. Almost everyone reading this will remember this magnificent object, which reached first magnitude and stayed there for weeks during March and April 1997. Although exact comparisons are difficult, Hale-Bopp is probably the most impressive comet seen in the northern hemisphere since the 1880s, edging out even the

great Halley in 1910. Far from hosting a comet drought, the 1990s laid claim to two of the outstanding comets of all time.

Five years after Hale-Bopp, in the spring of 2002, Comet Ikeya-Zhang gave northern-hemisphere observers another fine comet with stamina. Ikeya-Zhang remained at magnitude 3.5 as it approached Earth and receded from the Sun in late March and early April. In dark rural skies, it was a wonderful binocular sight, with a classic 5-to-10-degree tail. Even though it equaled or exceeded Halley in every respect for northern-hemisphere observers, the comet drew little popular interest, proving once again that name recognition is an all-powerful cultural phenomenon.

In compiling this dossier of recent comets, we can't forget Comet Shoemaker-Levy 9, the comet that was torn apart by Jupiter's immense gravity. The pieces plunged into the giant planet in July 1994, leaving black bruises on Jupiter's cloudy face. Visible in a 60mm refractor, the impact scars on Jupiter were the most prominent features ever seen on a planet. Placing the event in perspective, codiscoverer and comet expert Gene Shoemaker estimated that a comet of this size (the original nucleus was perhaps two kilometers wide) collides with Jupiter on average about once every 1,000 years.

## DISCOVERING A COMET

A generation ago, most comets were discovered by backyard astronomers scanning the night skies with modest-sized telescopes. Some of the hobbyists even turned the activity into a part-time career, gaining worldwide renown and literally notching their telescopes with each new find.

Those days are over. Apart from ever-increasing competition from their peers, the main change for comet hunters occurred in 1998. That's when the U.S. Air Force, at the request of NASA and other scientific agencies, began using a one-meter robotically controlled telescope to search for asteroids that might smash into Earth at some time in the future. The telescope turned out to be a superb comet finder as well. Known by the acronym LINEAR (LIncoln, Near Earth Asteroid Research), the New Mexico-based instrument has found more than three-quarters of all comets visible from the northern hemisphere since it became operational.

The LINEAR telescope's success has virtually put amateur comet hunters out of business. The pre-LINEAR rate of discovery by amateurs was four to eight a year. Now it is only one or two per year. Even rarer is the accidental visual find—someone happening upon a previously unobserved comet.

But that's exactly what happened at the 2001 Saskatchewan Summer Star Party, an annual gathering of about 200 astronomy enthusiasts at a dark provincial park in the southwest corner of the province. As amateur astronomer Vance Petriew was enjoying the views through his new 20-inch Obsession reflector telescope around 3:30 a.m. on August 18, he took a wrong turn on the way to aiming his telescope at the famous Crab Nebula in Taurus. The guide star he thought was Gamma Tauri was in fact Beta Tauri. Although that mistake meant he didn't find the Crab Nebula, Petriew did stumble upon a faint fuzzy object that proved to be a rare accidental visual comet discovery.

Comet Petriew was a 10th-magnitude comet, which means it would have taken at least an 8-inch telescope to detect. The comet never got much brighter, but the idea that an accidental comet discovery is still possible caused quite a stir in stargazing circles around the world.

▼ **Hale-Bopp the Magnificent**
A truly great comet is one that reaches first magnitude or brighter while at the same time remaining plainly visible for weeks. Such wonders occur only once or twice in a lifetime. This 10-minute exposure was taken by Terence Dickinson using Fuji 400 print film on a 7.5-inch f/2.3 Ceravolo Maksutov-Newtonian.

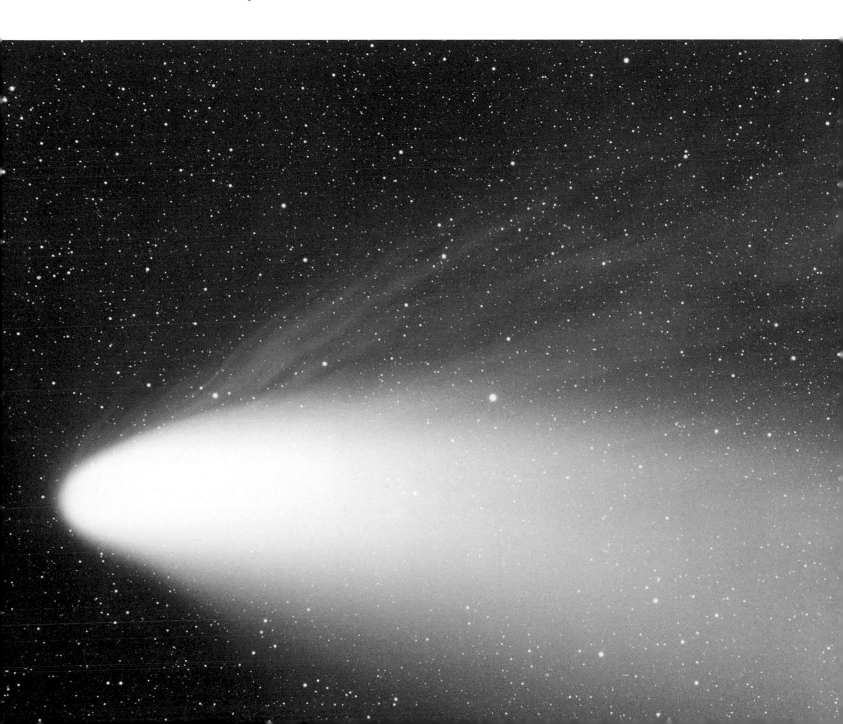

# Observing the Planets

Tens of thousands of years ago, prehistory's first Galileo would have noticed a few bright stars moving against the stellar background. This early planetary astronomer was probably also the first astrologer, because the natural question arises, Why do a small, select group of bright stars move, while the others remain fixed?

Astrological ruminations aside, planet watching remains a major element of today's recreational astronomy. On most nights of the year, at least one of the five naked-eye planets is visible. Quite often, it is the brightest object in the sky. When two or more planets cluster together, the sight can be so striking that it turns the heads of people who normally do not look skyward. And if the crescent Moon joins the scene, the result is one of nature's most alluring spectacles: a silvery sickle adorned by one or more jewels.

The stunningly beautiful planet Saturn is seen in its full glory in this Hubble Space Telescope portrait taken November 2000.

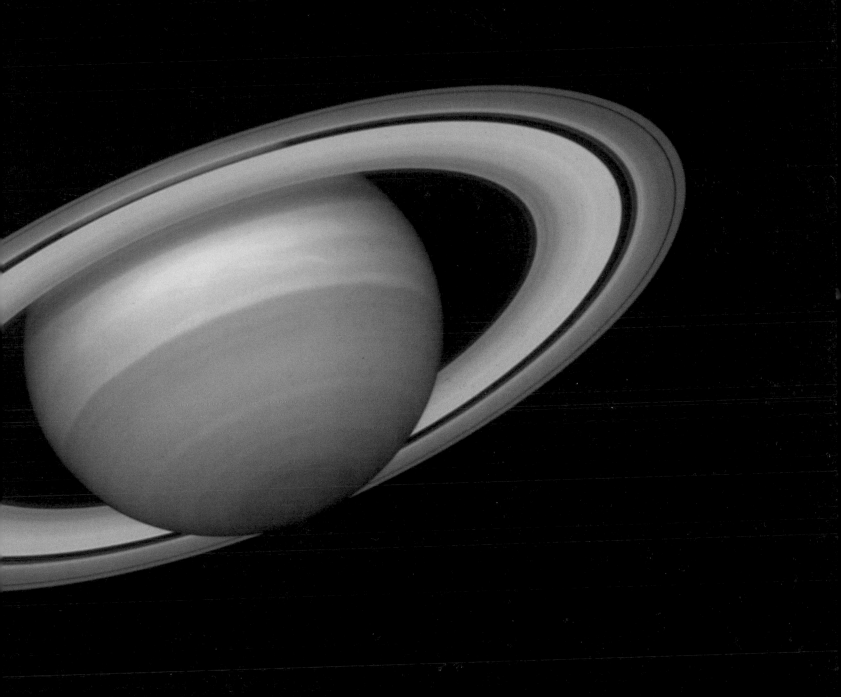

# Mercury

In many ways, Mercury is similar to our next-door world, the Moon. At only 1.4 times the Moon's diameter, it is a small, airless globe plastered with craters. But, at best, Mercury is more than 300 times farther away from us than the Moon. For a telescope user on Earth, that means the small world is typically six or seven arc seconds in diameter, only slightly larger than the apparent size of Uranus.

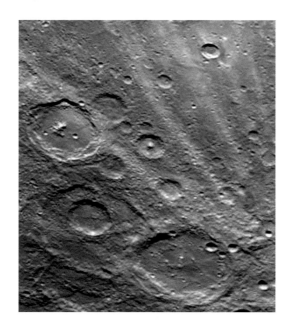

Mercury's surface material has approximately the same reflectivity and color as does the Moon's, and future explorers will undoubtedly find the landscapes of both worlds to be almost identical. Most of our knowledge of Mercury's physical characteristics is a result of the Mariner 10 close fly-bys of the planet in 1974, which provided excellent high-resolution images of one hemisphere. It takes an expert to distinguish some of the close-up images of Mercury from high-resolution photographs of the Moon taken with Earth-based telescopes.

Prior to Mariner 10, the finest Mercury observations were made by Eugène Antoniadi, a Greek-born French astronomer who became one of the greatest planetary observers of all time. Antoniadi did his best work during the 1920s with refractors ranging from 12 to 33 inches in aperture. He studied Mercury almost exclusively during the day and bright twilight, noting dusky patches and light regions on the planet's pale, creamy gray surface. Eventually, he was able to produce a map showing a rotation period of 88 days, the same time it takes Mercury to orbit the Sun. Antoniadi concluded that one hemisphere must constantly face the Sun.

Antoniadi was partly right. Through a quirk of celestial mechanics, Mercury's rotation is locked to its parent star, but not in the way he had thought. The planet actually makes half a rotation during one orbit around the Sun, so the same face returns to a sunward position after two orbits. It is a complex bit of celestial clockwork that is largely irrelevant to backyard astronomers, because seeing anything at all on Mercury is one of the toughest assignments in amateur astronomy. But the problem illustrates how difficult telescopic observation of Mercury is, regardless of the equipment used.

## IDENTIFYING MERCURY

As the solar system's innermost planet, orbiting the Sun at about one-third of the Earth's distance, Mercury never appears more than 28 degrees from the Sun. From the northern hemisphere, Mercury is seen in a dark sky for only a few spring evenings and fall mornings each year. Even at such times, its relative proximity to the horizon keeps the small planet mired in the absorption and distortions induced by the Earth's atmosphere at low celestial altitudes.

Unless your scope is equipped with precise setting circles or a permanently installed "Go To" mount with a hibernation mode to remember its position from day to day, just finding Mercury can be a task in itself. Binoculars or a good finderscope of at least 50mm aperture will allow you to spot the planet in a deepening blue sky about half an hour after sunset, well before it is visible to the unaided eye. There is usually little confusion; Mercury outshines all but the brightest stars found near the ecliptic—such as Aldebaran, Betelgeuse, Procyon, Regulus, Spica and Antares—and is occasionally a full magnitude brighter. If a star is available for comparison, the change in Mercury's brightness is noticeable over a few days, as is its shift in position from one night to the next.

## OBSERVING MERCURY

Simply from a standpoint of convenience, planet observers typically make Mercury identifications in the evening sky. This is fine for naked-eye or binocular observing, but telescopically, conditions could hardly be worse. Once Mercury is finally located, it is generally less than 15 degrees from the horizon and is swimming in hopelessly bad seeing. At such times, the planet looks yellow or ocher. In reality, it is a grayish white, like the Moon.

Moreover, false color is usually apparent, especially blue, green or red fringes— an effect caused by dispersion associated with atmospheric refraction, which causes celestial objects to appear to be displaced to higher altitudes. The effect is wavelength-dependent (being less for longer wavelengths), so for planets at low altitudes, this creates a red fringe on the lower side of the planet and a green (or blue) fringe on the upper. The effect also produces chromatic smearing to the whole image.

Observers with telescopes who hope to duplicate Antoniadi's feat of drawing planetary features will surely be disappointed when they look at the tiny remote world— unless they follow his regime of daylight or bright-twilight observing. Instead, watch Mercury's phase, which, during a typical three-week evening observing window, changes from nearly full to a slender crescent similar to a three-day-old Moon.

## OBSERVING MERCURY BY DAY

For a good telescopic view of Mercury, you must do what Antoniadi did: observe during the day. (The same applies to Venus, which can also be observed using the following technique.) The challenge is to find the planet in a bright blue sky. Here is the simplest way to do it.

Start by selecting a date when Mercury is a morning "star," and locate it in morning twilight with the unaided eye or binoculars. In the northern hemisphere, late August, September, October and November are the best months because of the higher angle of the ecliptic relative to the horizon, which brings Mercury to its highest altitude. The actual observing window will be just two to three weeks once or twice during the four-month span.

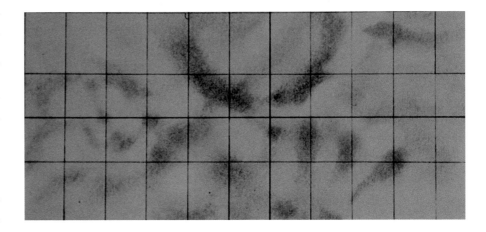

▲ 'Map' of Mercury
Based on visual observations in the 1950s and early 1960s with a 24-inch refractor and a 40-inch reflector, this chart of Mercury's surface represents all we knew about the surface features of the innermost planet before the space age.

Using a telescope with a polar-aligned, motor-driven equatorial mount, begin observing the planet, ensuring that the telescope is tracking properly as Mercury climbs from the twilight sky into full daylight. Ninety minutes after sunrise, Mercury will be about 30 degrees above the eastern horizon, high enough to allow a sharp view if the seeing is good. Daytime seeing is usually best around sunrise, before the Sun has warmed the air and the ground and produced the convection currents that have long plagued solar astronomers. Because the Sun heats the telescope as well, convection currents build up both outside and inside the instrument, and seeing inevitably deteriorates toward noon.

## DAYTIME PLANETARY OBSERVING PROTECTION

An important accessory for daytime planetary observing is an extended dewcap made of black construction paper or something similar. The paper is wrapped around the telescope tube and taped in position so that it extends in front of the tube at least a foot beyond the length of the dewcap. This prevents direct sunlight from falling onto a refractor's lens, a Schmidt-Cassegrain's corrector plate or a Newtonian's secondary mirror. Some observers erect a patio umbrella or a makeshift shade for the entire telescope once they have locked onto Mer-

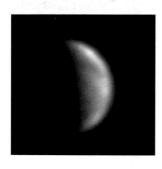

cury. Windy days are out, because the makeshift shade can be blown over or the extended dewcap ripped off. In any case, seeing is usually worsened under windy conditions. And wind transports dust. One blustery day a few years ago, we set up two telescopes, a Schmidt-Cassegrain and a refractor, to observe Mercury and Venus. After only a few hours, the front of the refractor's objective and the Schmidt-Cassegrain's corrector plate were covered with more dust than the instruments had collected in a year's worth of night sessions.

Under the best conditions of daytime or early-twilight viewing, Mercury is a sharply defined disk with hints of dark and light splotches just at the threshold of vision. During conditions of perfect seeing, however, the planet takes on a unique guise. Paler than Venus, its creamy surface has an appearance of being vaguely textured, like fine sandpaper. It is questionable whether observers are actually seeing evidence of craters on Mercury, but the planet is certainly different from Venus.

# Venus

As the brightest luminary in the night sky after the Moon, Venus holds a preeminent position among celestial objects. It is so dominant when it is in the sky that there is no mistaking it. Nothing approaching it in brightness is seen in the west at dusk or in the east just before dawn. The planet's brightness can be attributed to two factors: Venus comes closer to Earth than does any other planet and is often the nearest celestial object in the sky beyond the Moon; and the haze at the highest levels in Venus's atmosphere—the "surface" that we see from Earth—is extremely reflective. Sixty-five percent of the sunlight that falls on Venus is reflected back into space, the highest percentage of any planet in the solar system. And because Venus is closer to the Sun than any planet except Mercury, it receives a high dose of sunlight per unit surface area. Thus the observer is presented with a dazzling disk.

Although unrivaled in splendor when viewed with the unaided eye, Venus is generally a disappointment when seen through

a telescope—it is as featureless as a cue ball. But it passes through phases, as does Mercury, and for the same reason. Its orbit is interior to the Earth's, but Venus is twice Mercury's distance from the Sun. Consequently, it takes longer to orbit the Sun and appears, at best, twice as far away from the Sun in our sky. The results are longer viewing cycles and more favorable sky positions than Mercury ever attains.

The thick layer of cloud and haze blanketing Venus is extremely uniform. Yet sometimes when the planet is viewed through a telescope, there are dusky patches and lighter poles at the limit of visibility. Cloud features and circulation motion do exist but become obvious only in ultraviolet-light wavelengths that are invisible to the human eye. Nonetheless, for more than a century, visual observers have reported these features, and occasionally, their drawings have coincided with markings in ultraviolet photographs taken with Earth-based telescopes. Although elusive, the dark patches were drawn long before they were photographed.

## OBSERVING VENUS

When Venus is examined telescopically against the blackness of deep twilight, the contrast between the planet and the sky is so enormous that what little detail might be seen is lost. As well, a number of telescope aberrations that are commonly suppressed below detectable levels loom into view. The slightest amount of chromatic aberration in

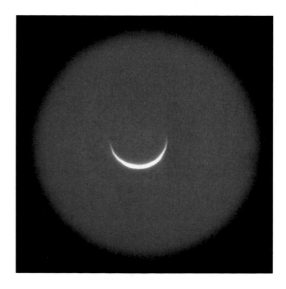

a refractor can produce a blue or purple halo around Venus. The secondary-mirror supports in a Newtonian generate spikes extending from the image. Both of these effects impair resolution of detail and detract from the aesthetic appearance of the planet's pure white hue and symmetric phase. The worst aberration does not occur in the telescope but, rather, in the Earth's atmosphere. It is dispersion associated with atmospheric refraction, as explained in the previous section on Mercury.

The combination of ultrahigh contrast, annoying optical effects and atmospheric dispersion means that the least desirable condition for observing Venus is a dark sky at low altitudes. As with Mercury, daytime viewing is the preferred mode. And, like Mercury, conditions are optimum during an autumn-morning appearance, shortly before or after sunrise. However, because Venus is much brighter and roams farther from the Sun than its smaller neighbor, conditions for seeing it reasonably well are less restrictive.

At its brightest, Venus can be located with the unaided eye in a clear, deep blue sky any time after 3 p.m. (near eastern elongation) or before 10 a.m. (near western elongation). Even at hours when the planet cannot be found in full daylight with the unaided eye, binoculars will readily reveal it.

## TELESCOPIC APPEARANCE

When observed by telescope in a rich blue daytime sky, Venus is a beautiful suspended pearl, its phase instantly evident and its sunward edge dazzling and distinct. Gone are the overwhelming contrast and the optical effects that come into play when it is low in the sky against a dark background.

The lower-contrast daytime observation of Venus reveals the surface's gradation in brightness—the difference between the sunward edge (the limb) and the day/night line (the terminator). Because the terminator receives only grazing sunlight compared with the direct rays on the areas of the planet turned toward the Sun, it is more subdued than the brilliant limb.

The visible terminator is actually inside the geometric, or true, terminator, which is 90 degrees from the point on Venus's

disk that lies directly beneath the Sun. Thus when the planet is half illuminated (a phase called dichotomy), according to references such as The Royal Astronomical Society of Canada's *Observer's Handbook*, it usually appears less than half illuminated in a telescope.

Because Venus passes from full phase toward crescent phase during its seven-month cycle through the evening sky, visual dichotomy occurs a few days prior to geometric dichotomy. The reverse happens when Venus is in the morning sky, passing from crescent to full phase. For decades, backyard observers have recorded these variances. Curiously, the morning-apparition figures do not match the evening ones. The difference between visual and geometric dichotomy is 4 to 6 days in the evening sky and 8 to 12 days in the morning sky. No reasonable explanation has ever been proposed, yet the discrepancy seems to be real. In view of these differences, begin watching for dichotomy far enough ahead of the predicted dates (given in the *Observer's Handbook*; listed as "Venus at greatest elongation east" for evening sky or "Venus at greatest elongation west" for morning sky).

▲ **Mercury and Venus**
Although considerably fainter than Venus, Mercury is still bright and easy to see, as this morning-sky portrait attests. Photo by Alan Dyer.

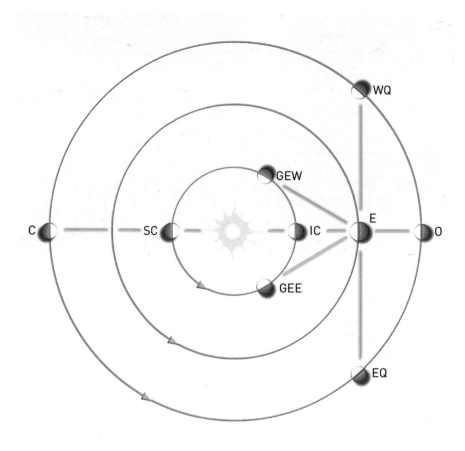

## INFERIOR CONJUNCTION

The celestial clockwork brings Venus to inferior conjunction every 19½ months. The most interesting time to observe Venus is the two months either side of inferior conjunction, when it is nearest Earth and less than 40 percent illuminated. When Venus is near inferior conjunction, its slender crescent can expand to 60 arc seconds across as the planet glides a few degrees above or below the Sun in the Earth's sky.

The days leading up to and following inferior conjunction are prime observing periods. No other celestial object except the Moon is seen at such thin phase angles. When viewing the slender crescent of Venus, the chief objective is a sighting of an extension of the cusps (the points of the crescent) into the planet's night hemisphere. This is not an illusion but sunlight illuminating Venus's upper atmosphere. Occasionally, the extension is seen as a complete ring, the extremely thin crescent and a very tenuous and ghostlike extension of the cusps making a full sphere. Conditions must be exceptional for such an ob-

servation, but it is far from impossible and does not require a large telescope.

Perhaps the most elusive Venusian feature is the planet's ashen light, a vague illumination of Venus's nighttime side when it is seen as a thin crescent. According to those who have reported it, the light is fainter than, but otherwise similar to, Earthshine on the Moon. However, there is no satellite to illuminate Venus's nightside, and Earth is too far away to do the job. We have never observed the ashen light, and it may well be an eye-brain contrast effect. If it is a real occurrence (unlikely in our opinion), some mechanism for lighting the nighttime side of Venus must be proposed.

# Mars

Apart from the Moon, Mars is the only object in the universe whose solid surface is seen in reasonable detail through Earth-based telescopes. Yet it is just far enough away from Earth that its features are difficult to observe with clarity. Mars tantalizes.

Once Mars has been viewed through a telescope with fine optics when the planet is fairly close to Earth, it is impossible not to be impressed by the feeling that it is a real world, a globe with obvious similarities to our own: polar caps, dark continents in the global salmon-hued deserts, dust storms and the occasional clouds—all have their terrestrial counterparts. What is missing on Mars, of course, are the oceans of water that characterize most of the surface of our planet. Nevertheless, nowhere else can so many comparable features be seen, including a 24.6-hour rotation period and an axis tilt only two degrees different from the Earth's. All of these together make Mars the most Earth-like planet in the solar system. When the seeing is good and the telescope's optics are equal to the task, it is easy to understand why some late-19th-century observers were convinced that they were gazing at a habitable world.

The pale, pinkish orange desert world suspended in the black void of space is an enchanting and unforgettable sight, but memorable views of Mars are not common. For only two to four months every two years, when the planet is within 0.8 AU

(1 AU, astronomical unit, is the Earth-Sun distance), is the disk large enough to reveal rich detail. Even then, Mars is a supreme challenge for both eye and telescope.

Large telescopes are limited by turbulence in the atmosphere of our own planet, which only rarely permits detection of features below 0.3 arc second (sometimes stated as 0.3 second of arc, or 0.3"). This is a tiny angle, equal to 80-kilometer resolution on Mars when the planet is at its closest. However, it can be reached by comparatively small telescopes. The normal good-seeing limit is 0.5 arc second, while 1.0 arc second is typical of many nights of the year at most sites. One arc second is the resolution of a 4-inch telescope; half an arc second is the theoretical capability of an 8-inch telescope; and 0.3 arc second is a 15-incher.

Rarely does the Earth's atmospheric turbulence settle enough for a 15-inch-class telescope to show full resolution. But when near-perfect conditions do occur, the overwhelming detail visible on Mars in a high-quality, large-aperture scope is astonishing. What large optics will do anytime, though, is make the planet appear brighter and enhance the color differences over the disk.

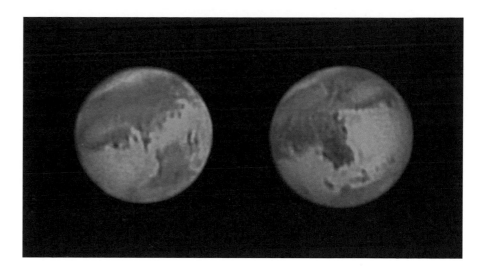

## OBSERVING MARS

Any good-quality telescope with more than 70mm aperture should reveal surface features on Mars during the weeks around the biennial closest approaches, including changes in the outlines of Martian dark zones from one opposition to the next. Once thought to be evidence of tracts of vegetation changing with the seasons, the alterations are caused by winds of up to 400 kilometers per hour transporting vast

▲ **Two Sides of Mars** These fine 2001 Mars portraits by expert CCD planetary imager Ed Grafton were taken with a 14-inch Schmidt-Cassegrain.

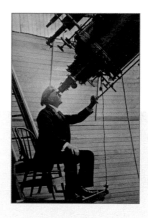

# Life on Mars and Percival Lowell (1855-1916)

In 1894, Percival Lowell, an American diplomat and aristocrat turned astronomer, built an observatory in Flagstaff, Arizona, to study thin linear features ("canals") reported on Mars by Italian astronomer Giovanni Schiaparelli a generation earlier. Schiaparelli had used an 8.5-inch refractor, but Lowell's new instrument was a 24-inch refractor.

Examining Mars with the new telescope, Lowell soon became convinced that the canals he observed were constructed by a Martian civilization to preserve a dwindling water supply on a desert planet. With the publication of a book about his theories and observations in 1895, Lowell quickly became the most famous astronomer of his day.

For the next two decades, Lowell continued his studies of Mars, making regular announcements about what he and his staff observed. The newspapers loved it. So did science fiction writer H.G. Wells, who, soon after reading about Lowell's theories, penned *The War of the Worlds*, about Martians invading Earth.

The canals ultimately proved to be the product of the brain's tendency to connect fine detail at the threshold of vision into linear features. There really are no straight lines on Mars. But the seeds had been planted, and Martians have been part of our collective psyche ever since.

 *An imposing figure, Percival Lowell sits in a reinforced wooden kitchen chair at the eyepiece of the 24-inch refractor at Lowell Observatory in Flagstaff, Arizona. Today, the telescope still looks much as it did then. Photo courtesy Lowell Observatory.*

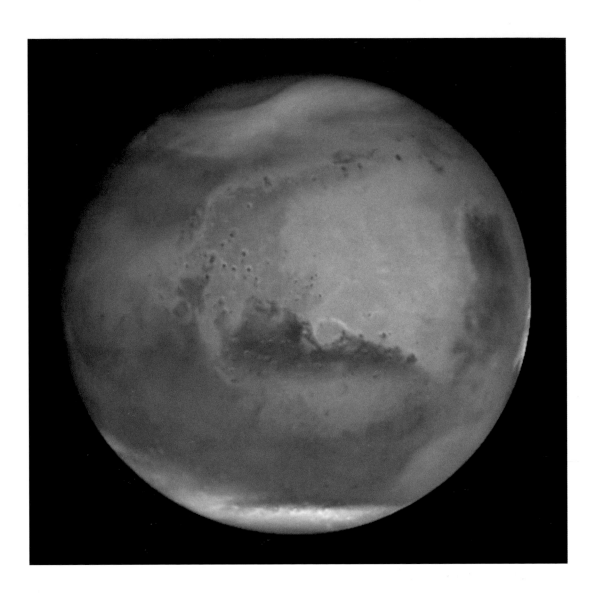

quantities of dark and light dust across the desert planet. Mapping the changes is therefore a study in Martian meteorology and geography. It is a challenging but rewarding area of planetary observation.

The clarity of Mars' surface features—white polar caps and dark, irregular patches on a largely peach-colored sphere—depends to a great extent on the transparency of the Martian atmosphere. Dust storms can lower contrast across large sectors of the disk within days of a storm's onset. The planet may remain partly or completely shrouded for many weeks thereafter (as happened in late June 2001, spoiling Mars viewing for the following five months). Experienced observers can detect emerging storms when familiar desert areas of the planet brighten and encroach on nearby dark features. Global dust storms are both rare and unpredictable. Only a few are well documented: 1956, 1971, 1973, 1977 (two) and the 2001 storm just mentioned.

The Martian south pole is tipped Earthward at oppositions favorable for northern-hemisphere observers. The brilliant white polar cap shrinks during the prime observing window (90 days centered on opposition), sometimes revealing detached patches or notches in the main cap. Summer is then beginning in the Martian southern hemisphere. When the southern polar cap is reduced to a tiny white button, part of the larger northern polar cap may be in view. A bluish white atmospheric haze known as the North Polar Hood often masks the northern cap itself. Both polar caps have a residual water-ice core that never disappears, but the rapid changes seen in the more extensive caps are due to the seasonal sublimation (gas to ice, or vice versa) of atmospheric carbon dioxide.

In winter, the polar regions are a brutal minus 140 degrees Celsius.

The best magnification for Mars is about 35x per inch of aperture up to 7 inches and 25x to 30x for larger telescopes. This yields a pleasing image while avoiding the effects of irradiation, a contrast phenomenon that originates in the eye and causes brighter areas to encroach on dark adjacent areas. In the case of Mars, irradiation produces the apparent enlargement of the polar caps and the apparent loss of fine, darker details next to the desert areas. The effect is most troublesome at magnifications below 25x per inch of aperture. Filters are essential in large telescopes used below 25x per inch.

Regardless of the instrumentation, experience is the key to detecting the wealth of detail that Mars can present to backyard observers. At first glance, the planet appears so small that you may wonder how anyone sees anything on it. The trick is to start observing Mars at least two months before opposition to train your eye to detect the ever-so-subtle features that abound on our neighbor world. Then, around opposition, when the best views are available, you will be ready to squeeze the most out of eye and telescope, rather than discovering at the time of optimum conditions how challenging it is to observe the red planet.

## PLANETARY FILTERS

Inexpensive color filters (about $15 each) that screw into the bottom of 1.25-inch eyepieces often improve the visibility of Martian surface features by reducing the effects of chromatic aberration in refractors and increasing contrast in all types of telescopes. A blue filter (Wratten No. 80A) reveals the few nondust clouds that float in the atmosphere of Mars. Other filters are orange (No. 21) and red (No. 23A), which enhance the contrast of the dark areas. The red filter may be a bit dark for instruments with apertures smaller than 8 inches, but try both to get the most out of observing Mars.

Larger telescopes also benefit from the filter's reduction of the brilliance of the Martian disk. In addition, a red filter improves seeing by cutting out the shorter wavelengths that are most affected by atmospheric turbulence. For telescopes greater than 8 inches in aperture, also try the deep red No. 25 filter. Generally, telescopes with more than 5 inches of aperture respond better to color filters than do smaller instruments. For aesthetic appeal, however, nothing matches an unfiltered view of the coral deserts, pure white poles and gray-green dark regions.

During the past few years, Sirius Optics of Kirkland, Washington, developed three new filters for planetary observers. The minus-violet interference filter, model MV-1 ($75), is exclusively for suppressing the effects of chromatic aberration (false color) in achromatic refractors. Chromatic aberration is apparent as a bluish purple halo around planets—Jupiter and Venus, in particular. The halo is just the obvious part of a wash of unfocused light that is actually smeared across the whole planet. False color has been eliminated in modern apochromatic refractors, but at significant expense. A typical apo refractor costs 5 to 10 times as much as an achromatic scope. In recent years, Chinese-made achromatic refractors have reduced the cost of this type of telescope even more than before, widening the price difference between them and apos. The result is that today, more achromatic refractors are being used for planetary observation than ever before.

▼ **Mars Oppositions**
Every 26 months, on average, Mars and Earth are nearest each other, an event known as Mars opposition. Favorable oppositions (from an Earth observer's point of view) occur around Martian perihelion. The 2003 perihelion opposition is the closest Mars has come to Earth since biblical times, appearing 25 arc seconds in diameter. Compare that with the 14-arc-second diameter at aphelion oppositions, such as the one in 2010.

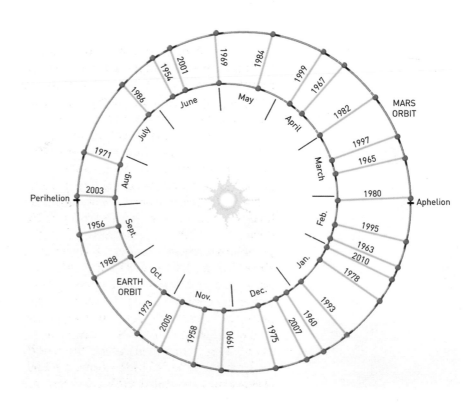

This is the background behind the MV-1 filter, which applies the same technology used in nebula filters and markedly suppresses the bluish purple halo that is most obvious around Jupiter. (We used the popular 120mm f/8.3 SkyWatcher refractor for our tests on Jupiter.) However, the filter shifts the planet's color to a distinct yellowish hue—a natural consequence of filtering out much of the blue part of the spectrum. Contrast is improved in the filtered image, but the image is noticeably dimmer.

The more advanced Sirius planetary-contrast filter (PC-1; $75) offers further improvement by eliminating more false color while introducing a less jarring yellowish cast to the image. The PC-1 filter is said to increase planetary contrast on all types of telescopes, but we found the improvement most obvious on achromatic refractors over 4-inch aperture. However, we feel there is an important aesthetic issue here. As long-time planet observers, we enjoy the beauty of natural planetary hues. For us, this is one of the strong attractions of planetary observing with apo refractors.

The third Sirius filter, called the variable-filter system, is triple the price of the two filters just mentioned, but it is unique in astronomy. Again, it uses an interference (nebula-type) filter; the difference is that by using a thumbwheel, the filter is variable in color over a wide range. Unfortunately, this filter appeared just days before this book went to press, but in principle, it is a promising concept for both planetary and deep-sky applications.

# Jupiter

Unlike Mars, which requires superb optics and relatively high magnifications before much detail is seen, Jupiter has a number of features that are easily revealed by an 80mm refractor at about 100x: its main belts, the Red Spot (unless it is particularly faded) and the shadows of the four Galilean satellites. Jupiter's disk ranges from 4 to 100 times larger in area than the face of Mars, depending on Mars' distance from Earth. Major disturbances on the equatorial belts and the dark "barges" in the tropical and

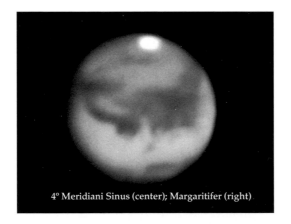
4° Meridiani Sinus (center); Margaritifer (right)

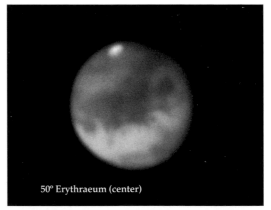
50° Erythraeum (center)

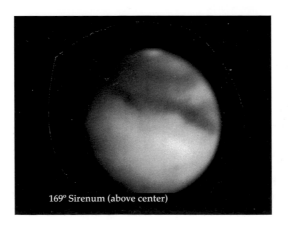
169° Sirenum (above center)

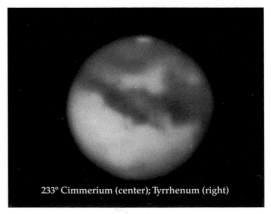
233° Cimmerium (center); Tyrrhenum (right)

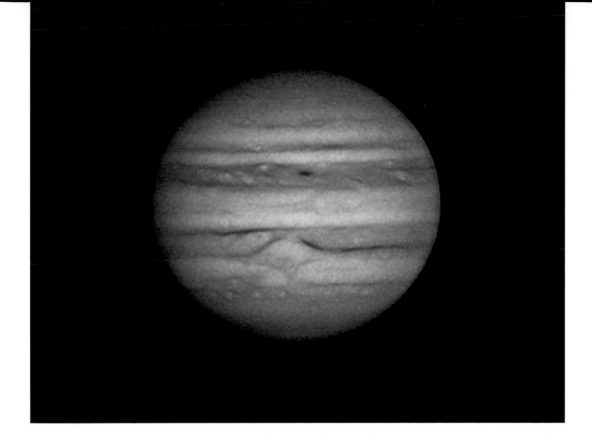

◀ **Backyard Jupiter**
This remarkable image of Jupiter's cloud-belted face shows the maximum amount of detail that anyone can expect to see through a telescope eyepiece. It is a composite of nine images taken in rapid succession with a digital camera through a 14.5-inch Starmaster Newtonian telescope on a night of perfect seeing in 2001. Photo by Ken Schmidt.

temperate belts of Jupiter are also detectable in an 80mm refractor at 100x to 130x.

Every increase in telescope size reveals more, but as with all planetary observing, telescope aperture is definitely secondary to the instrument's optical quality. Jupiter is a

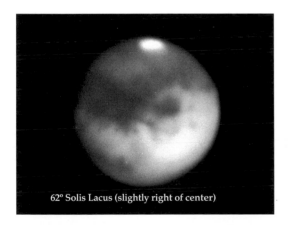

62° Solis Lacus (slightly right of center)

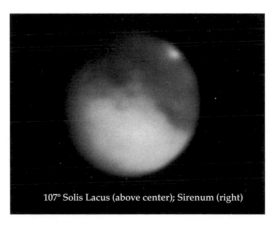

107° Solis Lacus (above center); Sirenum (right)

◀ **Mars: Full Rotation**
In autumn 1988, Mars was closer to Earth than it had been in a generation (a perihelion opposition). This sequence of eight images by Don Parker, taken through a 16-inch telescope, shows the same level of detail that a backyard observer can see under the best seeing conditions with an 8-inch or larger telescope. The degrees indicate the longitude on Mars at the central meridian. The major dark feature near the central meridian is named. Following long-standing tradition, south is up in these Mars pictures, as would be seen in a Newtonian reflector.

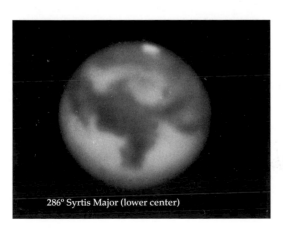

286° Syrtis Major (lower center)

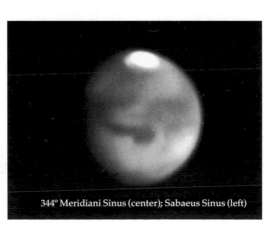

344° Meridiani Sinus (center); Sabaeus Sinus (left)

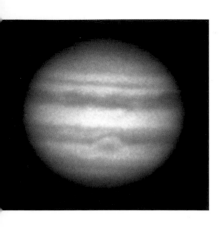

**Jupiter's Great Red Spot** ▲
In summer 2000, Gordon Bulger used a Nikon 995 digital camera and a 12-inch Schmidt-Cassegrain telescope to take this image.

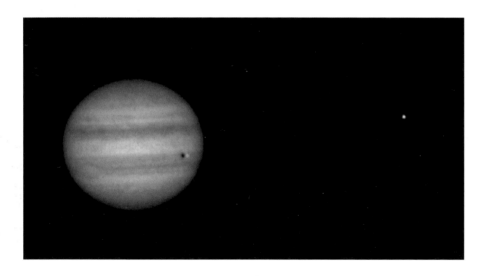

**A Moon and Its Shadow** ▲
When Jupiter is near opposition, its moons can appear next to the shadows they cast. Io is seen just to the right of its own shadow in this 10-inch Newtonian image by Luis Eguren, taken November 2001. Europa is at far right. Two dozen images gathered by an inexpensive webcam were digitally stacked to produce the final image.

bright object; there is plenty of light. Focusing that light into a sharp, high-resolution image of the disk is the key. In fairly good seeing, 25x to 35x per inch of aperture should be adequate for observing Jupiter. The image should remain sharp-edged and well defined.

Selecting something typical—for example, a 4-inch refractor or an 8-inch Schmidt-Cassegrain—what can an observer expect to see? Because of Jupiter's constantly changing cloud-covered surface, the amount of detail varies from one observing season to the next. However, three or four dark belts are always visible and sometimes as many as 10, depending on how the wind circulation is divvying up the clouds. Bumps, projections, loops and general turbulence would be evident along the edges of the dark belts. Even if the Great Red Spot has faded, its

home—the Red Spot Hollow—should be visible as an indentation in the south equatorial belt. In late 1989, for instance, the south equatorial belt itself disappeared over a period of just a few weeks, and the Red Spot, after years of near invisibility, regained some prominence. In 1990, the Red Spot faded again as the belt returned.

For many decades, several large, white ovals, one-quarter to one-third the size of the Great Red Spot, roamed in the next belt south from the one occupied by the Red Spot. During the 1990s, the white ovals began merging, and by 2002, only one, almost half the size of the Red Spot, remained.

The Red Spot is a swirling maelstrom of

clouds pumped up from a vortex that penetrates to a lower level in the planet. Its coral or pinkish color (it rarely appears red) is usually quite distinct from the hue of any other feature on the planet, indicating that its source material is probably deeper than that of other features. The Red Spot completes a counterclockwise rotation once every six days, but the anticyclonic motion is usually below the detection threshold with amateur equipment. What can be seen is vague texture within the Red Spot and variations in the color over time. As the south equatorial belt moves past the Red Spot and is disturbed by it, a churning swath of clouds is heaved into the wake (the region behind it, in terms of the planet's rotation). This wake feature is usually more obvious than details within the spot.

In the early 1960s, the two equatorial belts almost merged, and the activity between them was extraordinary, with thick, twisting bridges of dark material crossing the lighter equatorial zone. Such activity is simply cloud and circulation phenomena. The light zones are at a higher altitude than the dark belts and consist mostly of ammonia haze; the dark belts are chiefly ammonium hydrosulfide. Incursions of one belt into another are constantly appearing and disappearing as the belts and zones slip by one another. The equatorial zone and the adjoining parts of the equatorial belts are known as System I, while the rest of the planet (except the polar regions) is System II. The polar regions are referred to as System III, but for amateur-observation purposes, only Systems I and II are of interest.

System I rotates in approximately 9 hours 50 minutes; System II has a general rotation speed about five minutes longer. The two systems continually slide by each other inside the equatorial belts, making them the most active areas of the visible surface of Jupiter. System II contains the Great Red Spot. Using tables and a calculation method outlined in the *Observer's Handbook*, you can determine the System II longitude. *Sky & Telescope* usually lists the Red Spot's longitude when Jupiter is well placed for viewing. If the System II longitude is within 50 degrees of the Red Spot, the spot should be visible. During the 1980s, the Red Spot stayed fairly constant with respect to System II, not straying too far from 10 to 20 degrees longitude. During the 1990s, it moved several degrees a year and, by 2002, was near 80 degrees. In times past, the Red Spot has been known to shift by up to 30 degrees in a single year, so it is impossible to predict where it will be. Once seen, though, it will be visible in the same spot two days and one hour later (five Jupiter rotations).

Overall, Jupiter's disk is creamy white, bright and distinct, yet there is an aspect to it called limb darkening—the edge of Jupiter's disk is about one-tenth as bright as its center—that becomes apparent only when sought out. It is not obvious to the eye, because the disk edge abuts the blackness of the sky. Limb darkening is caused by solar illumination that is absorbed by a thin haze in Jupiter's upper atmosphere, above the highly reflective clouds. The limb-darkening phenomenon subdues the visibility of the cloud features near the disk's edge. The Great Red Spot, for example, is not visible at the edge of the disk. Rather, it is seen clearly only when it has rotated one-quarter of the way around. The same is true of all the Jovian features, which are seldom visible for more than 2½ hours before they rotate out of view.

If you have any inclination to draw what you see, Jupiter is probably the best planet on which to try it. Beautiful strip charts of Jupiter, sketched over a few hours as the planet rotates, have been made by absolute beginners with only a few nights' training at the telescope. "The more you look, the more you'll see" is an axiom that is never more apt than in the case of Jupiter.

## TRACKING THE FOUR MOONS

Among the most dramatic sights in astronomy is the transit of one of Jupiter's moons across the planet's cloudy face. As one of the four large satellites appears to touch the Jovian disk, it is transformed from a dazzling spot against the dark sky to a tiny, fragile disk etched in the limb. Each moon has its own characteristic appearance as it crosses the planet.

Io, the innermost of the four big moons, has a light pinkish hue and high surface brightness. When it enters the disk, it is always a brilliant dot on the darker limb. As it begins its transit, Io can become lost

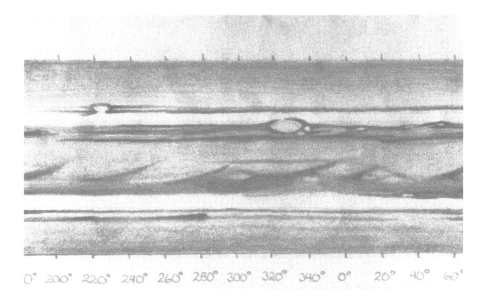

0°  200°  220°  240°  260°  280°  300°  320°  340°  0°  20°  40°  60°

in the white zones, which have similar reflectivity. When in front of the darker belts, the moon is usually a minute but distinct bright spot.

Europa is the smallest of Jupiter's four major moons but has the most reflective surface. It is intensely white, especially as it enters the limb—a tiny, white dot against the edge of the cloud-strewn giant. As Europa marches in front of the big globe, it usually encounters the white clouds of the zones and disappears from view. It is probably the most difficult moon to see throughout a complete transit. On the rare occasion when it tracks across a dark belt, Europa can be observed for its entire journey.

Ganymede, the largest satellite, is easier to follow across Jupiter's face because of its size and because its color is duller than the white clouds and brighter than the dark ones. When transiting a white zone, Ganymede is light brown in color and resembles a washed-out satellite shadow. Superimposed on the limb, the reverse is the case: The limb is darker than Gany-

▲ **Jupiter Strip Map**
In the mid-1980s, members of the Hamilton Centre of The Royal Astronomical Society of Canada mapped Jupiter by combining their sketches of the planet. Courtesy Derek Baker.

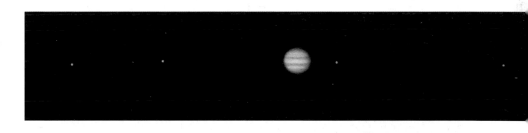

▲ **Family Portrait**
Jupiter and its four largest moons are seen in this digital-camera image just as they appear at about 100x in a small backyard telescope. Jupiter and the moons are bright enough to overwhelm any light pollution and will look like this even from a balcony "observatory" in the middle of a major metropolis. Photo by Jean Guimond.

# Jupiter's Four Major Satellites

| Satellite | Diameter (km) | Visual Magnitude | Orbital Period (days) | Average Maximum Distance From Planet Center (arc seconds) | Apparent Diameter (arc sec.) | Shadow Diameter (arc sec.) | Effective Shadow Diameter* (arc sec.) |
|---|---|---|---|---|---|---|---|
| Io | 3,630 | 5.0 | 1.77 | 138 | 1.2 | 1.0 | 1.1 |
| Europa | 3,140 | 5.3 | 3.55 | 220 | 1.0 | 0.6 | 0.8 |
| Ganymede | 5,260 | 4.6 | 7.16 | 351 | 1.7 | 1.1 | 1.4 |
| Callisto | 4,800 | 5.6 | 16.69 | 618 | 1.6 | 0.5 | 0.9 |

*Includes darker part of penumbral shadow.

mede, so the moon appears as a bright dot on the darker background. As with Io, there is a transition zone in which Ganymede almost always disappears as its intensity matches the background between the limb and the more brightly illuminated central part of the disk.

Callisto is probably the easiest moon to follow, because its dull surface material makes it darker than almost anything it

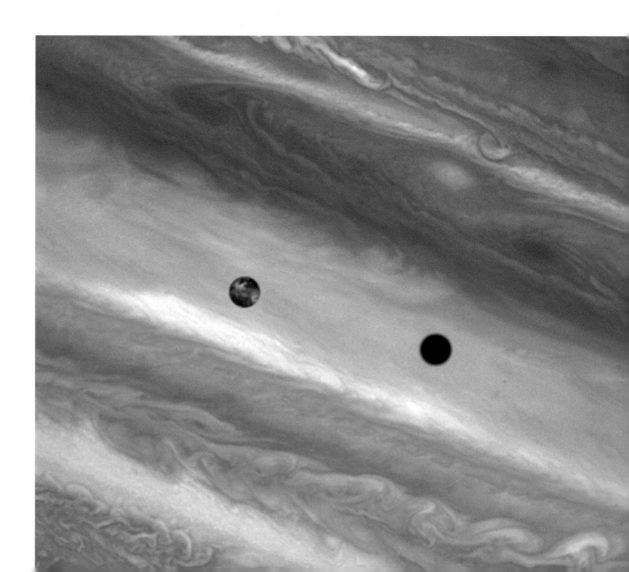

**Shadow Trek** ▶
In this Hubble Space Telescope image, Jupiter's moon Io casts its shadow on the giant planet's clouds. Note the shadow's penumbral ring, referred to in the table above.

encounters, except at the very edge of the disk when it enters or exits its transit. Callisto's crossings are by far the rarest of the four Galilean moons: It orbits Jupiter only once in 17 days, compared with 7 days for Ganymede, 3½ days for Europa and just 42 hours for Io. Also, for more than half of Jupiter's 12-year solar orbit, the planet's satellite system is angled so that Callisto passes above or below Jupiter as seen from Earth and misses transiting the disk altogether.

The shadows of Jupiter's moons are much easier to observe than a transit, because they are black dots, far darker than any of the planet's surface features. But the shadows vary in size, with Ganymede's the largest, Io's the second largest and Europa's and Callisto's the smallest. Ganymede's shadow is visible in a 60mm refractor; the others usually require a 70mm refractor for a definite sighting.

# Saturn

No photograph or description can adequately portray the astonishing beauty of the ringed planet floating against the black-velvet backdrop of the sky. Of all the celestial sights available through backyard telescopes, only Saturn and the Moon are sure to elicit exclamations of delight from first-time observers.

Almost any telescope will reveal Saturn's ring structure. A 60mm refractor at 30x to 60x clearly shows it. The view is outstanding in 4-inch or bigger telescopes. Such instruments also reveal several of Saturn's large family of satellites, which appear as tiny "stars" beside, above and below the planet.

Although spacecraft have discovered hundreds of identifiable rings, only three components can be distinguished visually through Earth-based telescopes. They are known simply as rings A, B and C. Rings A and B are bright and easily visible in any telescope. They are separated by Cassini's division, a gap about as wide as the United States. Cassini's division looks as black as the sky around Saturn but is actually a region of less densely packed particles rather than a true blank space. The division is largely the result of the gravitational per-

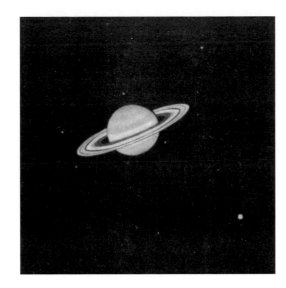

**◄ Magnificent Planet** This illustration of the ringed planet appeared in Richard Proctor's **Saturn and Its System,** published in 1865.

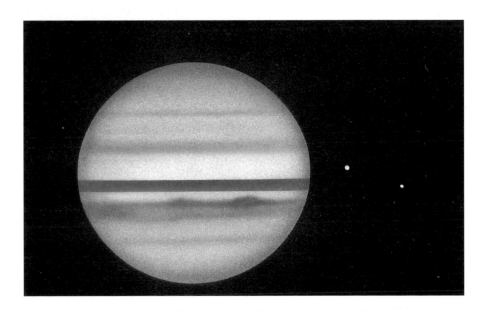

**▲ Saturn Without Rings** In 1966, Saturn's rings were edge-on to Earth and invisible for months. Only the ring shadow (darkest "belt") remained. Three of Saturn's moons are at right. Illustration by Terence Dickinson using a 7-inch achromatic refractor.

turbations of Saturn's moon Mimas. Ring particles orbiting in the gap are in resonance with Mimas and, over time, are gravitationally nudged into different orbits, thus largely clearing out the section. The other gaps that produce the many rings seen by the Voyager spacecraft are probably generated in much the same manner but involve far more complex interactions with other moons and large ring particles.

Ring A is less than half the width of ring B and is not as bright, although the difference between the two is subtle. Ring C, the innermost ring, is so dim that an experienced eye and at least a 6-inch telescope are needed to distinguish it. Also known as the crepe ring, C is a phantomlike structure extending roughly halfway toward

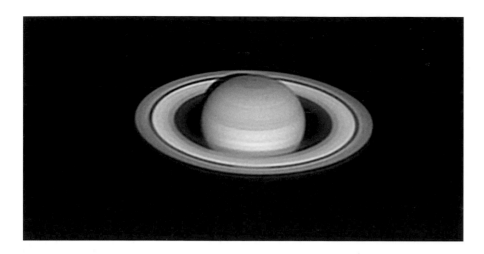

**Lord of the Rings** ▲
One of the finest portraits of Saturn ever obtained with backyard equipment, this CCD image was captured by Ed Grafton using a Celestron 14-inch Schmidt-Cassegrain telescope.

**Glorious Saturn** ▶
This superb illustration of Saturn by English amateur astronomer and artist Paul Doherty is based on his observations with a 16-inch Newtonian reflector during the 1980s.

the planet from the inner edge of ring B.

One other division besides Cassini's is visible in Earth-based equipment: the Keeler gap, which is generally (and incorrectly) referred to as the Encke division. Located near the outer edge of ring A, the Keeler gap is extremely difficult to detect. The first person credited with seeing a gap in that position was James Edward

gesting some variation in intensity over time.

## OBSERVING SATURN

When observing conditions permit, you can spend hours at the eyepiece looking at Saturn, and the planet certainly deserves attention on nights of good seeing when it is well placed. Here are some features to look for, in order of increasing difficulty:

• The rings themselves. It usually takes only 30x to see them clearly and 60x to show that they really do resemble a washer surrounding a marble.

• The shadow cast by the planet on the rings. It can be quite small around opposition but rapidly increases when the planet moves away from opposition.

• Cassini's division. An 80mm refractor will reveal it, but a good 5-inch instrument is usually required to detect it clearly all the way around.

• The dusky belt(s) of Saturn separating the creamy yellow equatorial region from the beige temperate zone.

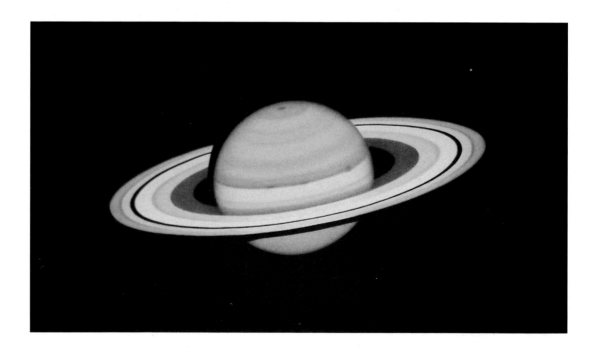

Keeler, who was using the Lick Observatory's 36-inch refractor at 1500x. The actual Encke division is not a true division but a shaded zone in the middle of ring A that has been seen with telescopes as small as a 6-inch refractor. Sometimes, it is invisible in much larger instruments, sug-

• The shadow of the rings on the planet. This is usually narrow but is not that difficult to see if you specifically look for it. Depending on the geometry between Earth, the Sun and Saturn, the shadow can appear on the disk either above or below the rings as they pass in front of the planet.

- At a much higher level of difficulty are the gentle cloud features in the planet's atmosphere. Saturn seems to have a high-level haze of ammonia ice crystals that is largely absent on Jupiter, and this tends to subdue the contrast of surface features. Very rarely, a white spot will erupt to disturb the scene for a few weeks. It happened in 1933, 1960 and 1990. At maximum intensity, the white spots were visible in 4-inch telescopes.

- Finally, there's the Keeler gap, the most difficult Saturnian feature. The gap is so thin—a meager 320 kilometers across—that it is detectable only by experienced observers using excellent scopes with serious aperture. Neither of us has ever seen it with certainty, not even on one excellent night with the 26-inch U.S. Naval Observatory refractor when, at 330x, the planet appeared perfectly steady, like a Voyager picture taken from a few million kilometers away.

## SATURN'S SATELLITE FAMILY

Seven of Saturn's moons are visible in 8-inch telescopes. Although their number surpasses the four big moons of Jupiter, the Saturn family is much more difficult to observe.

Titan, an eighth-magnitude object orbiting Saturn in approximately 16 days, is by far the largest of Saturn's moons and is easily seen in any telescope. When at its maximum distance from the planet, it appears to be five ring diameters from Saturn's center. Titan is the only satellite in the solar system known to have a substantial atmosphere.

Instruments of more than 60mm aperture should reveal Saturn's 10th-magnitude moon Rhea less than two ring diameters from the planet. Iapetus is next on the list of visibility, but only when it is in one part of its orbit. It has the peculiar property of being five times brighter when it is to the west of Saturn than when it is to the east. Iapetus ranges in brightness from 10th to 12th magnitude. One side of the moon has the reflectivity of snow, while the other resembles dark rock. When at its brightest, Iapetus is located about 12 ring diameters west of its parent planet. Because stars may appear at a similar distance from Saturn, several observations are necessary for a confirmed sighting.

The next two moons inward from Rhea are Dione and Tethys, each of which is magnitude 10.4 and readily seen in 6-inch or larger telescopes. Inward from Dione and speeding around the edge of the rings are Enceladus and Mimas, both more than a magnitude dimmer and significantly more difficult to detect.

# Uranus

Uranus was discovered in 1781 by English astronomer William Herschel, probably the greatest observational astronomer of all time. In the course of a systematic program to examine every object visible in his 6.2-inch Newtonian, Herschel observed a sixth-magnitude "star" that did not look like a point of light. Herschel was in the habit of using high magnification to study celestial objects, and when he came across Uranus, he was using 227x.

In his report of the discovery, published in *Philosophical Transactions* in 1781, Herschel stated: "From experience, I knew that the diameters of the fixed stars are not proportionally magnified with higher powers, as the planets are. Therefore, I now put on powers of 460 and 932 and found the diameter of the comet increased in proportion to the power, while the diameters of the stars to which I compared it were not increased in the same ratio."

Herschel thought he had discovered a comet, but it soon became clear that his find was a planet orbiting the Sun beyond Saturn. It had been seen before and had even been plotted in a star atlas. At magnitude 5.7, Uranus is barely visible to the unaided eye and is an easy target for binoculars. However, it takes about 100x before its 3.9-arc-second disk ceases to resemble a star. Apparently, prior to Herschel, nobody had used enough magnification.

A modern telescope similar in size to Herschel's easily reveals Uranus's pale, bluish green disk, but that's all. No surface features were ever clearly seen on the planet prior to the Voyager spacecraft's encounter with Uranus in 1986. And Voyager showed that there is nothing to be seen anyway, just featureless, aquamarine haze—the top layer of the planet's thick atmosphere.

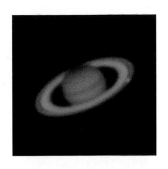

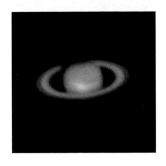

▲ **Rare Saturn Events**
Top: An extremely rare passage of Saturn in front of a bright star was captured on film on July 3, 1988. Above: Saturn's seldom-seen white spot was observed in 1990. Both photos by Don Parker.

## Saturn's Most-Observed Satellites

| Satellite | Diameter (km) | Visual Magnitude | Orbital Period (days) | Average Maximum Distance From Planet Center (arc seconds) | Apparent Diameter (arc sec.) | Shadow Diameter* (arc sec.) |
|---|---|---|---|---|---|---|
| Titan | 5,150 | 8.4 | 15.95 | 197 | 0.85 | 0.7 |
| Rhea | 1,530 | 9.7 | 4.52 | 85 | 0.25 | 0.2 |
| Dione | 1,120 | 10.4 | 2.74 | 61 | 0.17 | 0.15 |
| Tethys | 1,060 | 10.3 | 1.89 | 48 | 0.16 | 0.15 |

*Titan's shadow is seen only when the rings are nearly edge-on. Other satellite shadows are exceedingly difficult to see.

Five of Uranus's moons were known before the Voyager flyby, and 10 more were uncovered by the spacecraft's cameras, but the brightest are only 14th magnitude. Uranus is simply too far away to be of much interest to the backyard astronomer beyond mere identification.

# Neptune

In some ways, Neptune is more interesting than Uranus for the backyard astronomer; it is certainly more challenging. For the beginning observer, Neptune is a fairly demanding target in binoculars, although moderately experienced observers should have no trouble picking it out from among the stars of Capricornus, where it will be until 2010. In most standard references, such as the *Observer's Handbook*, charts of Neptune's position include stars down to eighth magnitude; Neptune is magnitude 7.7.

Although 100x will show Neptune as a disk, only powers close to 200 unmistakably reveal it as one. Its 2.5-arc-second disk is definitely blue in 6-inch or larger telescopes. In smaller instruments, the planet generally looks pale gray. Neptune has one large and seven small satellites. Six of the small ones were discovered by Voyager 2 when it encountered the planet in August 1989. The biggest moon, Triton, is 13th magnitude, making it visible in moderately large

scopes. On the night of May 28/29, 2009, Jupiter will be less than one-half degree from Neptune, offering an interesting pairing in a medium-power telescope field.

# Pluto

A tiny world smaller than the Earth's Moon, floating on the rim of the solar system, is a challenging target for owners of 6-inch or larger telescopes. The challenge is simply to see it. Pluto looks identical to a 13.7-magnitude star. It has been seen in 4-inch telescopes but is a tough assignment; 8-inch or bigger telescopes are recommended. Not only does Pluto have to be identified using the charts published in the *Observer's Handbook*, *Astronomical Calendar* or *Sky & Telescope*, but it must be identified on two nights within a few days of each other so that its motion among the stars can be plotted and the identification confirmed. A single sighting is usually not good enough, because much of the time, at least one faint star with a magnitude similar to Pluto's is nearby.

For a definite Pluto sighting, plot the stars in the immediate vicinity of the planet in a notebook, and be as precise as possible in the positioning of Pluto. Return to the eyepiece a few nights later, check the same field, and see whether the object identified as Pluto has moved.

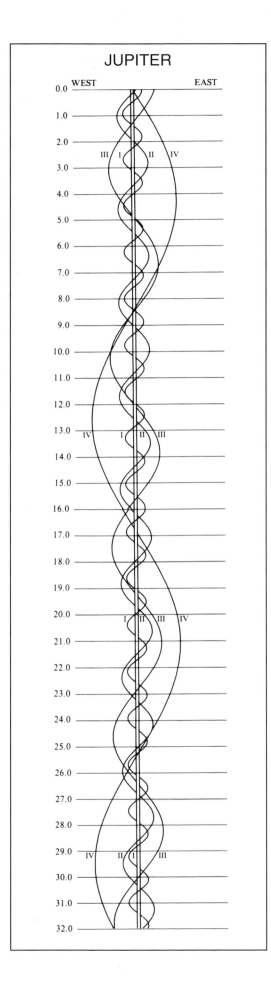

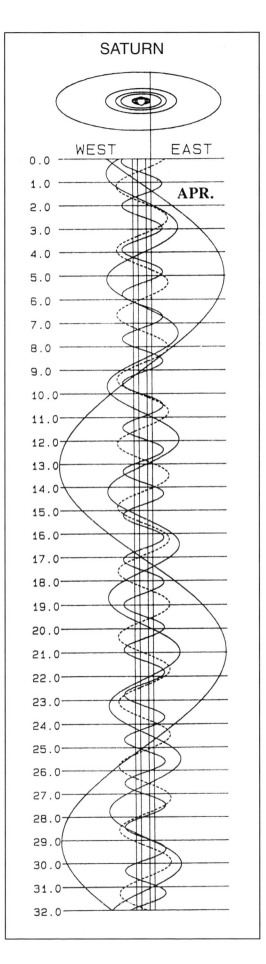

◀ **Follow Those Moons**
Using corkscrew diagrams like these for the current month, you can identify the four brightest satellites of Jupiter and Saturn. The horizontal lines represent 0 hours Universal time (UT) on the date indicated (0 hours UT = 7 p.m., EST, the previous day). The two straight vertical lines in the Jupiter diagram represent the disk of the planet. The wavy lines are the orbiting moons' positions at any time. The four vertical straight lines in the Saturn diagram are Saturn's disk (inner two) and its rings (outer two). The four satellites shown for Saturn are, in order outward, Tethys, Dione, Rhea and Titan. Saturn satellite charts for the current year are published in the **Observer's Handbook**. Jupiter charts are published in the **Observer's Handbook** and in all the major astronomy magazines. Courtesy U.S. Naval Observatory (Jupiter) and Larry Bogan (Saturn).

# Finding Your Way Around th

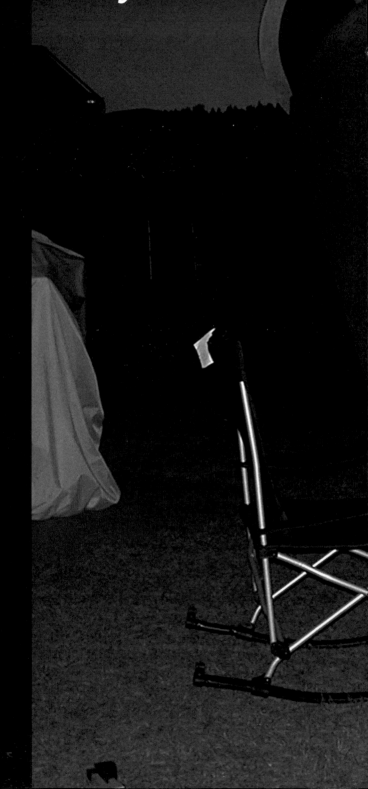

Familiarity with the night sky's geography and motions is what being an amateur astronomer is all about. While books like ours can serve as a guide, such knowledge can be gained only through the practical experience of observing the sky and understanding what you see. Without this knowledge, the most expensive telescope will add little to the hobby.

We have seen this advice ignored many times. People buy telescopes with fancy-looking mounts and setting circles (numbered dials on the mount) or computerized motors that can point automatically to thousands of objects, yet they do not know a single constellation. Or they know Orion and Andromeda but not the best season of the year to see them. Without a proper understanding of what the sky has to show or how the sky works, the observer will probably spend a few frustrating nights outside with the new telescope, then neglect it.

......................................

Remarkably, owners of some of the largest and most elaborate telescopes rely on no more than star charts, finderscopes and an ever-improving knowledge of the sky for locating deep-sky objects.

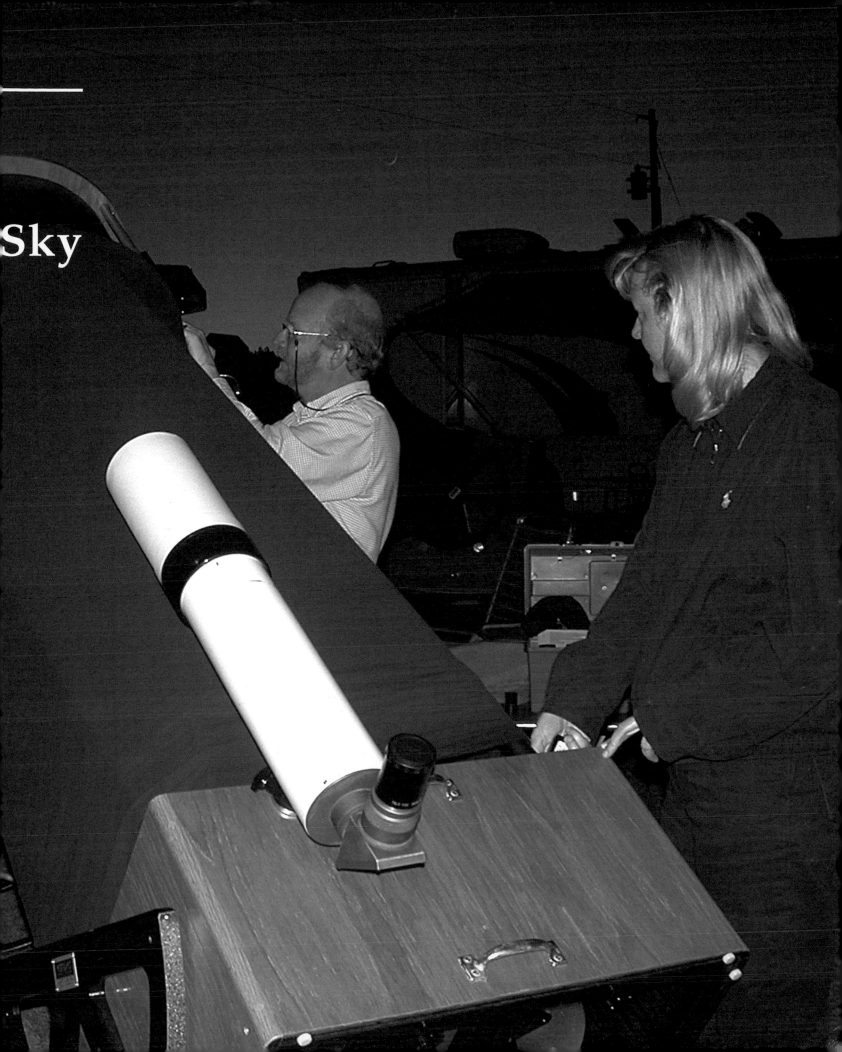

Sky

# How the Sky Works

Before we dive into the tools and techniques for finding your way around the sky, we'll explore the mechanics of the sky: how it moves and why stars and planets appear where they do. This emphasis stems from our observing philosophy.

We feel that the point of backyard astronomy is not just to peek into a telescope eyepiece; rather, it is the total experience of a personal exploration of the cosmos, a process that begins with the first identification of the Big Dipper, Orion and the bright planets. Recognition of the less obvious constellations follows, as well as a growing appreciation of the sky's motion due to the Earth's rotation and its revolution around the Sun. The quest can then extend thousands or millions of light-years via binocular sightings of the brighter star clusters, nebulas and a galaxy or two. As the months pass, the seasonal shift of the celestial panorama elicits a sense of cyclical change within a chamber of immense proportions. You see and understand the celestial sphere that turns above us each night.

Before this stage, a telescope of any description tends to be a distraction rather than an aid, diverting attention from the big picture. Until a novice stargazer understands the basics of how the sky is arranged and how it moves over time, attempting to navigate a telescope around the sky will only invite confusion. After you learn the basics, the next step is to learn to use star charts to find specific objects.

## NIGHT MOVES

Feeling at home under the sky won't happen until a mental picture clicks in: You are standing under a large dome that appears to be perpetually spinning around an axis. The sky dome is tilted at an angle that depends on where you live on Earth. The sky's rotation axis lies not overhead but to the north (assuming you live in the northern hemisphere). Think of the stars and constellations as being fixed to that spinning dome, as ancient astronomers did. Riding on that dome, however, are several moving targets—the Sun, Moon and planets—that travel among the stars along set and predictable paths.

## AS THE WORLD TURNS

The daily motion of the sky from east to west is an obvious fact of life—at least during the day. Everyone sees the Sun rise in the east and set in the west and knows that this motion comes not from the Sun but from Earth spinning on its axis. Although not so apparent, the same motion occurs at night, causing the starry dome to rotate above our heads. Stars, too, move from east to west, something that often surprises people unfamiliar with the sky. (We get UFO reports of objects hovering near the horizon that are no longer there an hour later!)

## UNDER THE CELESTIAL SPHERE

To understand this motion, imagine Earth surrounded by a large dome studded with stars. The Earth's North and South Poles point to the poles of the celestial sphere. The sphere is divided into northern and southern halves by the celestial equator, a projection of the Earth's equator into space. Earth spins on its axis within that sphere, from west to east. Living on the surface, we don't sense the Earth's movement. Instead, we see the sky turn above us in the opposite direction, from east to west. The entire sky appears to rotate around its two poles.

Now here's where it takes a little mental gymnastics. If we lived at the North Pole, the pole of the sky would lie directly overhead. The sky would turn parallel to the horizon. But most of us live at the mid-latitudes of the northern hemisphere (we picked 40 degrees north for the illustration at top right). Our local horizon, a line tangent to the Earth's surface at that latitude, cuts through the celestial sphere as indicated. Anything in the sky below that horizon line is out of our sight.

From our midlatitude viewpoint, the celestial sphere seems to be tipped over, with the north celestial pole and the North Star due north, 40 degrees above the northern horizon. The celestial equator arcs across the southern part of our local sky.

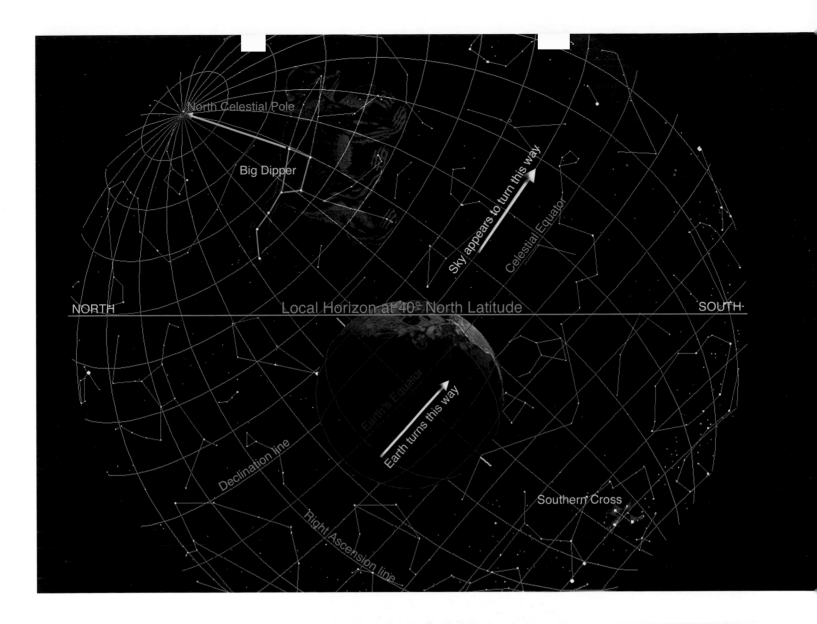

It is here that we see the Sun, Moon and planets—in the south. The location of the poles and equator in your local sky does not change through the year unless you move north or south on Earth. But the whole sky does turn through the night.

## WHERE'S THE NORTH STAR?

Most stargazers know the trick: The two stars in the pouring edge of the Big Dipper's bowl point to Polaris, the North Star. Locate that star, and you have found true north—Polaris never moves off its position in the sky as the night hours go by. But *where* you find Polaris depends on how far north or south you are on Earth. The altitude of Polaris above your northern horizon equals your latitude.

### Looking North ▶

When we look north at night, we see the sky rotating about the celestial pole, marked in our northern sky by Polaris, the North Star. This star barely moves through the night, and the celestial sphere appears to spin counterclockwise around it. Like the Big Dipper, stars and constellations near the North Star are circumpolar—they never set below the horizon but travel in endless circles about the pole.

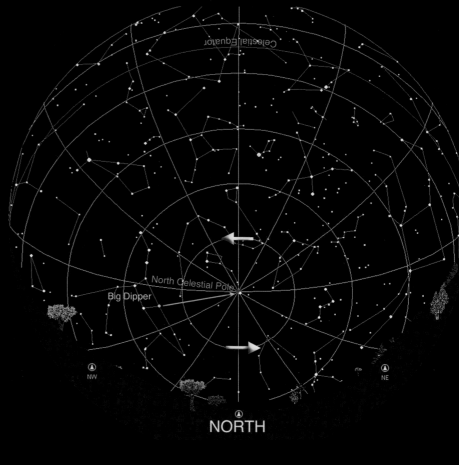

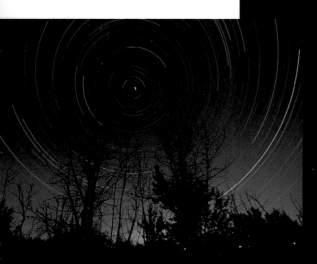

### Circumpolar Trails ▲

A 6-hour time exposure looking due north turns the stars into streaks as they spin around Polaris (the shortest bright streak near the rotation point), the north celestial pole.

### Looking East ▶

From northern latitudes, stars in the east appear to rise at an angle to the horizon, moving to the right as they climb higher.

All charts adapted from Starry Night Pro™/Space.Com

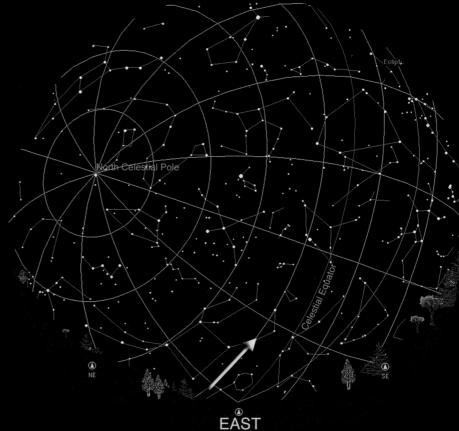

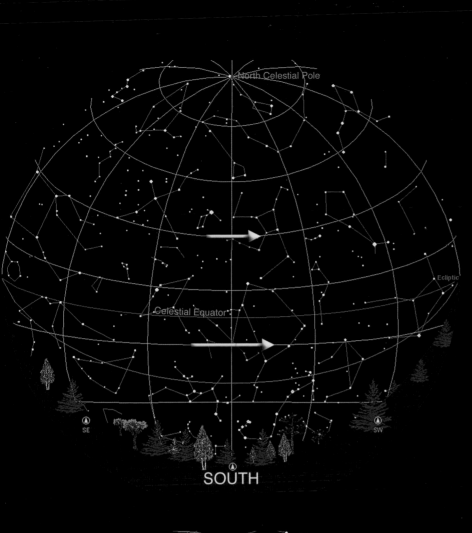

North Celestial Pole

Ecliptic

Celestial Equator

SE

SW

SOUTH

◀ **Looking South**
As we gaze due south, stars drift from left to right (east to west) across the sky. This region of sky contains the so-called seasonal constellations—those which change through the year, unlike the circum-polar patterns that can be seen all year long.

▲ **South-Sky Trails**
This 1-hour time exposure looking due south shows how those stars appear to move in horizontal paths parallel to the horizon.

◀ **Looking West**
As Earth turns, celestial objects eventually set in the west. The stars appear to move to the right as they sink toward the western horizon.

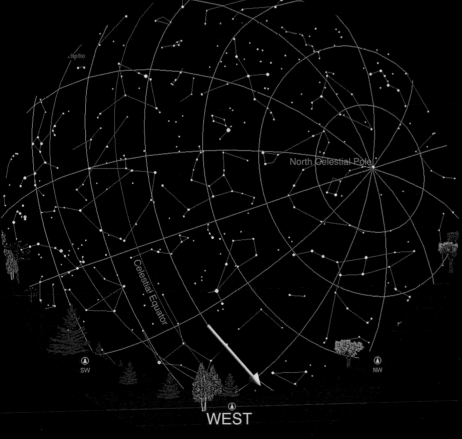

ecliptic

North Celestial Pole

Celestial Equator

SW

NW

WEST

211

**From 50 Degrees North** ▶

From western Canada, for example, at a latitude of 50 degrees north, Polaris shines high in the north, 50 degrees above the northern horizon. The Big Dipper remains visible all night and all year, even as it swings under the pole.

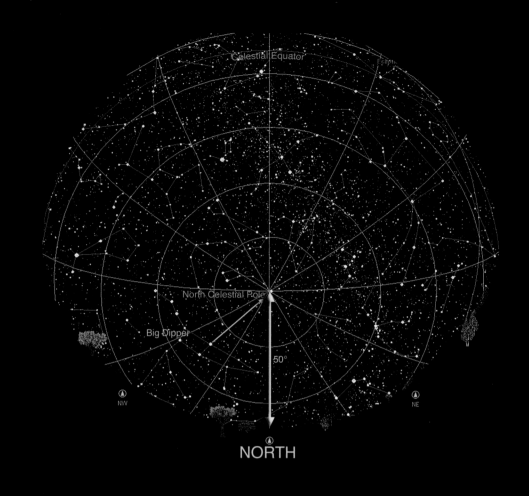

**From the Equator** ▶

For anyone used to northern skies, the view from the equator (perhaps from a game park in Kenya) is strange indeed. Stars rise straight up, perpendicular to the eastern horizon, and set straight down into the west. This makes for rapid sunrises and short-lived sunsets—the Sun drops below the horizon quickly, creating the fast onset of darkness typical of the tropics. From this location on Earth, the celestial equator passes directly overhead. The two celestial poles lie on the horizon, opposite each other, due north and south.

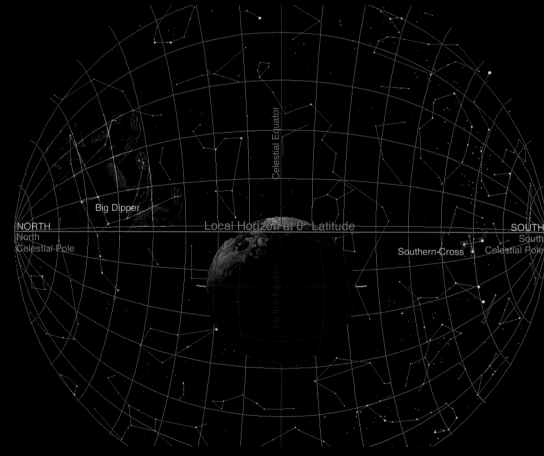

All charts adapted from Starry Night Pro™/Space.Com

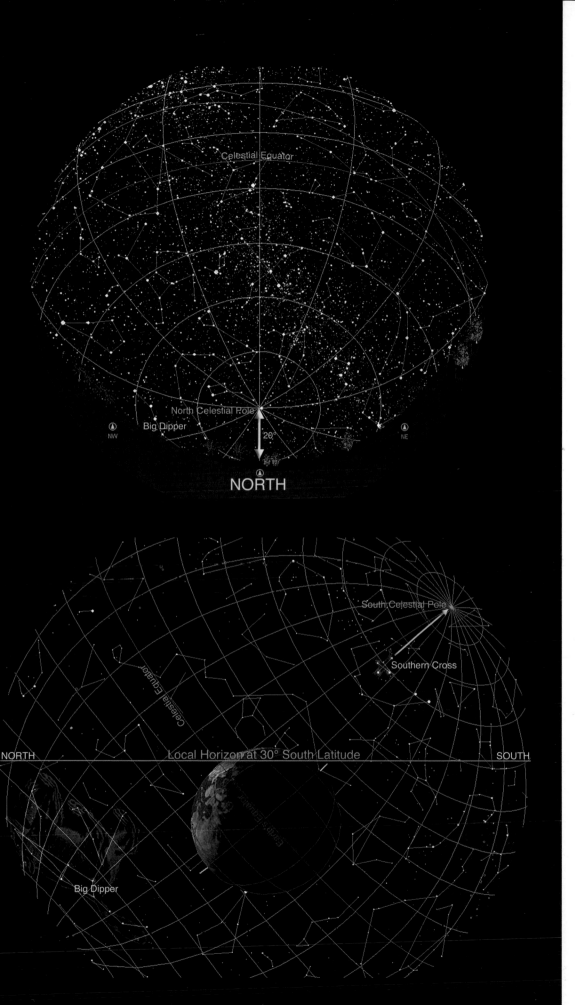

Celestial Equator

North Celestial Pole

Big Dipper

NW

NE

20°

**NORTH**

South Celestial Pole

Southern Cross

Celestial Equator

NORTH

Local Horizon at 30° South Latitude

SOUTH

Big Dipper

**◀ From 20 Degrees North**
From the latitudes of
Hawaii and Mexico, at
20 degrees north, Polaris
shines low in the sky, just
20 degrees above the horizon. At certain times of the
year, the Big Dipper slips
below the horizon, as it is
doing here.

**◀ From 30 Degrees South**
From the latitude of Aus-
tralia, southern Africa or
South America, the sky
seems turned upside down
compared with a northern
viewpoint. Of course, to
southerners visiting the
north, the North American
or European sky is upside
down. From 30 degrees
south, the north celestial
pole is forever below the
horizon. The sky turns
clockwise around the south
celestial pole, which lies
due south (the Southern
Cross points to it), at an
angle above the horizon
equal to the site's latitude
below the equator. The
celestial equator now
arcs across the northern
sky. Southerners look
north to see the Sun, Moon
and planets.

## THE EARTH'S ORBITAL TREK

Earth is said to *rotate* around its axis, taking 24 hours to complete one rotation with respect to the Sun. Earth also *revolves* around the Sun, taking a year to complete one orbit. The Earth's orbital period defines our 365-day year (more or less). The Earth's motion around the Sun gives rise to the sky's other great annual motion: the seasonal parade of constellations. We can't see Orion in June or Sagittarius in Decem-

**Path of the Sun ▶**
As Earth revolves around it, the Sun appears to move from west to east against the background stars. The imaginary line the Sun follows is called the ecliptic. This is the view we would see from space on December 21.

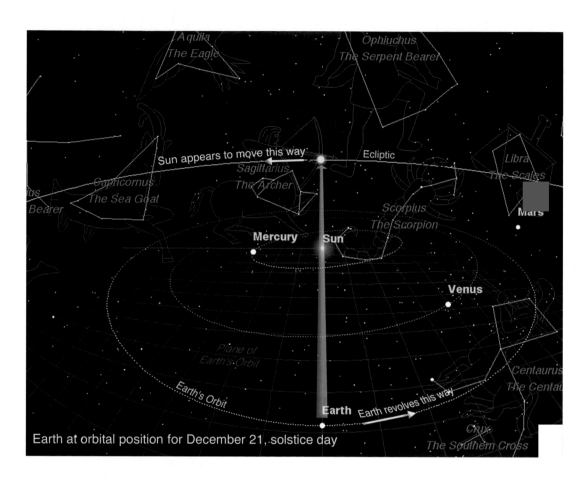

Earth at orbital position for December 21, solstice day

**The Sun and Stars ▶**
Taken by the orbiting SOHO satellite, this image of the Sun captures a view we can never see from the Earth's surface: the Sun (represented by the yellow disk) surrounded by stars, in this case the star clouds of Sagittarius the archer. Courtesy NASA/ESA.

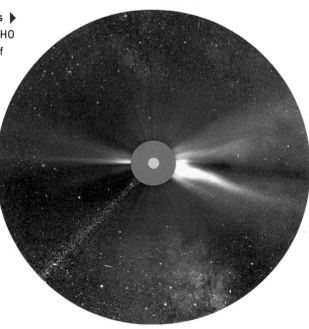

ber. Every constellation has its season, and here's why.

## THE VIEW FROM SPACE

From an omniscient viewpoint, we gaze down on Earth at its place in its orbit on December 21, winter solstice for the northern hemisphere. Looking from Earth toward the Sun, we see the Sun apparently sitting in the constellation Sagittarius. As Earth revolves around the Sun, the Sun appears to shift eastward against the background stars. From Sagittarius, the Sun moves into Capricornus. The Sun's path in the sky is the ecliptic line. Over the course of a year, the Sun travels along the ecliptic, moving through 12 constellations, the familiar signs of the zodiac. The Sun spends approximately one month in each constellation of the zodiac. However, as you can see below, the Sun also spends some time in a thirteenth constellation, Ophiuchus, whose foot long ago was placed between the scorpion and the archer.

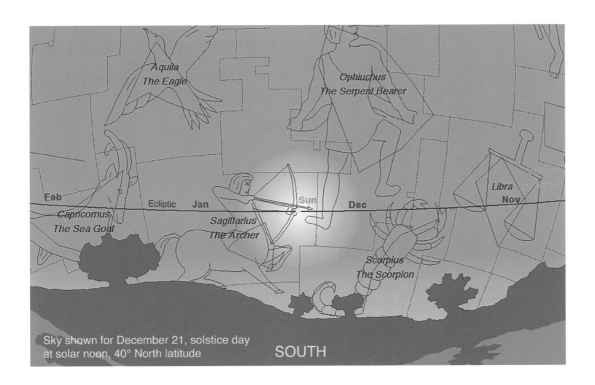

Sky shown for December 21, solstice day at solar noon, 40° North latitude

SOUTH

**◀ View From Earth by Day**
Shift your viewpoint from high in space to the Earth's surface at a midnorthern latitude. The date remains the same, December 21. At high noon, the Sun lies due south in the constellation Sagittarius. It and the surrounding constellations are, of course, invisible in the day sky. We'll see these constellations due south on summer nights six months later, when the Sun has moved 180 degrees along the ecliptic and lies in Gemini. In June, Gemini, Orion and the winter constellations will occupy the daytime sky.

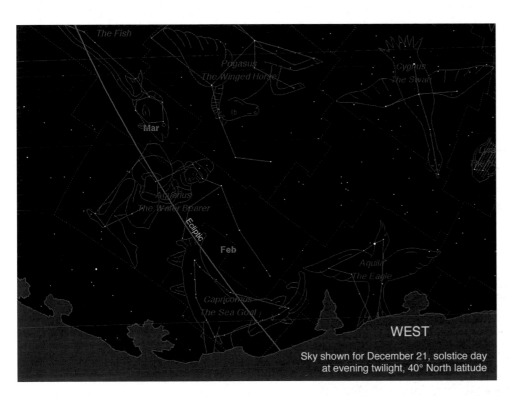

WEST

Sky shown for December 21, solstice day at evening twilight, 40° North latitude

**◀ View at Nightfall**
Wait until dark on December 21, then look west. Sagittarius has set with the Sun, and the stars of Capricornus and Aquarius shine low in the evening twilight sky—but not for long. As the Sun continues its trek along the ecliptic in the days to come (of course, Earth is actually doing the moving), these constellations disappear behind the Sun. By January, they are gone from the evening sky, and the zodiac constellations of Pisces and Aries are low in the west. So it goes through the seasons—a parade of changing constellations that repeats like clockwork every year.

All charts adapted from Voyager III™/Carina Software

## MARCH OF THE CONSTELLATIONS

The annual motion of Earth around the Sun creates the shifting array of constellations we see each year. Like the birds of spring and the leaves of autumn, constellations are familiar signs of the seasons. People are always amazed that backyard astronomers can simply look up and name a star. But the stars and constellations return to the same place in the sky each year. Learn them one year, and you'll know their habits for every year to come. If it is February, for example, that bright star twinkling in the south must be Sirius. But two months later, Sirius is low in the southwest. To the south, new stars have taken its place.

The progression of Earth around the Sun makes stars rise and set about four minutes earlier each night. (To be precise, the difference is 3 minutes 56 seconds.) This adds up to two hours per month. For example, Orion sets two hours earlier in March than it does in February. After 12 months, this advance of the constellations adds up to 24 hours, and we have gone full cycle—the same constellations that were due south in February this year will be there again next February.

**February: Orion ▶ Shines Due South**
In this series of three diagrams (right and facing page), the sky is shown at the same time (8 p.m., standard time) each night, looking south. Each successive view is one month later. In mid-February, the stars of Orion shine due south at 8 p.m., as high as they get for the night. Sirius sparkles in the south-southeast. Leo the lion is just rising in the east.

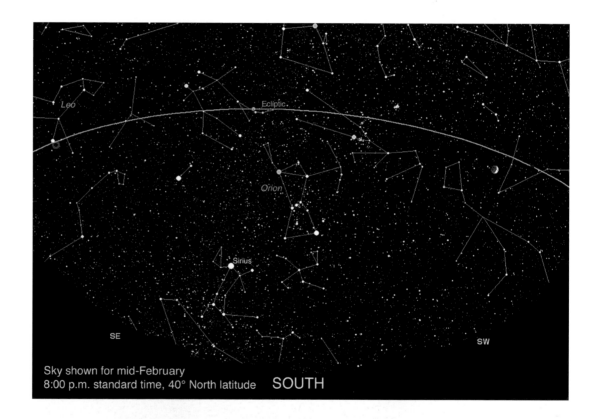

Sky shown for mid-February
8:00 p.m. standard time, 40° North latitude  SOUTH

**Winter Sky Setting ▶**
Looking west in April, we see the stars of Orion and the winter sky sinking into the combined glow of twilight and light pollution from a nearby city. April nights are the last chance to see Orion until it emerges from behind the Sun in the dawn hours of August.

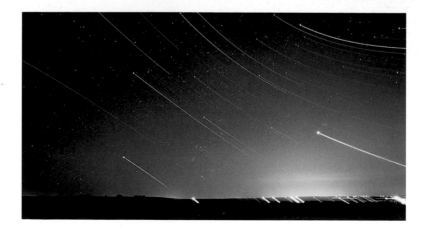

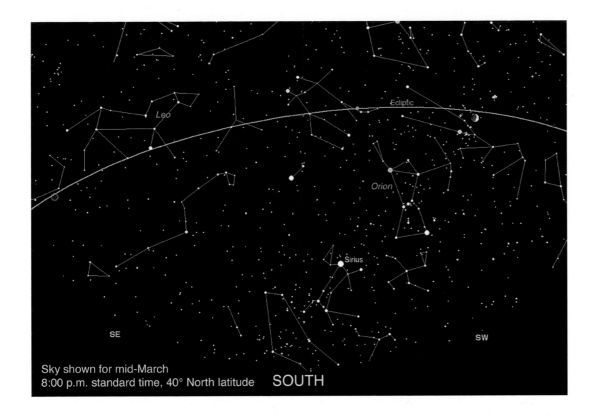

Sky shown for mid-March
8:00 p.m. standard time, 40° North latitude    **SOUTH**

### March: Orion in the Southwest

Thirty days later, at the same time of night, Orion is sauntering over to the southwest, two hours of sky rotation past the point where the celestial hunter stood a month earlier. Sirius is just past the meridian, the line that runs due south to north and bisects the sky. Leo is now well up in the east.

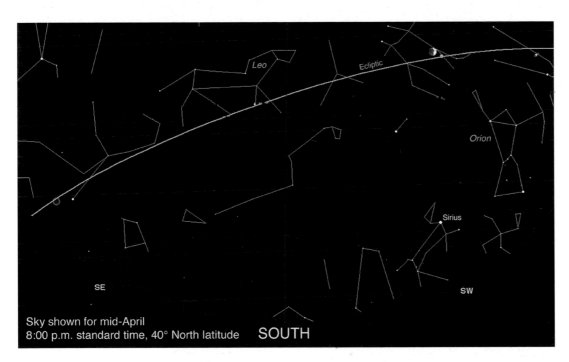

Sky shown for mid-April
8:00 p.m. standard time, 40° North latitude    **SOUTH**

### April: Orion Sinking into the West

By mid-April at 8 p.m. (9 p.m., daylight time), Orion is sinking into the western sky and Sirius is shining low in the south-west. The spring stars of Leo are due south now. And so the winter stars surrounding Orion give way to the spring constellations. Note that the sky is now deep blue—by April, the Sun sets later, and darkness doesn't come as early as it did in February. By the time the night is fully dark, Orion is gone.

All charts courtesy
TheSky™/Software Bisque

## WANDERING PLANETS

The stars and constellations return to the same place in the sky each year, but not the planets. Just as Earth revolves around the Sun, so do the other planets—each at its own rate. Of the bright naked-eye planets, Mercury travels the fastest around the Sun, while Saturn travels the slowest. The orbital motion of the planets carries them against the background stars, a motion that can be obvious even after a few days or weeks. Jupiter, for example, moves eastward from one zodiac constellation to the next over the course of a year, taking 12 years to make one cycle around the sky.

## PATH OF THE PLANETS

The celestial sphere contains several fundamental lines and points, the celestial poles and equator among them. But another key line is the ecliptic—the path the Sun apparently describes as it travels around the sky. It can also be pictured as the plane of the Earth's orbit around the Sun.

It is in this same plane, give or take a few degrees, that the planets orbit. The solar system is like a flattened disk, an artifact of its formation billions of years ago from a spinning disk of gas and dust.

planets are tipped over on their rotation axes. Earth is one of them. Our planet is tilted 23.5 degrees off the vertical. For this reason, the ecliptic does not coincide with the celestial equator.

## WOBBLING EARTH

Earth moves through another long-term motion whose effect is much more subtle than its daily rotation and annual revolution. Over 26,000 years, the Earth's spin axis (defined by the red arrows below) wobbles like a top around a radius of 23.5

**Ecliptic Plane ▶**
When viewing the solar system from space, we tend to depict it with the ecliptic as the standard horizontal plane. In this view, Earth is then tipped 23.5 degrees from a line perpendicular to the ecliptic. This tilt of the Earth's axis gives rise to the seasons. When the northern hemisphere is tipped away from the Sun, as it is here, we experience winter in the north, while the southern hemisphere, tipped toward the Sun, enjoys hot summer days. Six months later, with Earth on the opposite side of the Sun, the northern hemisphere is tipped toward the Sun, bringing summer to the north and winter to the southern hemisphere.

Adapted from Starry Night Pro™/Space.Com

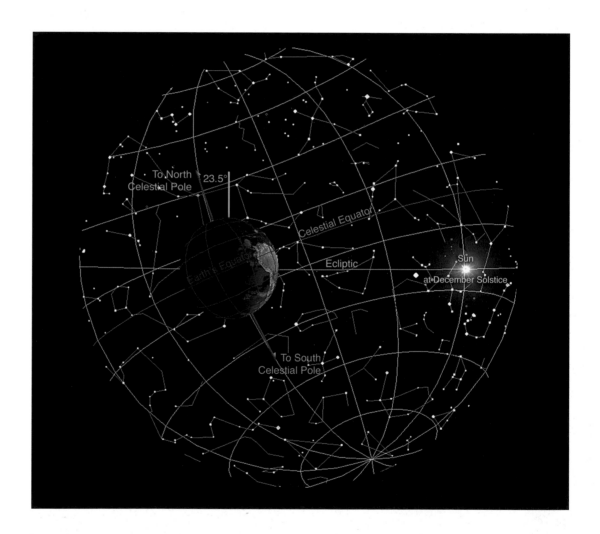

It is along the ecliptic line in our sky that we find the Moon and planets. Only oddball Pluto, some asteroids and many comets appear far off the ecliptic.

Their common origin explains why all the planets revolve around the Sun in the same direction, counterclockwise as seen looking down from the north. But some

degrees. The center of that wobbling motion is the ecliptic pole, the vertical green line protruding from Earth in the diagram above. This slow precession motion will gradually cause the north pole of the sky (the north celestial pole) to shift away from Polaris. In 12,000 years, Vega will be the pole star.

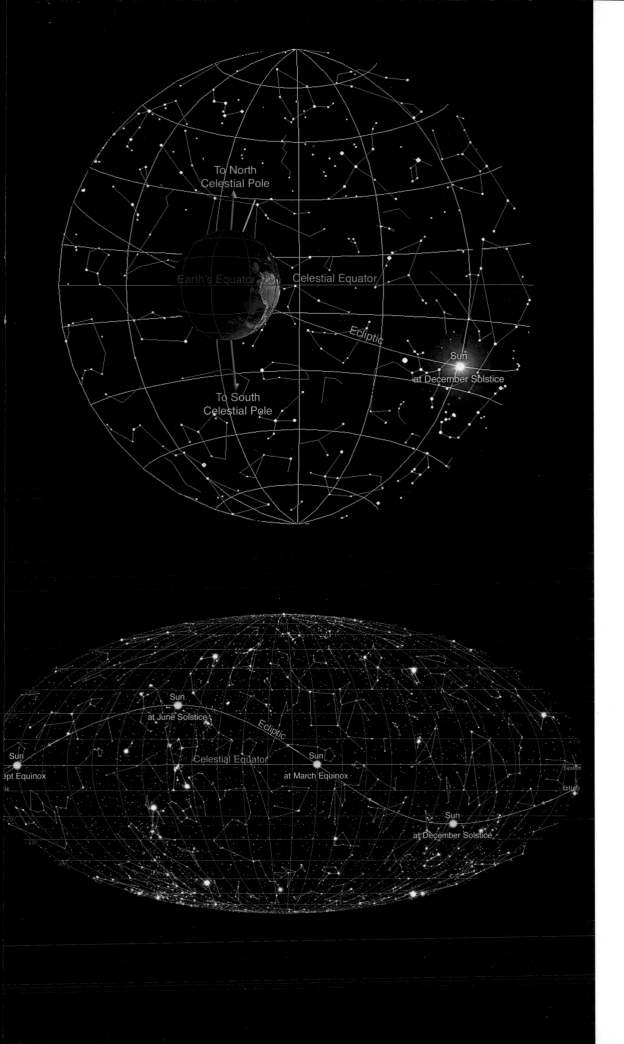

**To North Celestial Pole**

Earth's Equator  Celestial Equator

Ecliptic

Sun
at December Solstice

**To South Celestial Pole**

Sun
at June Solstice

Ecliptic

Sun
at September Equinox

Celestial Equator

Sun
at March Equinox

Sun
at December Solstice

Equator

Ecliptic

### ◀ Tilting Earth Upright

To understand why we see the ecliptic where we do in our sky, rotate the previous scene so that the Earth's rotation axis stands upright. The celestial equator becomes the horizontal plane. The ecliptic now swings above and below the equator at an angle of 23.5 degrees. From our geocentric point of view on the Earth's surface, this is how we prefer to depict the sky.

Adapted from Starry Night Pro™/Space.Com

### ◀ The Sky Unwrapped

Now let's unwrap the entire heavens we see from Earth into one all-sky chart. The yellow lines are the grid work of the celestial coordinate system of right ascension and declination that maps the sky. The ecliptic, the plane of the solar system, is the wavy line swinging above and below the equator. As the Sun moves along the ecliptic, it reaches four key points each year. Two are the solstices, when the Sun is at its maximum distance (23.5 degrees) above and below the celestial equator. The other two points are the equinoxes, when the Sun crosses the equator, heading north or south.

Adapted from SkyChart III™/ Southern Stars Systems

## ECLIPTIC HIGHS AND LOWS

It's a common misconception that the seasons are caused by the Earth's changing distance from the Sun. Although the Earth's orbit is slightly elliptical, the change in distance through the year has little effect on climate. It's the change in the Sun's altitude (due to the Earth's tilt), from high in summer to low in winter, that produces seasons. To know where to look for the Moon and planets, it's important to understand the wobbling ecliptic and to develop a mental picture of where the ecliptic sits each season. The following illustrations depict the sky from a latitude of 40 degrees north.

**Wintertime Highs ▶**
In winter, the ecliptic swings low across the daytime sky (see "View From Earth by Day" on page 215 for an illustration). But at night, just the opposite occurs. The ecliptic we see on a chilly winter evening is where the Sun sits on a summer day. As depicted in this scene from December 2001, wintertime planets always ride high in the sky. Winter full Moons also shine down from the highest altitude of any of the year's full Moons.

All charts adapted from SkyChart III™/ Southern Stars Systems

**Springtime Swing ▶**
As we look west on a spring evening, the Sun lies near the vernal equinox and has just set. The section of the ecliptic that rises above the celestial equator is now in the west. This swings the evening ecliptic to its highest angle above the horizon for the year, placing twilight planets, such as Mercury and Venus, at their highest altitude above the horizon. The crescent Moon also rides higher in the sky, making it easier to sight thin young Moons.

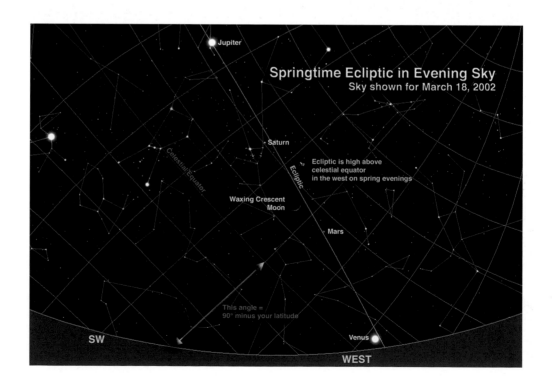

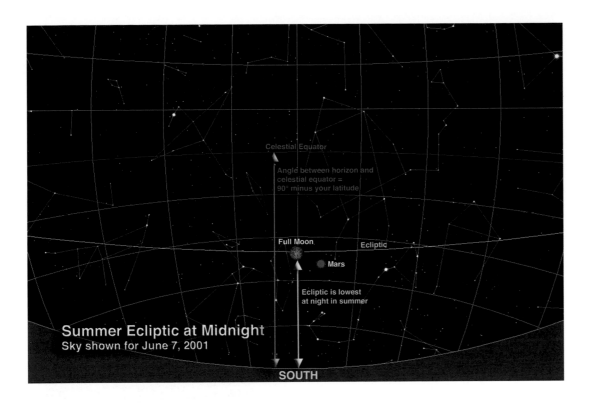

Celestial Equator

Angle between horizon and
celestial equator =
90° minus your latitude

Full Moon

Ecliptic

● Mars

Ecliptic is lowest
at night in summer

**Summer Ecliptic at Midnight**
Sky shown for June 7, 2001

SOUTH

The evenings may be warm
and pleasant, but summer
is not a good season for
planet watching. At night,
the ecliptic swings low
across the southern sky,
as do the summer planets.
This was the situation in
June 2001, when Mars was
at its biennial closest approach to Earth. The red
planet's low altitude put it
amid the murk and turbu-
lence of our atmosphere,
making for blurry tele-
scopic views. The same
occurs when Jupiter and
Saturn reach this realm
of the ecliptic in Scorpius
and Sagittarius.

## A REAL SCENE

The predawn image at left, taken on a mid-
August morning in 2001, shows how the
planets and Moon lie along a line. We can
also see Orion and the winter stars emerging from the Sun's glare.

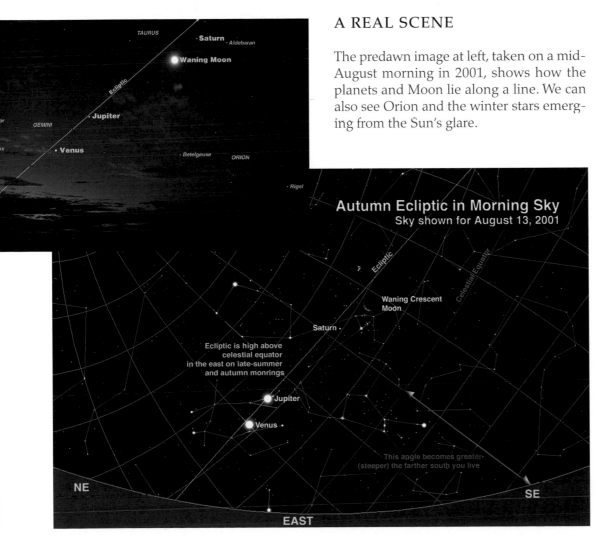

TAURUS

·Saturn
·Aldebaran

● Waning Moon

Ecliptic

·Jupiter

GEMINI

· Venus

· Betelgeuse    ORION

· Rigel

**Autumn Ecliptic in Morning Sky**
Sky shown for August 13, 2001

Ecliptic

Celestial Equator

Waning Crescent
Moon

Saturn ·

Ecliptic is high above
celestial equator
in the east on late-summer
and autumn monrings

·Jupiter

● Venus ·

This angle becomes greater
(steeper) the farther south you live

NE

SE

EAST

◀ **Autumn Angles**
In autumn, the ecliptic
arcs low across the evening
sky, but just the opposite
occurs at dawn. Late sum-
mer and autumn are the
best times for seeing twilight planets and thin waning Moons in the morning
sky. The ecliptic is then
angled at its highest for
the year in the eastern sky.
Autumn is also the best
season for morning sight-
ings of the faint glow of
zodiacal light (which follows the ecliptic), while
spring is the best season
for evening sightings.

## GOING THROUGH A PHASE

Moon phases remain the most misunderstood natural phenomena. They are not caused by the Earth's shadow moving across the disk of the Moon, as is commonly believed. That's a lunar eclipse. The real explanation becomes obvious once you form a mental picture of Earth sitting in space, with the Moon revolving about us once a month in its own orbit.

Both Earth and the Moon bask in sunlight. The sides of our two worlds that face the Sun are in daylight, while the sides facing away from the Sun are in darkness. Now let's place ourselves away from Earth, facing the Sun and looking down on the night side of our planet. As the Moon revolves around Earth, what do we see?

## VIEW FROM SPACE

When the Moon comes between us and the Sun, the nightside of the Moon faces us.

This is new Moon, a phase that is invisible to us except when the Moon happens to cross directly in front of the Sun in a solar eclipse. As the Moon continues to revolve around Earth (from right to left in the illustration below), we see more and more of the lit dayside of the Moon. First, we see a thin crescent followed by first-quarter phase, about seven days after new. The Moon is then 90 degrees away from the Sun. After another week of so-called gibbous phases, the Moon reaches the point in its orbit opposite the Sun. The side of the Moon facing us is now entirely lit by the Sun, creating a full Moon 14.5 days after new.

## VIEW FROM EARTH: EVENING

Dropping back to Earth, here is what we see through that same cycle. The scene at top right depicts the evening sky at the same time each night for a two-week period. We chose a scene early in 2002, but a similar

**Synodic and Sidereal ▼**
The Moon revolves around Earth once every 27.3 days, its sidereal period. However, because Earth has also revolved one-thirteenth of its orbit around the Sun in that time, it takes the Moon another two days to return back to the same phase. So the interval between new Moon and new Moon is 29.5 days, the so-called synodic period that forms the basis of the calendar month.

Adapted from Starry Night Pro™/Space.Com

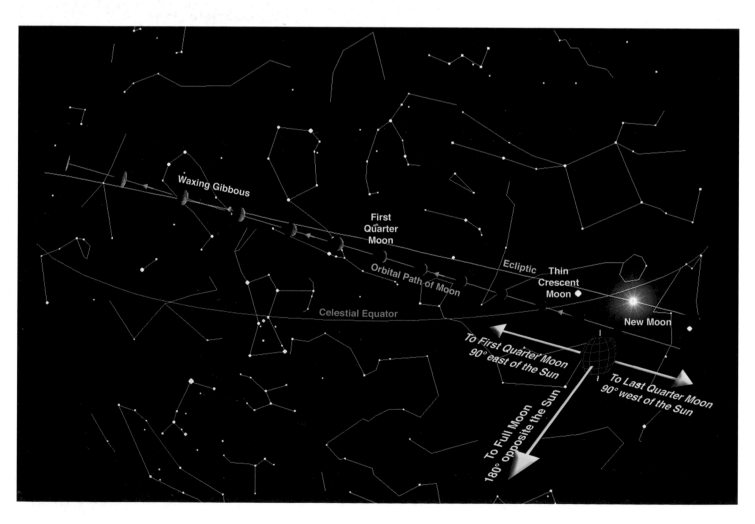

scene applies to any month in any year, with only the angle of the Moon's path shifting with the seasonal change in the angle of the ecliptic.

During the waxing (growing) cycle, we first sight the Moon as a thin crescent low in the evening twilight. This is possible about two days after new (anything less is tough to see). The Moon's motion around us carries it farther to the east each night—we see it higher and higher in the evening sky as the week progresses. By seven days after new (the number on each little moon indicates its age in days), the Moon lies due south at sunset. This is first-quarter Moon, so called because the Moon has traveled one-quarter of the way around its orbit. Half the disk of the Moon we see is now sunlit.

The Moon continues to wax through gibbous phases until it reaches full. Being opposite the Sun, any full Moon rises in the east as the Sun sets in the west. A full Moon sits due south at midnight and shines in the sky all night.

## VIEW FROM EARTH: MORNING

Now let's switch to a morning-sky view. We're still looking south, but the Sun is just about to rise in the east. This places the full Moon over in the west, ready to set. With each passing night, the Moon wanes in phase as it moves back toward the Sun. Each night, the Moon is roughly 12 degrees farther to the east than it was the previous night. The Moon therefore moves its own diameter, 0.5 degree, in about an hour. This motion is due to the Moon's revolution around us.

On clear mornings after sunrise, as the waning cycle proceeds, we see a gibbous Moon in the western sky. About 21 days after new, the Moon once again lies 90 degrees away from the Sun, creating a last-quarter Moon shining due south at sunrise. This wanes to an old, thin crescent Moon rising just before the Sun, low in the dawn sky. The cycle completes itself 29.5 days after the previous new Moon, when the Moon comes between us and the Sun. The week or so around new Moon, a period called dark-of-the-Moon, is cherished by deep-sky observers wanting dark moonless skies.

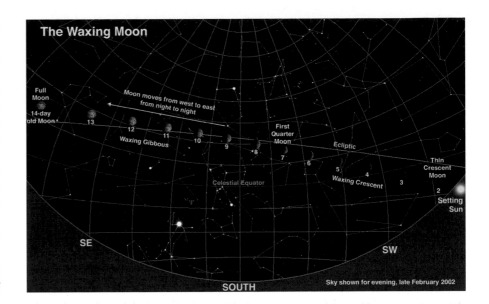

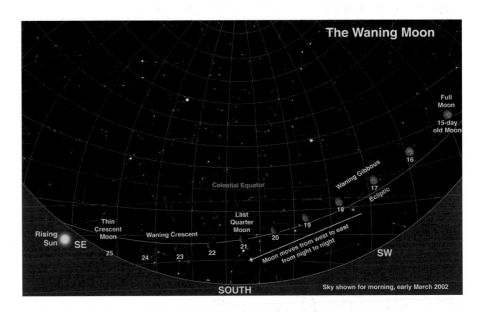

## WHY NO MONTHLY ECLIPSES?

In these illustrations, notice how the Moon follows the ecliptic, but not quite. Its orbit is tilted five degrees off the ecliptic, which is why we don't get an eclipse each month—because the Moon passes above or below the Sun's disk or the Earth's shadow. But when the Moon crosses the ecliptic at its new or full phase, we experience some type of eclipse. Such an alignment occurs from four to seven times each calendar year. Eclipses also come in pairs. Two weeks before or after every solar eclipse, when the Moon passes in front of the Sun, we usually get a lunar eclipse as the Moon moves through the Earth's shadow.

▲ **Wax On, Wane Off**
When the Moon is increasing in phase each night, from new to full, the cycle is called a waxing Moon. After full phase, the Moon wanes, or decreases in illumination, each night.

Adapted from SkyChart III™/ Southern Stars Systems (both)

# Star-Hopping

A prime question that every new telescope owner asks is, How do I find things? The quick answer is star-hopping. This technique, one every amateur astronomer masters, requires learning to read star charts, which are the maps of the night sky.

To become comfortable reading star charts, we recommend you spend a few months stargazing with binoculars first. Besides acquainting you with the capabilities of simple optics, binoculars will help you learn to use star charts to hop to bright binocular targets. After this, interpreting charts at the telescope (with its narrow field and inverted view) becomes a small step instead of the giant leap many newcomers find it to be.

## LEARNING TO STAR-HOP

The key to successful star-hopping is pattern recognition. First locate your target on a suitable star chart, one deep enough to show enough stars to act as steppingstones across the sky. One of the sixth-magnitude atlases described in the next section is a minimum. Pick a bright star near your target, and identify a pathway of stars leading from that guide star to your destination. Look for identifiable chains or triangles of stars that can serve as signposts.

The next trick is to transfer that mental picture of your route to the sky. This is where familiarity with the sky, the size of the constellations and angular distances is important, as is the basic skill of knowing how to turn the chart so that its orientation matches the sky. This comes with an understanding of how the dome of the sky is oriented above your head—for example, knowing that when you look east, the north celestial pole is to the left, so star charts with north up must be twisted counterclockwise to match the naked-eye sky. A further trick is the simple act of turning the star chart upside down to match the view in an inverting finderscope.

A polar-aligned equatorially mounted telescope (as opposed to a Dobsonian scope) can simplify star-hopping. No matter where it is pointed, the telescope moves only parallel to the lines of right ascension and declination, making it easier to track along east-west or north-south routes that match gridded star charts.

Star-hopping may sound like work, but star-hop routes will soon become as familiar as the backstreets of your hometown. With practice, you will soon be centering the telescope on any number of objects with no more than a couple of quick glances through the finderscope.

## FINDER AIDS

A good finderscope (at least 6x30, but preferably 7x50) is essential for star-hopping. As a supplement or an alternative, we also recommend one of the reflex-style finders, such as the Telrad or the Rigel QuikFinder (see Chapter 5 for details). These are generally superior to the red-dot style of sighting aids, which are fine for homing in on bright stars but are not up to more rigorous star-hops to fainter targets (it is hard to see faint stars through the small sighting windows). In a dark sky, a Telrad or QuikFinder is all you need to locate most targets. Many computer software programs allow you to overlay Telrad or QuikFinder reticle patterns onto the star field, a great feature for printing out your own custom star-hop charts at home.

**Hopping Tools ▶** Even owners of large-aperture telescopes routinely equip their scopes with a good optical finder and a reflex window-style finder, such as a Telrad (on this telescope) or a Rigel QuikFinder. Add star charts, and you're all set for a night of star-hopping.

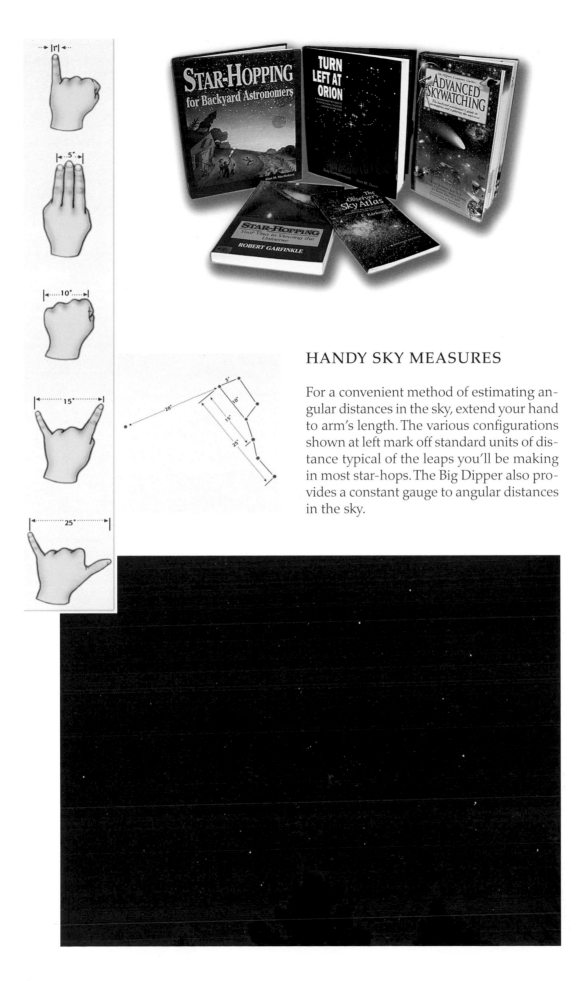

◀ Star-Hopping Guides
Specialized books offer a great selection of packaged star tours, including maps with star-hop routes marked. In their respective books, both titled **Star-Hopping,** Alan MacRobert (Sky Publishing) and Robert Garfinkle (Cambridge) conduct informative tours of selected regions of the sky. **Advanced Skywatching** (Time-Life; also sold in softcover under the title **Backyard Astronomy**), with contributions by Alan Dyer, contains standard background material complemented by monthly star charts and detailed star-hopping maps. **Turn Left at Orion** (Cambridge), a fine beginner-oriented guidebook by Guy Consolmagno, provides maps for hunting down a hundred or so of the best deep-sky objects. A more advanced work is Erich Karkoschka's **The Observer's Star Atlas** (Springer), a small book of 50 charts, each with wide-angle constellation views and close-up charts that depict the view through a finderscope.

## HANDY SKY MEASURES

For a convenient method of estimating angular distances in the sky, extend your hand to arm's length. The various configurations shown at left mark off standard units of distance typical of the leaps you'll be making in most star-hops. The Big Dipper also provides a constant gauge to angular distances in the sky.

◀ **Big Dipper and Polaris**
Compare the Big Dipper diagram at the center of this page with the real thing in the photo at left. The pointers from the bowl direct you to Polaris, the North Star, which is one dipper length away.

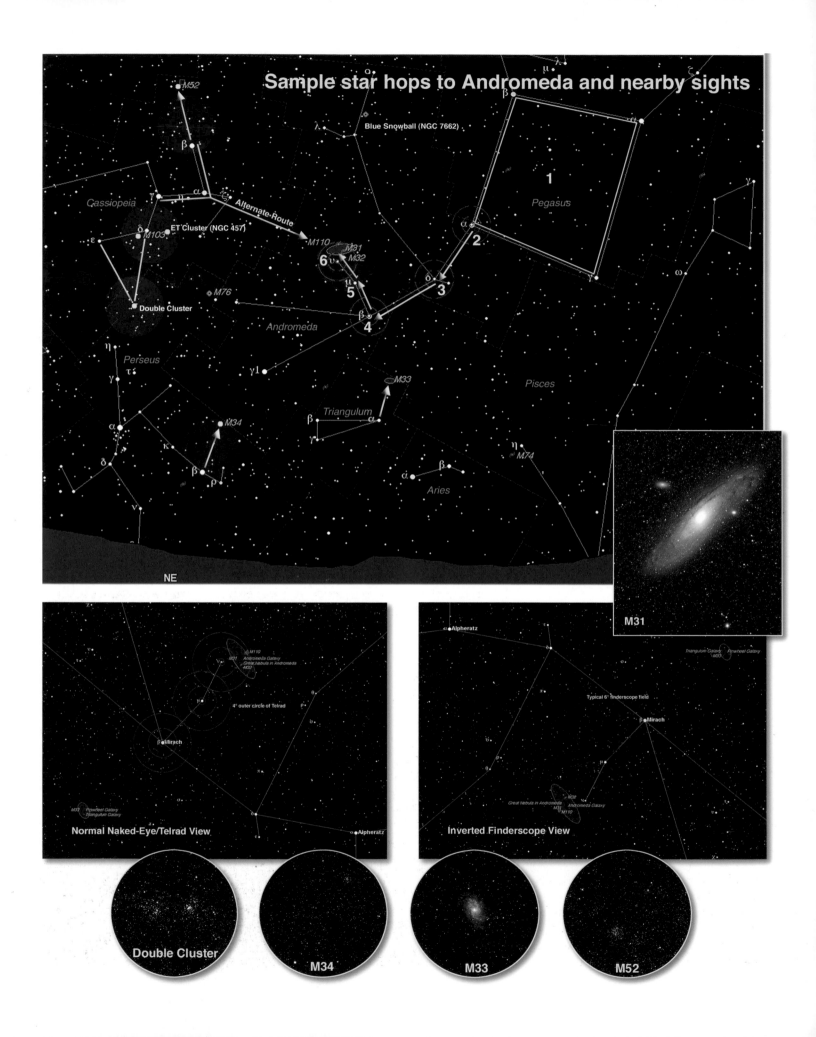

**Sample star hops to Andromeda and nearby sights**

M52

Blue Snowball (NGC 7662)

Cassiopeia

Pegasus

*1*

Alternate Route

ET Cluster (NGC 457)

M103

M110

M31

M32

Double Cluster

M76

Andromeda

Perseus

Pisces

M34

M33

Triangulum

Aries

M74

NE

M31

**Normal Naked-Eye/Telrad View**

M110
Andromeda Galaxy
Great Nebula in Andromeda
M32

4° outer circle of Telrad

β Mirach

Pinwheel Galaxy
Triangulum Galaxy

**Inverted Finderscope View**

Alpheratz

Triangulum Galaxy   Pinwheel Galaxy
M33

Typical 6° finderscope field

β Mirach

M32
Great Nebula in Andromeda   Andromeda Galaxy
M31
M110

**Double Cluster**

**M34**

**M33**

**M52**

# A SAMPLE STAR-HOP

It is a warm August evening on a dark-of-the-Moon weekend. The Milky Way shines overhead, and the stars of autumn are rising in the east. Among them is one of the most popular destinations for stargazers: the Andromeda Galaxy. Located some 2.5 million light-years away, Messier 31 (M31), as it is also known, is the most distant object you can see easily with the unaided eye. It is the largest and brightest galaxy in the northern sky, providing a thought-provoking target for binoculars or any telescope. But how do you get there?

**1.** First locate the nearest distinctive constellation or star pattern, in this case the well-known Square of Pegasus. In our August evening view, the Square is turned on its side, rising in the east. Once you have identified the Square, turn your star chart so that its orientation matches the sky.

**2.** Find a bright star to serve as a jumping-off point, in this case α Andromedae, or Alpheratz. The red circles indicate what a Telrad finder would show.

**3.** Get hopping! Swing away from the Square to the next brightest star, δ Andromedae, a couple of Telrad circles, or finder fields, to the east.

**4.** Keep going. Travel an equal distance again to β Andromedae, also called Mirach. Here's where you turn a corner.

**5.** Swing north about one Telrad circle, or finder field, until you come to a slightly dimmer star, μ Andromedae.

**6.** Keep going north for the last leg until you reach the next bright star, the slightly fainter υ Andromedae. M31 lies within a low-power eyepiece field of that star. You've arrived!

For an alternate route to Andromeda, imagine the three westernmost stars of Cassiopeia's W-shape forming an arrowhead pointing down to M31. Though it is a big leap for a telescopic star-hop, you can use this imaginary pointer line to confirm that you are in the right region.

While you are in the area and primed for star-hopping, try to track down these fine sights:

• The Double Cluster sits off the first side of Cassiopeia's W, just far enough away to be a challenge to find in light-polluted skies. But the view of hundreds of stars sprinkling the field in two bright clusters is worth the hunt. Though it carries no Messier number, the Double Cluster exceeds most Messier object clusters for spectacle.

• M34, a bright, loose open cluster listed in the Messier catalog (see Chapter 12), sparkles within a finder field of β Persei, also known as Algol, the Demon Star.

• Similarly, M33, a spiral galaxy in Triangulum, is within a finder field of α Trianguli. It lies about the same distance below β Andromedae as M31 lies above it. Like M31, M33 is a spiral galaxy that is a member of the Local Group of neighboring galaxies. Note: M33 is a diffuse object hard to pick out in bright skies or with small-aperture telescopes in any sky.

• The line joining α and β Cassiopeiae points to M52, another rich open cluster of hundreds of stars. With M52 just beyond the finder field (with β in the field) and no bright chain of stars to guide you, it is easy to get lost on this star-hop.

◀ **Facing page, lower left:** Mirach (β Andromedae) is a naked-eye star that serves as a starting point for a Telrad hop to M31.

◀ **Facing page, lower right:** To start a hop with an inverting finderscope, place β Andromedae (Mirach) at the top of the finderscope's six-degree field.

▼ **Determine Field of View** When star-hopping, it is useful to know the field of view of your low-power eyepieces. Aim your telescope at a star on or near the celestial equator. Turn off the drive motor, and let the star drift across the field. An equatorial star takes four minutes to drift one angular degree. Divide the time it takes the star to cross the diameter of the field by four to get the eyepiece's actual field of view in degrees.

Illustrations adapted from TheSky™/Software Bisque

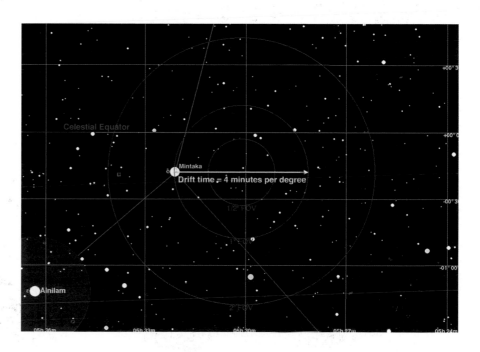

# Star Atlases

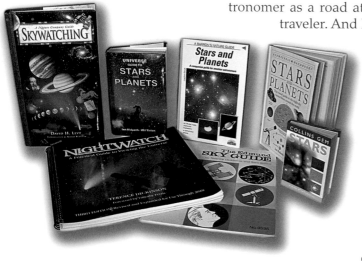

Even if you plan to use a computerized telescope or setting circles, you still need a star atlas. It is as essential to the backyard astronomer as a road atlas is to a highway traveler. And like the traveler, the astronomer can become lost by not selecting the right atlas for the situation.

To plan a cross-country trip requires, first, a national map for an overview, then state or provincial maps for more detail and, finally, regional or city maps for information about congested areas or sites of special interest. Astronomers use the same procedure when exploring the stars with binoculars or a telescope.

For the initial overview, the entire sky must be on one map. One such reference is the monthly circular charts in each issue of astronomy magazines. A more durable alternative is the plastic-laminated "Backyard Stars," a folding set of seasonal maps published in the Klutz series. Also popular are rotating sky charts that can be dialed to show what is above the horizon for any time and night of the year. Of these planispheres, or star wheels as they have become known, our favorite is "The Night Sky," designed by David Chandler and available for about $10 in durable plastic versions. Star wheels are configured for various latitude ranges. Buy the one suitable for your latitude or for the region where you might be traveling—planispheres are great aids to learning an unfamiliar tropical or southern-hemisphere sky.

Despite the proliferation of laptop and handheld computers, don't plan to use computer software in the field. A $2,000 computer isn't as convenient to use outside at night as a $10 planisphere.

Unless you are doing solely unaided-eye viewing, you need more than a single all-sky chart. Star atlases divide the sky into smaller areas that plot the sky's contents in greater detail. Star charts are categorized by their limiting magnitudes. Each increase of one magnitude more than doubles the number of stars and other celestial objects shown but produces a substantially bulkier atlas. More detail also demands a higher level of experience to use the atlas.

## FIFTH-MAGNITUDE STAR ATLASES

For those just beginning their tour of the night sky, several introductory books provide excellent fifth-magnitude star atlases along with plenty of support material. These include *NightWatch* by Terence Dickinson (Firefly Books), *The Edmund Sky Guide* by Terence Dickinson and Sam Brown (Edmund Scientific), *The Monthly Sky Guide* by Ian Ridpath and star-chart master cartographer Wil Tirion (Cambridge) and the *Universe Guide to Stars and Planets* by Ian Ridpath and Wil Tirion (Universe Books). A particularly attractive volume is *Skywatching* by David H. Levy (Nature Company).

Two of our favorite compact guidebooks are both named *Stars and Planets*, one in the Barron's Nature Guide series and one by publisher Dorling-Kindersley, the latter authored by Ian Ridpath. Both have excellent hemispherical monthly charts for learning the sky. A novel and truly shirt-pocket-sized guide is *Stars* in the Collins Gem series, a tiny but surprisingly complete constellation guide that can slip into any telescope or binocular case.

## SIXTH-MAGNITUDE STAR ATLASES

Every backyard astronomer needs a sixth-magnitude star atlas as a basic stargazing tool. In this category, there is none better than Wil Tirion's *Bright Star Atlas* (Willmann-Bell), a superbly practical sixth-magnitude star atlas with 10 large, full-page charts covering the entire sky to magnitude 6.5. Facing each chart are tables of nebulas, clusters, galaxies and double and variable stars found on that chart. All open and globular clusters to magnitude 7 and all galaxies to 10 are shown. Double stars and nebulas are limited to those visible in small

**Fifth-Magnitude Guides ▲**
Introductory guidebooks cost $30 or less and contain charts showing stars down to magnitude 5.0 or 5.5. Most include extensive introductory material as well as lists of hundreds of the brightest double stars and deep-sky objects. These titles count among our favorites.

telescopes. In softcover, it's a bargain at $10.

A step up in attractiveness is the hard-cover, full-color *Cambridge Star Atlas* (Cambridge; $25), with 20 charts to magnitude 6.5 and 900 deep-sky objects plotted. This atlas contains modestly more information than the slimmer *Bright Star Atlas*. Both are highly recommended.

Another choice is the *Mag 6 Star Atlas* by Terence Dickinson, Glenn F. Chaple and Victor Costanzo (Edmund Scientific; $20). The first half of the book provides introductory reference material on telescopes and techniques. The second half is an atlas of the sky to magnitude 6.2, divided among 12 charts reproduced on large, 12-inch-square pages. Three additional charts are blowups of congested areas.

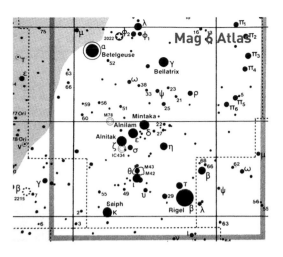

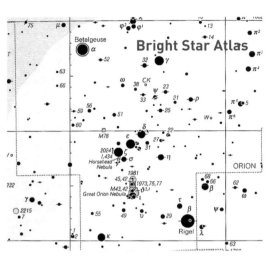

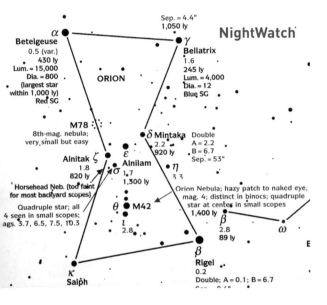

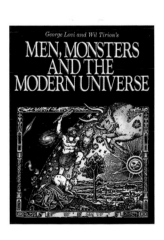

**Comparing Charts**
Star atlases are rated by their magnitude limit. Each increase in the limit corresponds to a significant boost in the amount of detail shown. The chart comparisons on pages 229-231 show a section of the constellation Orion from each of seven atlases, revealing the sky to progressively fainter limiting magnitudes.

The world's best-known sixth-magnitude star atlas is *Norton's Star Atlas*, first published in 1910. The recent edition, rewritten under the supervision of English astronomy writer and editor Ian Ridpath, is so heavily revised, its name has been changed to *Norton's 2000.0* (Longman; $32). There are 15 main maps to magnitude 6.5, each on a two-page, 11-by-17-inch spread, with reference tables on the previous two pages. Unfortunately, the tables still emulate the style established by the original author, Arthur P. Norton, which reflects the observing tastes in vogue early in the 20th century. Variable stars and double stars constitute 70 percent of the objects listed,

making this a good reference for fans of these objects. Clusters, nebulas and galaxies are relegated to the remaining 30 percent. The chart section is followed by 150 pages of tables and reference material.

## SEVENTH-MAGNITUDE STAR ATLAS

Only one volume falls into the seventh-magnitude category: *A Field Guide to the Stars and Planets* by Jay M. Pasachoff, with monthly sky maps and atlas charts by Wil Tirion (Houghton Mifflin). It is in the Peterson Field Guide series, the venerable and highly successful pocket-sized volumes for use by naturalists. Although the format may work for bird watchers, it fails in this instance. Tirion's beautiful seventh-magnitude star atlas is divided into 52 charts, but each is only four by five inches—too small for the detail they contain. The small-page format is unfortunate for a book this com-

▲ **Two Books in One**
A unique and highly regarded reference, *Men, Monsters and the Modern Universe*, by George Lovi and Wil Tirion (Willmann-Bell), contains both the *Bright Star Atlas* and one of the best sky-mythology texts in print.

prehensive. This is regrettable, because there is much useful information here, packed into 575 pages of small type. Priced at less than $20, though, it is an excellent value as a reference work, if not as a practical star atlas.

## EIGHTH-MAGNITUDE STAR ATLAS

Astrocartographic genius Wil Tirion produced the definitive eighth-magnitude atlas with his *Sky Atlas 2000.0* (Sky Publishing and Cambridge). The 26 charts are each 12 by 18 inches. This is a big atlas, but it works. Swatches of the sky, roughly 40 by 60 degrees, are presented. If they were smaller, they would not include enough of any individual constellation to provide a feel for the portion of the sky being examined.

The atlas is available in three formats: a deluxe spiral-bound edition with color-

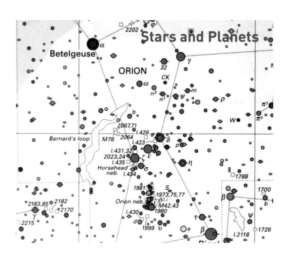

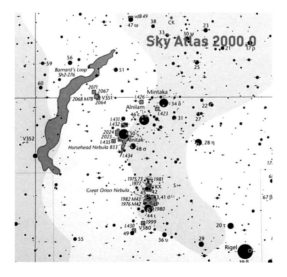

coded charts; smaller individual charts in a desk edition, with black stars on a white background; and the so-called field edition, with white stars on a black background to help preserve night vision at the telescope. At approximately $30, either the desk or the field edition is essential for owners of 8-inch or larger telescopes. The deluxe edition costs about $50. Plastic-laminated versions are available for just over double the price and are great for withstanding dew and abuse in the field—we recommend them. *Sky Atlas 2000.0* is the obvious step up from a sixth-magnitude atlas.

## NINTH-MAGNITUDE STAR ATLAS

Compiling an atlas of stars down to magnitude 9.75 was a monumental undertaking. To accommodate the observing agenda of serious amateur astronomers, more than 300,000 stars had to be plotted, along with tens of thousands of deep-sky objects, to 14th magnitude. The task was first completed in the late 1980s with the publication of *Uranometria 2000.0* by Wil Tirion, Barry Rappaport and George Lovi (Willmann-Bell). The scale of detail necessary meant that to show entire constellations on one chart would require a chart the size of a tablecloth. Obviously, that was impractical, so instead, the sky was divided into 220 double-page charts, printed in two bound volumes on 9-by-12-inch pages.

The package, greatly revised in 2001 for increased accuracy and ease of use, is a tour de force. The new edition adds improved wide-angle key charts plus 26 charts plotting crowded sky regions in close-up detail. Priced at $160 for the two-volume set and complementary one-volume Field Guide catalog, *Uranometria 2000.0* is probably the most advanced printed star atlas that backyard astronomers are likely to require.

## ELEVENTH-MAGNITUDE STAR ATLAS

It is hard to imagine a star atlas more advanced than *Uranometria*, but the *Millennium Star Atlas* is it. Produced by Sky Publishing and the European Space Agency, this three-volume set uses the Hipparcos satellite's superaccurate star-mapping data to plot more than one million stars down to magnitude 11. Depicting the whole sky requires 1,548 charts in three 9-by-13-inch volumes. The image scale is so large that each chart covers only 5.4 by 7.4 degrees, an area of the sky equivalent to the area of France on a map of the world. Deep-sky fanatics and lovers of star charts must have this set. But at $250, the *Millennium Star Atlas* is too luxurious to be used in the field and is likely to be bypassed by many deep-sky observers in favor of custom printouts from advanced software programs such as Guide, MegaStar or TheSky.

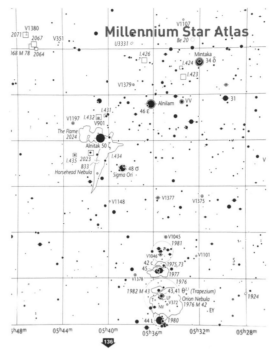

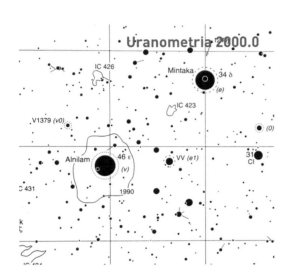

# Helpful Finder Cards

Two great aids make hunting deep-sky objects a snap yet go largely unappreciated by most observers. Brent Watson's "Sky Spot" books of charts (top right) feature one object per chart, highlighted with a Telrad reticle. The charts are plastic-laminated and spiral-bound for durability in the field. They are available from Sky Spot Publishing, 1263 East Beverly Way, Bountiful, UT 84010, for $25 to $50 each.

Covering many more objects, George Kepple's wonderful Astro Cards (bottom right) feature one main deep-sky object and several nearby objects on each three-by-five-inch index card. Each card has a wide-angle chart indicating where to look in the constellation, plus an adjacent chart covering a finderscope field. The cards are easy to handle at the eyepiece and can be flipped over to match a right-angle finderscope (by shining a flashlight through them). There is a set of cards for the Messier objects and two sets for fainter deep-sky targets. At $10 per set of 70 cards, they are a bargain. They can be ordered from Astro Cards, Box 35, Natrona Heights, PA 15065.

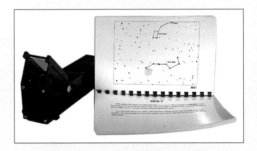

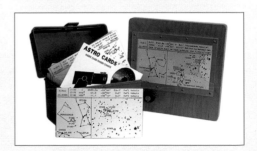

 *A recommended accessory for the Astro Cards is the backlit Cardlighter ($40), which holds and illuminates one card at a time.*

# Exploring the Deep Sky

Professional astronomers at mountaintop observatories do not look through telescopes. The instruments they use are giant cameras recording starlight onto electronic detectors. Seeing the subtle light from distant galaxies and nebulas "live" through an eyepiece is now the exclusive domain of amateur astronomers. As such, today's backyard stargazers share a common bond with the great visual observers of the 19th and early 20th centuries, who made their discoveries during all-night sessions at the eyepiece.

Through an amateur telescope, a cluster of galaxies might appear only as a field of faint fuzzies at the threshold of vision. Unimpressive at first glance, but not when you realize that each of those indistinct spots is another Milky Way, filled with stars, planets and, perhaps, curious minds like ours. Deep-sky observing is done as much with the mind as it is with the eye.

Deep-sky exploring takes us through the star- and nebula-packed spiral arms of our own galaxy, such as this rich region in Cygnus, then beyond, into the realm of the multitude of galaxies within reach of backyard telescopes. Photo by Alan Dyer.

**Glowing Nebula** ▶
The Trifid Nebula shines
in red and blue light.
CCD image by Chris Schur.

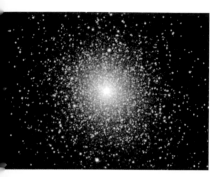

**Glowing Nebula** ▶
The Trifid Nebula shines
in red and blue light.
CCD image by Chris Schur.

# Geography of the Sky

The deep end of the sky officially starts at the edge of the solar system and extends out to clusters of galaxies and enigmatic quasars. Taken literally, it encompasses everything in the universe except our Sun and its family of worlds. The deep sky includes the many types of stars that populate the night sky. And yet, when amateur astronomers speak of deep-sky objects, they are usually referring to extended objects: the star clusters and nebulas of our own Milky Way Galaxy and the many types of galaxies that lie beyond the Milky Way.

## DEEP-SKY ZOO

Each of the thousands of objects within reach of backyard telescopes can be classified into one of half a dozen species in the deep-sky menagerie:

❖ Open Star Clusters
Open star clusters are congregations of stars bound together by their mutual gravity. Individual stars in an open cluster were all born about the same time from a nebula

similar to the Orion or Trifid Nebula. Roughly 1,200 open star clusters have been cataloged; most are easily accessible to backyard telescopes. About 10 to 25 light-years across in the prime of their lives, open clusters eventually disperse, scattering their member stars along the arms of the Milky Way.

❖ Globular Star Clusters
Like miniature spherical galaxies, globular star clusters contain hundreds of thousands of ancient suns packed into a space about 25 to 250 light-years wide. Approximately 150 have been found associated with the

**Globular Cluster** ▲
Omega Centauri rates
as the best globular by
far. Such clusters of
millions of stars are
stunning sights in large-
aperture telescopes.
Photo by Matt BenDaniel.

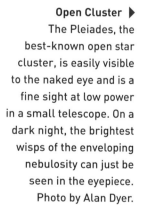

**Open Cluster** ▶
The Pleiades, the
best-known open star
cluster, is easily visible
to the naked eye and is a
fine sight at low power
in a small telescope. On a
dark night, the brightest
wisps of the enveloping
nebulosity can just be
seen in the eyepiece.
Photo by Alan Dyer.

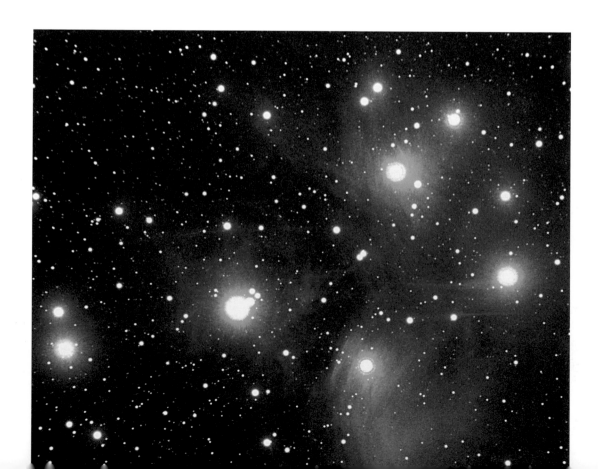

Milky Way Galaxy, the majority being suitable targets for amateur telescopes. Most are ancient, having formed 9 billion to 11 billion years ago as by-products of the creation of the galaxy itself. The nearest globulars lie several thousand light-years away, toward the core of the galaxy.

### ❖ Star-Formation Nebulas

Stars form from contracting clouds of interstellar dust and gas. These cold regions of hydrogen gas and complex molecules line the spiral arms of the Milky Way, where shock waves from nearby supernovas can trigger the initial collapse of the nebula. As stars begin to form in the densest regions of the nebulas, their ultraviolet starlight energizes the surrounding gas, creating the visible nebula. Most nebulas stretch a few dozen light-years across space, though the largest, the Tarantula Nebula, holds the record, at 900 light-years across.

### ❖ Planetary Nebulas

At the end of their lives, stars with one to six times the Sun's mass lose weight by blowing as much as one-quarter of their mass into space through a continuous stellar wind. The process takes thousands of years and often involves several episodes of stellar belching, which creates complex nebula structures as fast-moving gas shells run into older, slower-moving shells. The

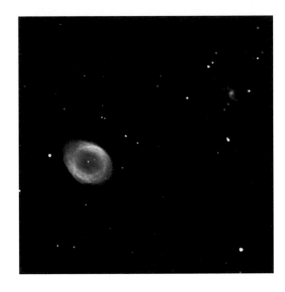

aging star then shrinks to a hot white dwarf the size of a small planet. About 1,500 planetary nebulas have been cataloged in our region of the galaxy.

### ❖ Supernova Remnants

Very massive stars end their lives abruptly. In just minutes, 90 percent of a star's mass is blasted into space, while the remaining core collapses into a superdense neutron star or, perhaps, a black hole. In the process, the star gives off as much energy as an entire galaxy—a spectacular finale to a star's life, but a rare one. Only a few of the most

◀ **Planetary Nebula**
The Ring Nebula may well be the first deep-sky object many stargazers see through a telescope. It is easy to find in Lyra the harp and is bright enough to pick out even in light-polluted skies. CCD image by Chris Schur.

◀ **Supernova Remnant**
Off the east wing of Cygnus the swan lie these feathery arcs of light, the remains of a star that exploded thousands of years ago. A nebula filter makes the Veil Nebula components an easy sight, even in an 80mm telescope. CCD image by Robert Gendler.

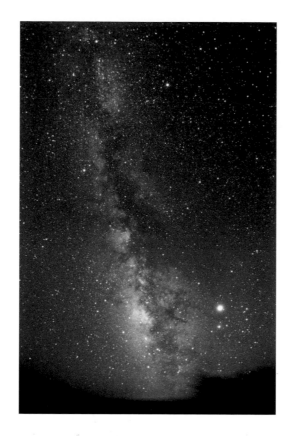

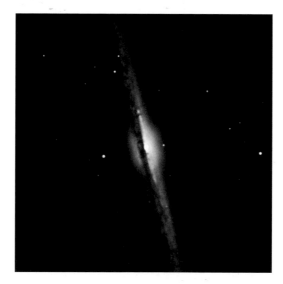

massive stars are supernova candidates. During historic times, just a handful of stars in our section of the galaxy have exploded as supernovas, and just a fraction of those have left visible nebulas.

❖ The Milky Way

The term Milky Way can be a little confusing. Everything we can see in the night sky easily with the unaided eyes (with the exception of the Andromeda Galaxy and the Magellanic Clouds) belongs to our spiral-shaped galaxy. We call the entire galaxy the Milky Way. But in its original sense, the term referred only to the gray band of light arching across the summer, fall and winter skies (*Via Lactea* in Latin). Greek mythology explains its origin in a tale of Hercules spilling milk over the sky. It took Galileo to discover that this milky band is actually made of stars. The stars we see with our unaided eyes all lie nearby, but the ones forming the Milky Way band are much more distant, their light blending together and glowing from thousands of light-years away in the spiral arms of the Milky Way.

❖ Galaxies

Our Milky Way, with its stars, clusters and nebulas, is but one of tens of billions of galaxies—spirals, ellipticals and odd-shaped irregulars—that come in all sizes, from dwarfs to giants. In fact, the stars we think of as countless are really just a scant foreground clutter between us and the real universe—a space tangled with galaxies that clump together into clusters, the clusters forming strandlike superclusters. Galactic superclusters are the largest gravitationally bound structures in the universe.

## MAKING SENSE OF THE SKY

Deep-sky objects aren't scattered at random about the heavens. Each species in the deep-sky zoo inhabits its own territory in the sky.

Nebulas line the spiral arms of our galaxy, so that's where we find them, almost exclusively along the Milky Way band. The vast majority of open star clusters, being the products of star-forming nebulas, also lie along the Milky Way.

Most globular clusters live in a halo thousands of light-years across that surrounds the distant core of the Milky Way. From our vantage point near the edge of the galaxy, we look toward the center and see a region of sky in Sagittarius and Scorpius populated with globulars, like bees buzzing around a distant hive.

Galaxies, objects that lie beyond our Milky Way, inhabit the parts of the sky where the Milky Way isn't. We see few galaxies embedded along the Milky Way band, not because they aren't there but because they are hidden or dimmed by the mass of star-stuff that makes up our galaxy.

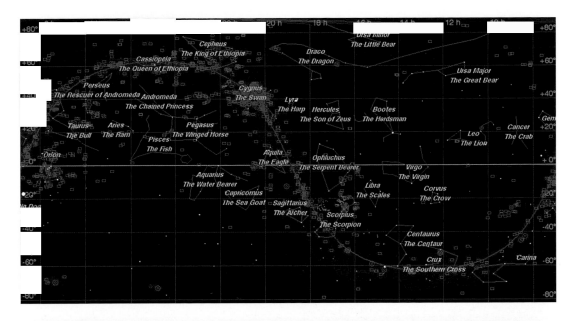

### Star-Forming Nebulas
Glowing regions of star formation, the pink objects, line the spiral arms of the Milky Way, the S-shaped band on this map. Nebulas populate the Milky Way constellations of Cygnus, Cassiopeia, Cepheus, Auriga, Orion and Carina and also the dense regions surrounding the center of the Milky Way in Scorpius and Sagittarius. Few nebulas are found off the Milky Way.

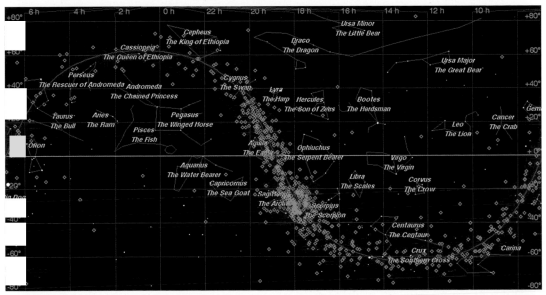

### Planetary Nebulas
Also residents of our galaxy, these stellar remnants, marked in green, primarily dot the Milky Way band, with a heavy concentration toward the center of the galaxy. However, nearby planetaries, such as the Helix Nebula and the Owl Nebula, can be found far off the Milky Way, in our local neighborhood of surrounding stars.

Diagrams on this page and on page 238 courtesy Voyager III®/ Carina Software

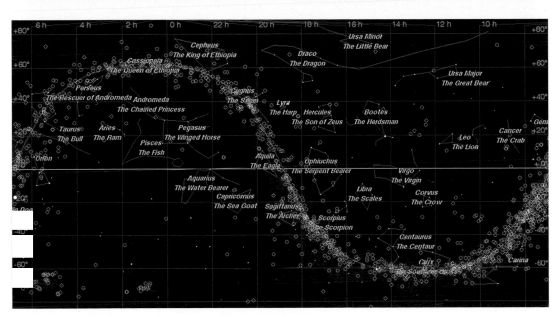

### Open Clusters
Like planetaries, most open clusters, marked in yellow, are confined to the Milky Way. However, some open clusters lie close by. These nearby clusters appear large enough to see with unaided eyes (for example, the Hyades, 150 light-years away; Coma Berenices, 260 light-years; the Pleiades, 410 light-years; and the Beehive, 525 light-years). These few nearby clusters sit well off the plane of the Milky Way.

**Globular Clusters ▶**

Because the galactic center lies far away in Sagittarius, most globulars (the blue dots) inhabit a wide circle centered on the Sagittarius and Scorpius region of the northern summer sky. Some lie superimposed on the Milky Way, but most float above or below the plane of our galaxy. We see few globulars in the northern winter sky and just a handful in the northern spring and autumn skies.

**Galaxies ▶**

In spring and autumn, we look out the top or bottom of the thin disk of our galaxy, rather than deep into its arms as we do in winter and summer. We look through the thinnest amount of obscuring galactic dust and gas, allowing us to see other galaxies in the distance. The rich concentration of galaxies on the right side of this map is the nearby Coma-Virgo galaxy cluster that dominates the northern spring sky.

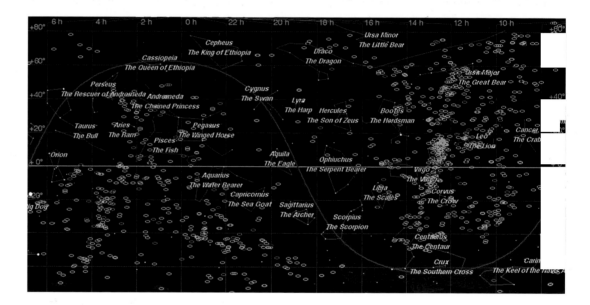

# Running the Messier Marathon

Seeing all 110 objects in Charles Messier's catalog in one night seems an impossible feat. And yet that's just what some avid observers attempt to do.

The period between March 5 and April 12 is prime season, with the ideal nights being March 30 to April 3 if the Moon is new. On marathon night, the autumn Messiers provide the targets in the early evening, with participants hastily picking M74, M77, M33 and M31 plus its companions out of the evening twilight. The race then turns to easier winter targets. After a midnight breather, the marathoners move on to Heartbreak Hill: the multitude of spring-sky galaxies. From there, it's a downhill run to the dawn sky and the congestion of the final mile in Sagittarius. M30 in Capricornus is the elusive prize, rising before dawn.

To plan a Messier marathon, the best reference is the superb observing guidebook *The Year-Round Messier Marathon Field Guide* by Harvard C. Pennington (Willmann-Bell, Inc., 1997). Its finder charts are helpful for Messier hunting at any pace.

# Diving Into the Deep Sky

You'll never run out of things to look at in the realm of the deep sky. From objects so large they can be seen with the naked eye to objects so small and faint they require a giant 24-inch telescope, the universe has much to offer. While deep-sky observing can be done with any instrument, it is hard to discount the need for sheer aperture.

Bigger telescopes resolve globular clusters into swarms of pinpoints. More aperture makes nebulas appear brighter and more like their photographs and turns galaxies from ill-defined spots into unmistakable spirals mottled with dust lanes. A field that barely shows one or two galaxies through a small telescope turns into a cluster of dozens of galaxies when viewed through a massive 24-inch reflector.

However, sheer brute-force aperture is not the sole requirement for deep-sky exploring. More important is the honing of some key observing skills.

## LEARNING TO SEE

No matter what telescope you use for exploring the deep-sky realm, a few techniques can help you see more than you might think possible.

❖ **Cultivate Night Vision**
Becoming dark-adapted is essential for seeing the faint fuzzies that make up the deep sky. Don't expect to see much through the eyepiece if you've just stepped outside from the bright indoors. Similarly, porch lights shining into your eyes will make it impossible to see faint objects. An initial dark adaptation, during which the pupil dilates to its maximum aperture, takes place in 10 to 15 minutes. But over another 15 to 20 minutes, a chemical reaction in the eye kicks in that further boosts its sensitivity. A single shot of white light can ruin all the hard-won improvement in night vision. One trick is to fashion a "monk's hood" to wear to block light.

❖ **See by Not Really Looking**
A faint object stands out better if you look just to one side of the object and not straight at it. Such averted vision places the object on the more sensitive rod cells around the periphery of the retina. Seeing an object by not really looking at it sounds odd, but it really works.

❖ **Practice Jiggle Vision**
Another trick is to jiggle the telescope. The slight motion in the field can reveal dim targets that might otherwise fade into the background. The trick works because our eyes and brain have been trained since the Pleistocene epoch to pick out nasty things moving in the night.

❖ **Apply Power as Needed**
For all the proscriptions against using high power on telescopes, sometimes power is just what you need. Boosting the magnification darkens the sky while enlarging the object enough to make it more obvious.

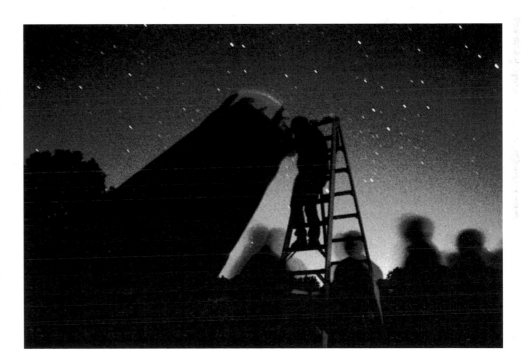

Small, faint galaxies invisible at 50x may pop into view at 150x. Magnifications of 100x to 150x are typical for most deep-sky viewing, with lower powers reserved for the biggest objects and for finding targets. On large-aperture scopes, even higher power is ideal for revealing faint planetary nebulas, dim open clusters and small globular clusters.

❖ **Head Out on the Highway**
The best accessory for any telescope is a dark sky. This is especially true of small telescopes. Well away from the interference

▲ **Getting High on the Deep Sky**
Big-aperture reflectors on Dobsonian mounts, with eyepieces often accessible only by ladder, are favored by deep-sky observers, but any telescope of any size can be used to explore the realm beyond our solar system.

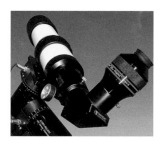

of city lights, even an 80mm refractor can show all the brightest members of the Coma-Virgo galaxy cluster.

❖ **Don't Give Up on the City**
Confined to the city? The brighter Messier objects, especially star clusters, are visible even through the murk of urban sky glow. Double stars also make suitable targets for the urban stargazer. "Go To" telescopes can locate the hard-to-find targets that are invisible in finderscopes and therefore difficult to find by star-hopping.

❖ **Drawing Upon Experience**
You may have already had this happen: At a star party, a veteran observer invites you, the novice, to look at this wonderful nebula. You look—and see nothing! The fault lies not in the stars or the equipment but in your inexperience. Seeing the faintest targets requires training the eye. As Gregg Thompson describes in his guest essay (page 265), the best way to learn to see is to draw what you see. Over time, you'll be amazed at the improvement in your visual acuity. Experience, often gained by sketching, and a dark sky are more important than aperture for seeing to the limits of the sky.

❖ **Log the Sky**
Even if you choose not to cultivate your artistic talents, start a deep-sky journal. Check off the objects you find, and note the ones you didn't locate. Record your visual impressions and the sky conditions. The act of recording your observations in any form helps you learn the sky.

## Low-Power Limit

Just as telescopes have a high-power limit, a boundary also exists at the low end of the power range. Use a reflecting telescope below its low-power limit, and you may see the dark shadow of the secondary mirror floating in the field. Use any telescope below its low-power limit, and you are effectively cutting down its aperture, as the cone of light leaving the eyepiece will be wider than the dark-adapted eye can accept. Not all the incoming light can make it into your eye.

To calculate the longest-focal-length eyepiece (i.e., the lowest power) you can use at night, multiply the telescope's focal ratio by 7mm (the diameter of a dark-adapted pupil) or, for older observers, 6mm.

| Telescope's focal ratio | Longest eyepiece usable |
|---|---|
| f/4 | 28mm |
| f/4.5 | 32mm |
| f/5 | 35mm |
| f/6 | 42mm |
| f/8 | 56mm |

# Inventories of the Sky

Using computerized telescopes, backyard stargazers can now call up the location and identity of thousands of deep-sky objects from a variety of digital catalogs. These inventories of the sky are the result of 200 years of meticulous observations. In the 18th and 19th centuries, explorers of the sky hunted down, charted and cataloged every-

thing within reach of their often crude telescopes. The result was a series of celestial catalogs, complete with baffling nomenclature, that are still very much in use today.

## MESSIER'S CATALOG

The most famous deep-sky catalog provides backyard astronomers with a ready list of the sky's best and brightest deep-sky targets. Ironically, it was compiled as a catalog of things *not* to look at.

In the late 1700s, Charles Messier did not set out to find deep-sky targets. To him, they were merely nuisance objects he kept bumping into during his searches for comets. He published lists of the fuzzy non-comets so that he and his comet-hunting colleagues would not be fooled by them. Today, Messier's comet discoveries are largely forgotten. It is his list of nuisance objects that is remembered.

Among the best-known Messier objects are M45, the Pleiades star cluster; M31, the Andromeda Galaxy; M13, a globular cluster in Hercules; and M42, the Orion Nebula. Modern versions of Messier's catalog contain 110 objects, providing a selection of the finest deep-sky targets for northern-hemisphere observers.

The identity of several Messier objects has often been questioned. Evidence exists that M91 and M102 are mistaken observations of M58 and M101, respectively. The objects M104 to M109 were discovered by an associate, Pierre Méchain, and reported to Messier but were never included in a published version of his catalog. These M objects are really "Méchain objects." NGC205, one of M31's companion galaxies, was apparently logged by Messier, though he never listed it in his catalog. Modern-day observers have dubbed this object M110.

Purists sometimes reduce the Messier list to 99 or 100 objects. A version compiled by coauthor Dyer for the annual RASC *Observer's Handbook* lists a full 110 entries, including two faint galaxies that some astronomers have suggested as substitute candidates for M91 and M102.

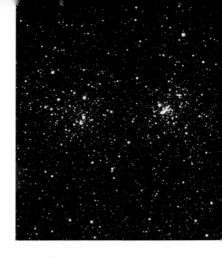

▲ **Twin Clusters**
One of the brightest and best-known deep-sky objects, the Double Cluster in Perseus, was missed or ignored by Charles Messier. Why that is so remains a mystery. Photo by Alan Dyer.

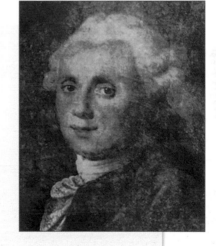

# Charles Messier, 'Ferret of Comets'

On September 12, 1758, French astronomer Charles Messier was tracking a comet he had found when he came across something unexpected in the sky. He called it "a nebulosity above the southern horn of Taurus. It contains no stars; it is a whitish light, elongated like the flame of a taper." Messier wasn't the first to see the object; English astronomer John Bevis had logged it 27 years earlier. But Messier's rediscovery of what eventually became known as the Crab Nebula inspired him to inventory more "comet masqueraders," objects he and others might mistake for the real celestial prize of the day: comets.

From a rooftop observatory at the Hotel de Cluny in Paris, Messier discovered 13 comets from 1760 to 1798, winning him the title of "ferret of comets" from King Louis XV. Messier's first "Catalogue of Nebulae and Star Clusters," published in 1771, contained 41 objects that he and his colleagues had discovered. For good measure, Messier added four well-known objects, the Orion Nebula complex (M42 and M43), the Beehive cluster (M44) and the Pleiades (M45), bringing his first list to a tidy 45 objects. Messier published revised versions of his catalog in 1783 and 1784, bringing the total inventory of published Messier objects to 103. Subsequent additions bring the total to 109 or 110.

The Messier numbers follow a haphazard sequence across the sky, because the objects are numbered in the order Messier located them or learned of them. Although he intended to, he never did publish a list with entries renumbered in order of right ascension, west to east, across the sky. Illness, old age and the French Revolution intervened.

........................................................................................................................

  *The largest telescopes used by Charles Messier were 190mm and 200mm reflectors. However, their mirrors of polished speculum metal would have had the light-gathering power of a modern 80mm-to-100mm reflector. Portrait courtesy Owen Gingerich.*

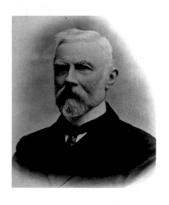

From a rural site, all the Messier objects can be seen with an 80mm telescope and many with only 7x50 binoculars. The largest telescope Messier himself used was a 6-inch reflector. Tracking down the Messiers over the course of a year or two is a rewarding endeavor. In the process, you will become familiar with the sky, learn how to see faint objects through the telescope and gain enough credits to graduate to the level of an experienced observer.

## THE NGC AND IC

Most deep-sky enthusiasts complete the Messier list. But then what? The next goal is collecting NGC objects.

NGC stands for New General Catalogue, now more than a century old. It was originally compiled by Danish-born astronomer John Louis Emil Dreyer under the aegis of the Royal Astronomical Society in England. Published in 1888, Dreyer's *New General Catalogue of Nebulae and Clusters of Stars* compiled observations of 7,840 objects from dozens of observers and replaced all previous lists and catalogs. Even the Messier objects were assigned NGC numbers. The NGC contained every nebula and cluster known in 1888. The fact that many of the "nebulas" were actually galaxies was unknown at the time; anything that could not be resolved into stars was called a nebula.

Unlike the random nature of the Messier catalog, NGC objects are all neatly ordered by right ascension. The numbering starts at 0 hours right ascension (or what was 0 hours right ascension in 1888) and increases from west to east across the sky. However, successively numbered NGC objects can be separated by many degrees north or south in declination.

Soon after the NGC was published, it required revision. Supplementary Index Catalogues were published in 1895 and 1908. Objects labeled IC (or simply "I") are

# The Herschel Dynasty

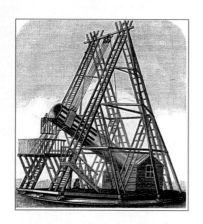

Few families have had an impact on science as significant as the Herschels' influence on astronomy. In the 1770s, William Herschel, a musician by profession, took up the craft of telescope making. His Newtonian reflectors, all with speculum metal mirrors, proved so superior to any of the day that Herschel became the equivalent of a millionaire just from telescope sales. On March 18, 1781, Herschel used his 6.5-inch reflector to sweep up what he at first thought was a comet. It proved to be Uranus, the first planet discovered in historic times. England's King George III appointed Herschel his private astronomer. Set for life, Herschel built ever larger reflectors, culminating in 1789 with a 48-inch telescope, which he used to survey the sky.

Herschel's principal assistant on his night shifts was his sister Caroline. In her own explorations of the sky, Caroline Herschel found eight comets and is credited with finding M110, one of the companions to the Andromeda Galaxy. Caroline was also instrumental in the publication of the catalogs of nebulas and clusters that the brother-and-sister team had found and logged.

William's only son, John, continued the dynasty, discovering more than 2,000 new deep-sky objects, many during a stint in the southern hemisphere. In 1864, the younger Herschel published a compilation of the family's life-work, the General Catalogue, an exhaustive listing of 5,000 nebulas and clusters. The master of visual astronomy, John Herschel became one of the pioneers of photography. The first photo ever taken on a glass plate is John's 1839 image of his father's soon-to-be-dismantled 48-inch telescope on the grounds of their British country home.

· · · · · · · · · · · · · · · · · · · · · · · · · · · · · · · · · · · · · · · · · · · · · · · · · · · · · · · · · · · · · · · · · · · · · · · · · · · · · · · · ·

*William Herschel referred to it as his 40-foot reflector, in keeping with the old tradition of describing a telescope by its focal length, rather than aperture. From his backyard in Slough, England, Herschel found hundreds of deep-sky objects with the 40-footer.*

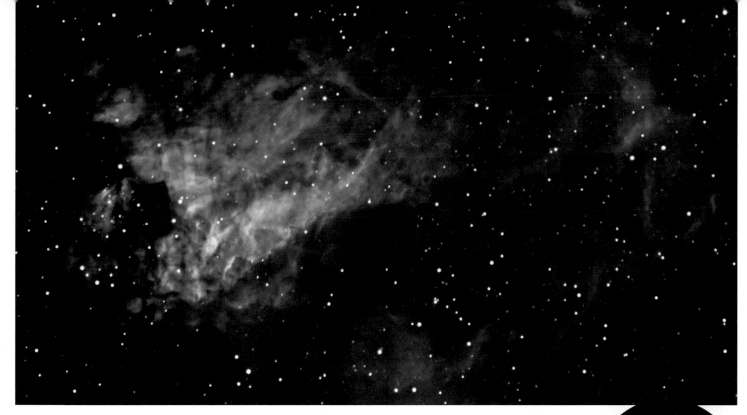

from one of those listings. The first IC contains 1,529 objects discovered visually between 1888 and 1894.

The second IC offers another 3,856 entries, many found between 1895 and 1907 through the new technique of photography. Most of these later IC objects (with numbers higher than 1529) are too faint to detect visually. They are I-don't-see objects!

The brightest NGCs are easy targets for an 80mm telescope. But a 5-to-8-inch telescope is a minimum to dive deeply into the NGC list.

## HERSCHEL'S CATALOG

Avid deep-sky observers wanting to venture beyond the Messier catalog can tackle hunting down 400 of the best Herschel objects. Prompted by a suggestion from observer and author James Mullaney in the 1970s, members of the Ancient City Astronomy Club in St. Augustine, Florida, embarked on a project to sort out the original core objects of the NGC, a total of 2,477 objects first observed by William Herschel in the late 1700s. From this, the club sifted out the 400 best, the so-called Herschel 400. All are known by NGC numbers but also carry Herschel numbers from his original catalogs, which predate the NGC by almost a century.

Herschel's catalog numbers and system for classifying deep-sky objects are now largely obsolete and disused, but a sampler of his original discoveries lives on for those who tackle the Herschel 400. Because William Herschel observed from England, this "best of" subset of the NGC contains only northern-sky objects. More details can be found at the Astronomical League's website and in its publications (for links, see www.backyardastronomy.com).

## CALDWELL CATALOG

In 1995, a new deep-sky best-of list appeared in the hobby, largely through the promotion of *Sky & Telescope* magazine. Patrick Moore, Britain's best-known astronomy author, presented his list of 109 of the best and most notable non-Messier objects, selected primarily from the NGC. Rather than have his list also contain M objects, he titled his list after his full last name, Caldwell-Moore. So we have Caldwell 1, or C1, better known as NGC188, a star cluster near the north celestial pole. Caldwell 2 is NGC40, a planetary nebula in Cepheus, and so on. The objects are arranged from north to south, in order of decreasing declination, and include southern-sky targets.

The Caldwell catalog sounds like a helpful idea for venturing beyond the Messiers.

▲ **Celestial Swan**
The Swan Nebula, aka the Omega Nebula, is one of the brightest nebulas in the Messier catalog. The CCD image by Robert Gendler (top) reveals its full extent and structure. The eyepiece sketch by John Bianchi (above) shows what you will see through a small 80mm telescope under dark skies.

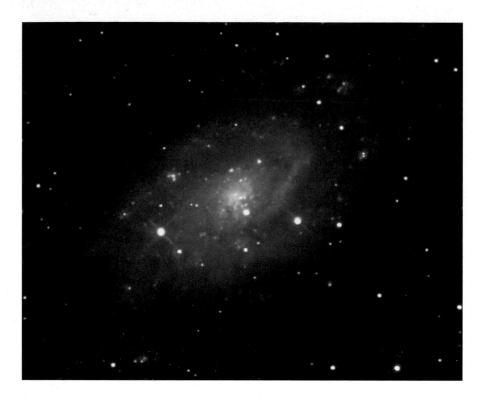

**Classic Spiral Galaxy** ▲
NGC2403 in the obscure northern constellation of Camelopardalis rivals most Messier galaxies for size and brightness and is an easy target for any telescope. This object is also known as Caldwell 7 in Patrick Moore's listing of the most notable non-Messier objects. CCD image by Chris Schur.

However, the list has rankled many deep-sky observers, who take exception to the relabeling of objects already well known by established catalog numbers. The other problem is that some of the Caldwell objects, though ostensibly the best in the sky, fall far short of that criterion. Caldwell 5 is IC342, a diffuse spiral galaxy that would challenge most small-scope observers. The same is true of Caldwell 17 (NGC147), a hard-to-see companion galaxy to M31; Caldwell 31 (IC405), a dim nebula in Auriga; and Caldwell 51 (IC1613), a faint Local Group galaxy in Cetus, to name a few. Errors in the positions, sizes and magnitudes of objects also plague many print and electronic versions of Moore's list.

Both of us prefer not to use this catalog. In the 1980s, coauthor Dyer prepared a list of the "110 Finest NGC Objects," published each year in the annual RASC *Observer's Handbook*. Another good source of selected objects is the Orion *DeepMap 600*, which plots and lists all the Messiers, plus 100 of the best variable, double and colored stars and about 400 superb non-Messier objects, all selected by Steve Gottlieb, one of the world's most experienced deep-sky observers.

Yet another fine source is the privately published *Celestial Harvest: 300-Plus Show-*

*pieces of the Heavens* by veteran skywatcher James Mullaney. Mullaney's choice of objects and the descriptions compiled from observers over the past century provide a wonder-filled introduction to the sky. Complement your computerized telescope with one of these recommended sky guides, and you'll have the information and tools to direct you to great deep-sky views.

## BEYOND THE NGC

Even with its several thousand entries, the New General Catalogue ignores an entire class of object: the dark nebula. William Herschel commented on dark "starless spots" he encountered as he swept the skies in the late 1700s, but it was left to American astronomer Edward Barnard to compile the first catalog of these regions, or B objects, as they are called. Barnard's "Catalogue of 349 Dark Objects in the Sky" was included in his 1927 work *A Photographic Atlas of Selected Regions of the Milky Way*.

In the class of emission nebulas, advanced amateur astronomers pursue elusive objects with prefixes such as Ced (from S. Cederblad's 1946 list), Mi (Minkowski), Sh2 (Sharpless), vdB (van den Bergh) and Gum (from Colin Gum's 1955 nebula survey). Most of the non-NGC nebulas are very small or very large. All are extremely faint.

With the advent of nebula filters, experienced amateur astronomers routinely pick off non-NGC planetary nebulas once thought to be invisible. Many come from Perek and Kohoutek's 1967 *Catalogue of Galactic Planetary Nebulae*, thus the PK designations. Most are 13th to 16th magnitude. A subset of this group is Abell planetaries, from George Abell's list of some 100 large, faint planetaries he discovered by examining photographs taken with the Palomar 48-inch Schmidt telescope in the 1950s.

Detailed star atlases plot open clusters with such prefixes as Be (Berkeley), Cr (Collinder), Do (Dolidze), H (Harvard), K (King), Mel (Melotte), Ru (Ruprecht), St (Stock) and Tr (Trumpler). Many of these non-NGC clusters were missed by earlier eyepiece surveys because they are either very large or so sparse they do not stand out from the background star field.

Over the past few decades, specialized surveys at research observatories have cata-

loged thousands of galaxies missed by the compilers of the NGC. Advanced star atlases and computer programs plot galaxies labeled UGC (from the 1973 Uppsala General Catalogue of Galaxies ), MCG (from the Morphological Catalogue of Galaxies compiled from 1962 to 1974) and ESO (from the European Southern Observatory's 1982 catalog of faint southern-sky galaxies). Many star-atlas programs for home computers extract their galaxy data from the PGC (Principal Galaxies Catalogue), a 1989 reworking of earlier galaxy catalogs.

There is even a catalog of galaxies not included in any catalog: the MAC, or Mitchell's Anonymous Catalogue. This is amateur astronomer Larry Mitchell's compilation of 27,000 officially unnumbered galaxies he has found on Palomar Observatory Sky Survey (POSS) images.

Large Dobsonian telescopes now allow amateur astronomers to sight entire clusters of galaxies whose existence was unknown until the advent of photographic surveys. The main catalog here was compiled by George Abell in the 1950s. The brighter Abell clusters (charted as A objects, but often with NGC or UGC galaxies as their brightest members) appear as fields of glowing spots through large reflectors.

# Deep-Sky Tour One: The Stars

Not all deep-sky targets need superdark skies. From the confines of a city, some of the best sights are the many types of stars.

▲ **Cluster Country**
Cassiopeia contains clusters of clusters. Here, M103, far right, is outsized by NGC663, far left, paired with NGC659, bottom. Trumpler 1 is the clump right of center. NGC663 is also known as H VI 31 in Herschel's catalog.

◀ **Can You See the Coathanger?**
A large and unique star cluster missed by the Messier and NGC catalogs lies just south of the star Beta Cygni. Officially listed as Collinder 399, this binocular object is also known as Brocchi's Cluster, or the Coathanger. Can you see it hanging upside down in a dark lane of the Milky Way? Both photos by Alan Dyer.

## DOUBLE STARS: STELLAR JEWELS

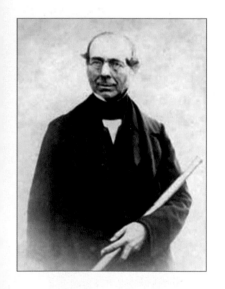

**Colorful Double** ▲
Albireo, also known as Beta Cygni, is one of the sky's best double stars. Its gold and blue components are easy to split in any telescope, even at low power. Eyepiece sketch by John Bianchi.

The majority of stars (estimates suggest at least 50 percent) aren't single but belong to systems of two or more stars orbiting each other, the result of a common origin in a spinning cloud of gas and dust.

Many double stars have measured orbital motions, though it may take centuries for the stellar partners to dance around each other. Some doubles are merely traveling together through space (dubbed CPM, or Common Proper Motion).

Some doubles don't belong together at all and are merely coincidental lineups in the sky. The best example of this type of optical double is Zeta (ζ) Ursae Majoris, better known as Mizar, the middle star in the handle of the Big Dipper. Its naked-eye companion, Alcor, lies three light-years farther away than Mizar, too far away to be in orbit. However, Mizar itself is a true physical double whose components, 14 arc seconds apart, are resolvable in any telescope.

In our opinion, the best double stars fall into two categories: strongly colored pairs (yellow and blue stars, for instance) and pairs with close but nearly equal-brightness stars (usually white or blue-white) that look like a pair of headlights in the eyepiece. Gamma (γ) Arietis and Gamma (γ) Virginis are classic "headlight" doubles. Doubles with too wide a range in magnitude or separation lose their attractive twin nature.

Double stars have been compiled into their own catalogs. One widely used list is the Aitken Double Star catalog, published by Robert Grant Aitken in 1932—thus the ADS numbers assigned to double stars in some "Go To" telescope databases. (The ADS contains 17,180 entries.) Famed double-star observers such as Wilhelm and Otto Struve and Sherburne Burnham compiled earlier lists in the 19th century. Some stars are still best known by their

# Eagle-Eyed Dawes

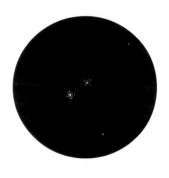

"Using a capital little refractor of only 1.6 inches aperture…I worked away almost every night, when uncertain health would permit, and found and distinctly made out…Castor, Rigel, ε¹ and ε² Lyrae, σ Orionis, ζ Aquarii and many others….The difficulty was often to get to bed in summer before the Sun extinguished the sight of the game."

Such dedication earned England's William Rutter Dawes the reputation of being one of the finest observers of the mid-1800s—that and his eagle eyes. Try splitting those double stars with a 1.6-inch refractor, and you'll appreciate Dawes' remarkable acuity. With a series of larger refractors, the Reverend Dawes (he was an ordained minister) performed measurements of double-star positions so accurate, they are still relied upon today. Seeking ever better telescopes, Dawes became the first important customer for refractors made in America by Alvan Clark. Dawes' endorsement established the Clark enterprise, which went on to create the largest refractors in the world.

In today's amateur circles, Dawes is best known for his rule on resolving power: "I thus determined…that a one-inch aperture would just separate a double star composed of two stars of the sixth magnitude, if their central distance was 4.56"….Hence, the separating power of any given aperture, a, will be expressed by the fraction 4.56"/a." We still use this rule of thumb today for calculating the resolving power of a telescope. But keep in mind that this empirical rule was devised for stars of equal and moderate brightness. Double stars whose components exhibit a wide range in brightness may be harder to split.

 *For all his eagle-eyed ability at the telescope, William Dawes was so nearsighted that legend has it he could pass his wife on the street and not recognize her. Portrait courtesy Royal Astronomical Society.*

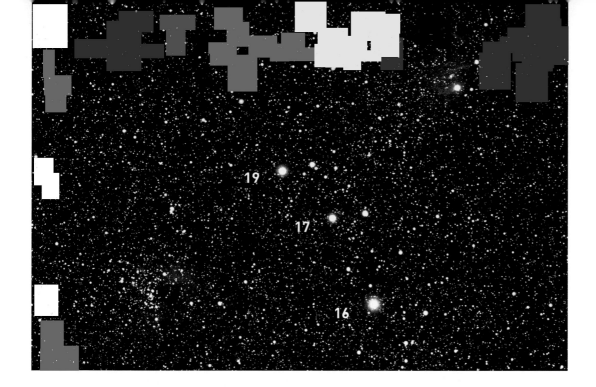

## Letters

Astronomers class stars by the arcane sequence of spectral letters, easily remembered from hot to cool by the mnemonic: Oh Be A Fine Guy/Girl Kiss Me. (All temperatures are in Kelvin, with 0°K equal to absolute zero at –273°C.)

**SPECTRAL CLASSES**

O = blue
Temp. = 25,000° & up
B = blue-white
Temp.=10,000°–25,000°
A = white
Temp. = 7,500°–10,000°
F = yellow
Temp. = 6,000°–7,500°
G = yellow-orange
Temp. = 5,000°–6,000°
K = orange
Temp. = 3,500°–5,000°
M = red
Temp. = 3,500° or less

Our Sun is an average G-type star with a surface temperature of 5,800°.

Struve (Σ or OΣ) or Burnham (β) numbers.

In historic guidebooks, the colors of double stars read like an artist's palette of pastel hues: azure, lilac, aquamarine, cerulean blue, rose; or a jeweler's tray of gems and minerals: golden, silvery, emerald, topaz, turquoise. Be warned, these tints are subtle and subjective. Only a few doubles, such as Albireo, or Beta (β) Cygni, Delta (δ) Cephei and Gamma (γ) Andromedae, show obvious and striking colors. Most exhibit pale tints at best.

In addition, the colors of fainter companion stars are often illusory, an effect of contrast with the brighter primary star. For example, the companion of Antares isn't really green; it just appears that way next to its bright yellow-orange primary.

The objective method of measuring a star's color subtracts a star's brightness in the visual yellow-green, or V, band of the spectrum from the star's brightness in the blue, or B, band. This B-V value, called the color index, has been set to 0 for blue-white stars such as Vega. Bluer stars have a slight negative color index. Moving the other way down the spectrum, our yellow Sun's color index is +0.65. Red giant stars have color indices around +1.5 to +2.0.

Double stars can put your telescope optics to the test. Stars with separations between one and two arc seconds will tax the resolving power of telescopes under 4 inches aperture, and stars less than one arc second apart will test 8-inch and larger telescopes. Because of pervasive atmospheric turbulence at most sites, telescopes of any size will seldom resolve stars closer than 0.5 arc second apart.

Very bright doubles with components two arc seconds apart, such as Castor and Alpha (α) Piscium, can be tough to split even in a large telescope, especially on nights of mediocre seeing. Doubles with faint companions orbiting brilliant primary stars (such as Antares, Rigel or the infamous white-dwarf star around Sirius) can be difficult to resolve in any backyard telescope.

Some stars show significant orbital motion over several years. Sirius is becoming easier to resolve after the 1993 minimum separation (2.5 arc seconds) between Sirius and its faint white-dwarf companion, Sirius B; maximum separation of 11 arc seconds comes in 2022. Gamma (γ) Virginis (aka Porrima) is closing up until 2008, when the stars reach periastron, swing around each other only 0.4 arc second apart, then begin to move apart.

### CARBON STARS: COSMIC COOL

Some stars pulsate in brightness over hours, days or weeks. Some, like Algol in Perseus, belong to pairs of stars that eclipse each other. Most variables are stars that actually physically pulsate in size and brightness.

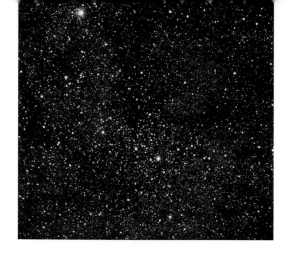

**Stellar Garnet ▶**
On the south edge of the large nebula IC1396 lies the strikingly red Garnet Star, a bloated red giant and one of the brightest carbon stars in the sky. Photo by Alan Dyer.

molecules that absorb blue light. The filtering effect of their atmospheres contributes to the stars' deep red tints.

Unlike so-called red-giant stars, such as Betelgeuse and Aldebaran, which look more yellow-orange, carbon stars really do look red. Observers have described them as blood- or ruby-red or like glowing coals. One of the best known, Mu ($\mu$) Cephei, was dubbed by John Herschel as the Garnet Star. Hind's Crimson Star, discovered in 1845 by John Hind and officially known as R Leporis, has a color index of 5.74, making it one of the reddest stars known.

The intensity of the color varies with aperture; carbon stars sometimes look more intensely red in small telescopes than in large ones. Being variable, carbon stars can slowly pulse in a range of up to five magnitudes over many hundreds of days. As a rule, carbon stars look redder when they are near minimum brightness.

Even with bright carbon stars, not all ob-

A colorful subset is the long-period variables, which take months to cycle up and down in brightness. Many are also classed as carbon stars, among the coolest known, with surface temperatures of less than 3,000 degrees C. This alone makes them among the reddest stars in the galaxy.

Their atmospheres are cool enough to support carbon compounds such as $C_2$, CN (cyanogen) and CO (carbon monoxide),

# It's All Greek to Me

In 1603, Johannes Bayer published the *Uranometria*, a set of star charts in which he labeled stars with Greek letters. Usually, he called the brightest star in a constellation Alpha ($\alpha$), the second brightest Beta ($\beta$), and so on, down the 24-letter Greek alphabet to Omega ($\omega$). We still use these Bayer letter designations today. For example, Sirius is Alpha ($\alpha$) Canis Majoris.

By tradition in such designations, the constellation name takes the Latin genitive, or possessive, case. Thus it is Gamma Arietis (not Gamma Aries), meaning Gamma of Aries. And it is Sigma Orionis, Delta Cephei and the awkward Alpha Canum Venaticorum.

A French edition of John Flamsteed's 1729 *Atlas Coelestis* introduced the system of numbering stars from west to east across a constellation, the so-called Flamsteed numbers. Most naked-eye stars carry Flamsteed numbers. Sirius is 9 Canis Majoris.

The first computer-generated star atlas, published by the Smithsonian Astrophysical Observatory in the 1960s, catalogs 260,000 stars down to ninth magnitude. Sirius, for instance, is designated SAO 151881. Most "Go To" telescopes call up stars by their SAO numbers.

The most recent and complete star catalog, the Tycho, compiles data on one million stars observed by the European Hipparcos satellite in the 1990s. In this classification, Sirius is HIP32349.

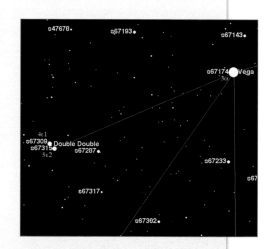

 *Stars can have several names. Vega is Alpha Lyrae (Bayer); 3 Lyrae (Flamsteed); and SAO 67174 (Smithsonian). The Double-Double star is Epsilon¹ and Epsilon² Lyrae, aka 4 and 5 Lyrae. Chart courtesy TheSky™/Software Bisque.*

servers see the stars as deeply red as they are often described, which can likely be attributed to differences in individual color sensitivity. In such cases, try slightly defocusing the image to bring out the subtle color.

# Deep-Sky Tour Two: Star Clusters

Astrophotographers can make a good case that the full glory of a nebula or a galaxy can be seen only on film or CCD chips, but when it comes to star clusters, they lose the argument. Few astrophotos capture a cluster's visual quality of glittering diamond dust.

## OGLING OPEN CLUSTERS

The word open, which is used to describe these clusters, refers to their resolvability: All the stars in them can be seen individually, in contrast with the haze of globular clusters. Open clusters range from objects that fill the eyepiece with brilliant star fields to clusters that appear as smudges resolvable only with moderate power.

How well a cluster shows up in the eyepiece depends on several factors. One is size. Large clusters (more than 30 arc minutes in apparent diameter) require low power and a wide field. For example, the Pleiades and Beehive star clusters can look better in a finderscope than they do in a long-focal-length telescope. To appreciate a cluster fully, you need a field of view twice the size of the cluster so that the cluster appears distinct from its background. Conversely, small clusters (less than five arc minutes) require high power to resolve.

Magnitudes of open star clusters range from 1.5 for the Pleiades to fainter than 12 for the dimmest ones. You would think that the brighter the cluster, the better it would look. That is not necessarily the case. A cluster's magnitude is just a measure of the total brightness of all its member stars. If it contains merely a handful of bright stars, a cluster's appearance may be disappointing despite a high magnitude rating.

What makes a cluster interesting is its richness (the number of member stars) and the contrast between it and the surrounding star field. The best clusters usually contain 100 or more member stars, earning them the official designation of rich, as opposed to moderate (50 to 100 stars) or poor (fewer than 50 stars).

Gorgeous NGC7789, in Cassiopeia, is one of the least known of the rich-cluster cream of the crop, while M11, in Scutum, is one of the best known and one of the finest clusters in the sky. Some clusters are duds—no more than scattered groupings of stars that look much like the surrounding field. A few otherwise poor clusters are intriguing because of some unusual trait. Gaze at NGC457 in Cassiopeia, and you'll see the

▲ **Wild Duck Cluster**
M11, seen here in an eyepiece sketch by John Bianchi, is one of the richest open clusters in the sky. Early observers likened it to a flight of wild ducks.

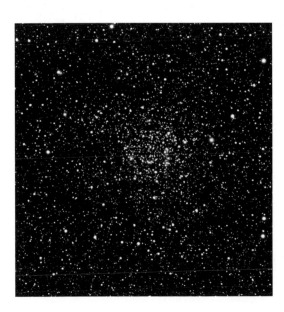

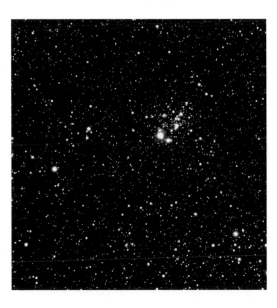

◀ **Contrast in Clusters**
Two Cassiopeia clusters display the range of cluster types. NGC7789 (far left) is a rich collection of faint stars. NGC457 (left) is a looser collection of brighter stars. Can you see the outline of E.T. the Extra-Terrestrial in this cluster? Both photos by Alan Dyer.

# Rating Open Clusters

To categorize the appearance of open clusters, astronomers use the Trumpler system ratings (e.g., II3r), which are often listed in deep-sky catalogs:

**Concentration of Stars**

| | |
|---|---|
| I | well separated, with a strong concentration to the center |
| II | well separated, with little concentration to the center |
| III | well separated, with no concentration to the center |
| IV | not well separated from the surrounding star field |

**Range in Brightness of Stars**

| | |
|---|---|
| 1 | small range in brightness |
| 2 | moderate range in brightness |
| 3 | large range in brightness |

**Richness of Cluster**

| | |
|---|---|
| p | poor, fewer than 50 stars |
| m | moderate, 50 to 100 stars |
| r | rich, more than 100 stars |

 *The best open clusters have a concentration of I or II and lots of stars (r). The Pleiades cluster, above, is a class I3r, while the larger Hyades, at bottom of frame (with Saturn nearby), is a looser, less rich class II3m. Photo by Alan Dyer.*

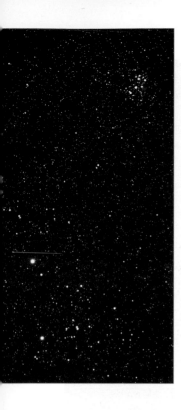

stellar outline of E.T. the Extra-Terrestrial. In Orion, NGC2169's pattern of stars resembles the number 37 or an XY, depending on how your mind interprets this celestial Rorschach test.

## GLORIOUS GLOBULARS

To see globulars in all their glory requires sharp, well-collimated optics as well as aperture. A good 4-inch telescope will begin to resolve the best globulars, such as M13 in Hercules, M3 in Canes Venatici, M5 in Serpens and M22 in Sagittarius. The legendary Omega Centauri explodes into stars with even smaller aperture. But through a 10-to-12-inch instrument, the best globulars are breathtaking sights, like sugar bowls of stars.

Yet not all globulars are as dazzling as those showpiece objects. Globulars vary in appearance because of their apparent size and concentration.

Globular clusters range in size from 1 to 20 arc minutes. The best are the largest ones. Small globulars tend to be more dif-

ficult to resolve, appearing as fuzzy-edged spheres. How well even large globulars can be resolved depends on their concentration. Some are so highly compressed that they are impossible to resolve in even a large-aperture telescope.

At the other end of the scale, a few globular clusters are so loosely concentrated, they take on the appearance of very rich, finely resolved open star clusters. NGC288 in Sculptor, NGC5466 in Boötes and NGC5897 in Libra are good examples of this type. All are best seen in large apertures; smaller telescopes at low power show them only as circular glows. One oddball globular in this class that is well worth a look is the bright M71 in Sagitta. For many years, it was considered to be an open star cluster. On the other hand, NGC2477 in Puppis is such a rich open star cluster, it borders on being a globular.

As might be expected, the most distant Milky Way globulars appear faint and small (one to two arc minutes across), making them difficult to resolve in all but the largest amateur telescopes. With intergalactic wan-

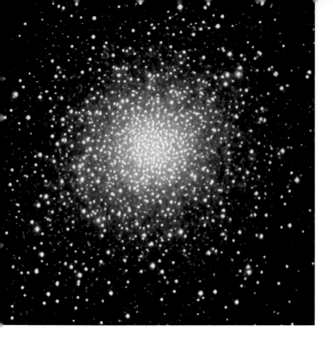

blur only 0.6 arc minute across that resides in a dwarf elliptical galaxy called the Fornax Dwarf. Although the galaxy is too faint to be seen in most amateur telescopes, this one globular stands out, shining from a distance of 300,000 light-years.

But NGC1049 is not the distance champ. Owners of large instruments can go after globular clusters 2.5 million light-years away, surrounding the Andromeda Galaxy. About 300 such globulars have been cataloged. The brightest can just be seen in a 12-inch or larger telescope. At about magnitude 15, these globulars are impossible to distinguish from faint foreground stars in our own galaxy. *The Night Sky Observer's Guide, Vol. 1* by Kepple and Sanner contains a good finder chart.

Another big-scope challenge is to hunt down all 15 of the Palomar globular clusters, most of which are buried in the Milky Way and heavily obscured by intervening dust. All of them were discovered on Palomar Observatory Sky Survey (POSS) photos and appear as no more than smudges in the eyepiece. Finish those, and you can then attempt the even more obscure (literally!) globulars in the Terzan catalog.

◀ **Class V Globular**
Often overlooked, M5 in the northern spring and summer sky is one of the best globulars. Its concentration class of "V" makes it richly endowed yet easy to resolve into myriad stars. CCD image by Chris Schur.

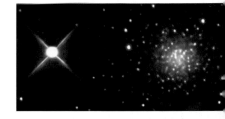

▲ **Distant Globular**
Dubbed the Intergalactic Wanderer, NGC2419 lies so far away that it appears as little more than a circular glow in most telescopes, impossible to resolve at the eyepiece. CCD image by Chris Schur.

derers, such as NGC2419 in Lynx (300,000 light-years away) and NGC7006 in Delphinus (185,000 light-years away), the reward is simply seeing them.

If reaching out to greater distances sounds appealing, try M55 in Sagittarius. This easily resolved globular belongs to another galaxy, the Sagittarius Dwarf, a newly discovered companion to the Milky Way. If you live below 40 degrees north latitude, try NGC1049, an 11th-magnitude

# Rating Globular Clusters

As with open clusters, there is a classification system for rating the appearance of globular clusters. Devised by Harlow Shapley, the globular-cluster rating system goes from Roman numeral I through XII.

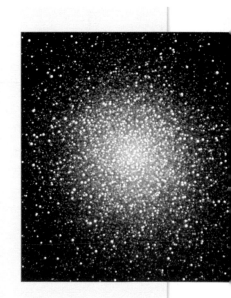

| | |
|---|---|
| I | very highly concentrated; very difficult to resolve |
| II – XI | decreasing degree of concentration |
| XII | least concentrated; very loose globular; easily resolved but not as richly spectacular |

The finest globulars fall in the middle of the range, about Class V to VII, which is the best compromise between richness and resolvability. Class I and II globulars seem star-poor, while Class XI and XII globulars are so loose, they resemble rich open clusters—pretty, but not the expected appearance.

 *The great Hercules globular star cluster (M13) is one of the northern summer sky's finest sights. A good 4-inch telescope begins to resolve it, but in a 10-inch or larger scope, it explodes into thousands of stars. CCD image by Robert Gendler.*

# Deep-Sky Tour Three: Where Stars Are Born

Images of colorful nebulas are icons of astronomy, and the anticipation of seeing their technicolor forms swirling in the eyepiece entices many newcomers to a telescope. When that's the case, the views are bound to disappoint. The human eye is simply not sensitive enough to pick up the colors that film or CCD chips record in long exposures. The live views of these puffs of celestial smoke are in monochrome gray. Only a few (Orion, the core of Eta Carinae and some planetaries) are bright enough to excite the eye's color receptors.

Despite this, a 6-to-12-inch telescope can show a wealth of detail in the brightest nebulas, producing views that look almost photographic, albeit in black and white. Some nebulas, such as the Orion Nebula, look better in real life than they do in many astrophotos. The eye can capture the full range of detail, from bright to dim, along with stars embedded within the nebula that are often washed out in long-exposure images.

## Asterisms, the Un-Clusters

Everyone comes across these when scanning around the sky: interesting alignments of stars that *look* like something—and certainly not just random patterns. Yet that's exactly what they are—chance arrangements of stars which we perceive as familiar shapes. These are not star clusters but asterisms, which are unlabeled on many star charts.

The best-known example lies in dim Camelopardalis. Kemble's Cascade is a two-degree-long chain of fifth-to-eighth-magnitude stars obvious in binoculars. Look for it at R.A. 03h 57m and Dec. +63°, with the small open cluster NGC1502 at its southern end. Here's a sampler of other notable asterisms:

• Nearby in Cassiopeia, within a binocular field of M103, lies the Kite (at 1h 40m; +58° 30'), a three-degree-long gathering of fifth-to-seventh-magnitude stars forming a diamond with a tail.

• A binocular field southwest of M52 in Cassiopeia (at 23h 07m; +60°) is a three-degree-long chain of stars that looks like a number 7 on its side.

• Polaris forms the sparkling jewel of a 45-arc-minute-wide semicircle of faint stars dubbed the Engagement Ring by Robert Burnham Jr.

• The one-degree-long Little Fish cluster of a dozen stars lies in a rich field in Auriga (at 05h 18m; +33° 30').

• Three degrees southwest of the cluster M50, near the Monoceros-Canis Major border (at 06h 53m; –10° 12'), lies the Number 3 cluster, a backwards "3" of stars half a degree across.

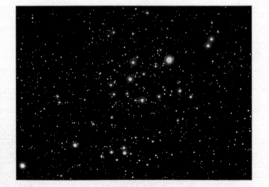

• The "W" cluster in Draco (at 18h 35m; +72° 18') looks like a miniature Cassiopeia only half a degree across.

• Finally, the bright, sprawling gathering of stars that surrounds Alpha (α) Persei, or Mirfak, looks as if it should be an official cluster. Indeed, it carries the designation Melotte 20. But this group is classed as an OB Association, a collection of hot young stars that formed together 50 million years ago and are now only loosely bound in the nearby Perseus Arm of the Milky Way.

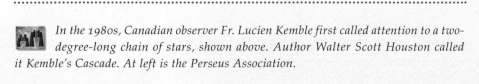

*In the 1980s, Canadian observer Fr. Lucien Kemble first called attention to a two-degree-long chain of stars, shown above. Author Walter Scott Houston called it Kemble's Cascade. At left is the Perseus Association.*

## GLOWING GAS CLOUDS

The Orion Nebula (M42) is one of the first deep-sky objects amateur astronomers observe, and it is the object everyone returns to time after time. It is an example of an emission nebula—a nebula that glows with its own unique light.

Embedded within every emission nebula is a very hot blue star (more often, a group of them, such as the four Trapezium stars at the heart of M42), newly formed out of the surrounding cloud. The star emits prodigious amounts of ultraviolet light into the heart of the nebula. The neutral hydrogen atoms in the nebula absorb the ultraviolet radiation and are pumped up by this shot of energy. As a result, the atoms are torn apart into a sea of free electrons and protons, a process called ionization, which turns the neutral hydrogen into singly ionized hydrogen atoms called H-II. Emission nebulas are often dubbed H-II regions.

The electrons and protons eventually recombine to form neutral hydrogen, but as the wayward electrons are recaptured, they give up their excess energy as visible light in a series of narrow wavelengths.

In photographs, emission nebulas look red, the result of light emitted at a wavelength of 656.3 nanometers, the hydrogen-alpha spectral line deep in the red end of the spectrum. In the eyepiece, however, emission nebulas, if they show any color at all, appear greenish. M42 is a case in point. Its green color is produced in part from the hydrogen-beta line at 486.1 nanometers, but it arises primarily from a pair of emission lines at 500.7 and 494.9 nanometers. These two lines come from oxygen that has lost two of its eight electrons. Doubly ionized oxygen is called O-III. The fact that nebulas emit light at these discrete wavelengths makes nebula filters possible—they allow the select wavelengths to pass through while rejecting all others, improving the contrast between object and sky.

The Eagle Nebula (M16) in southern Serpens is a good example. Spotting this nebula without a filter can be difficult even in a dark sky. A filter reveals it clearly as a field of grayish haze surrounding a cluster of stars. Other fainter nebulas that are normally barely visible in a dark sky stand out dramatically when seen through a filter.

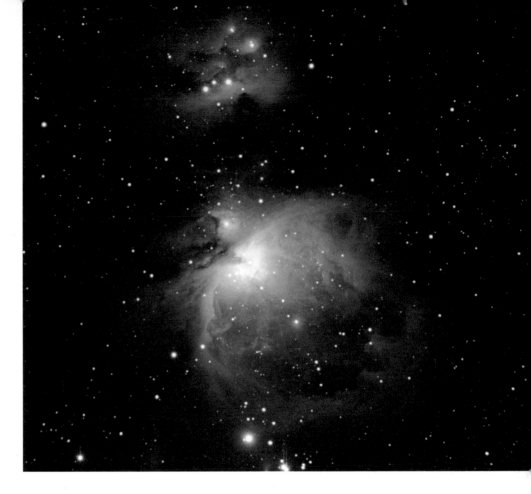

## MISTY REFLECTION NEBULAS

Most nebula filters do little to enhance the view of reflection nebulas. These gas and dust clouds do not emit their own light. They shine because the light of nearby stars scatters off the nebulas' clouds of minute dust particles. "Dust" is a catchall term for any interstellar particulate matter larger than molecules. The dust inside nebulas is thought to be graphite coated with ice. When starlight hits fields of dust, the light is simply reflected. The spectrum of a reflection nebula is the same as the broad continuous spectrum found in stars. Since newly formed stars are usually bluish, most reflection nebulas are blue as well.

Reflection nebulas are less common than emission types. Most are also fainter and more difficult to see, since they are often washed out by the glare of the nearby source star. For example, the only reflection nebula in the Messier catalog is M78.

One hazard in seeking reflection nebulas is that dew or a film of dirt on the eyepiece or the main optics can produce pale glows around bright stars. Seeing the nebulosity surrounding the Pleiades, for in-

▲ **Best Northern Nebula**
The Orion Nebula, consisting of M42 and M43, is the brightest nebula visible to northern-hemisphere observers. Its hydrogen atoms glow reddish pink. Above it is the blue glow of the reflection nebula NGC1973-5-7. Photo by Alan Dyer.

▼ **Best Reflection Nebula**
Not far from M42 lies M78, one of the sky's brightest reflection nebulas, paired with NGC2064 at right. CCD image by Robert Gendler.

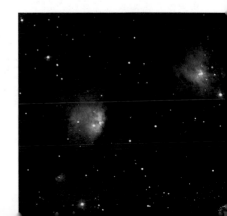

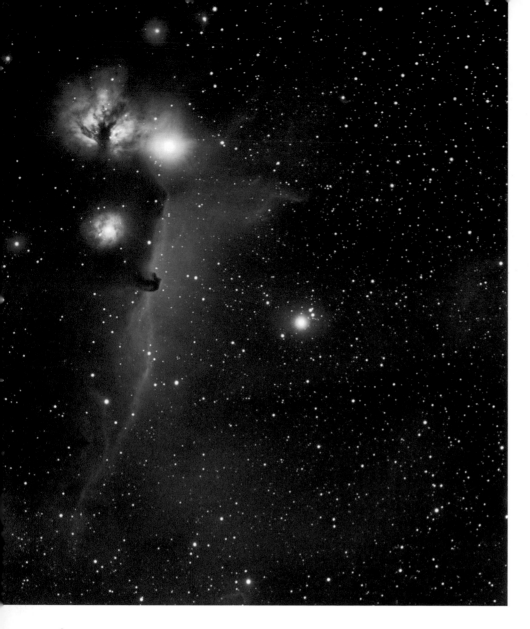

degree-wide Coal Sack near the Southern Cross is a mass of obscuring dust roughly 500 light-years away.

Not all dark nebulas are so large. Many fit nicely into the low-power field of a telescope. But how does one observe something that gives off no light? The trick is to use a wide field (one degree or more) to see the dark area framed by the surrounding bright star field. A large telescope is not needed—a short-focus 3.5-inch (90mm) refractor will do nicely, as will giant 10x70 or 11x80 binoculars.

Once you find dark nebulas, how do you identify them? Older star atlases either do not plot dark nebulas or do not label them. Exceptions are the *SkyAtlas 2000.0* (2nd Edition, 1998) and the *Uranometria 2000.0*. In computer programs, try Guide 8.0 or Megastar. With these, you can locate dark nebulas such as B86 and B92 in Sagittarius, both small, opaque patches nestled in rich Milky Way star fields. A good binocular object is Barnard's E, a three-pronged nebula formed by B142 and B143 within a binocular field of Altair, the brightest star in Aquila.

The most famous dark nebula, but perhaps the most elusive, is B33, a protrusion of dusty material that forms the Horsehead Nebula, south of Orion's belt. Transparent skies, a telescope larger than 8 inches and a nebula filter or, better yet, a Hydrogen-beta filter help bring out the faint emission nebula called IC434, which forms the glowing background to the Horsehead's surprisingly tiny silhouette.

**Infamous Horsehead ▲**
This dark nebula has the reputation of being well known but seldom seen. Visually, the horse's head barely shows as a dark bay in the dim band of IC434. Above the Horsehead is the blue reflection nebula NGC2023 and the glowing orange emission nebula NGC2024. CCD image by Robert Gendler.

**Cocoon Nebula ▶**
Easily visible in binoculars, this red nebula, IC5146, lies at the tip of the long dark nebula Barnard 168. Photo by Alan Dyer.

stance, requires clean optics. A humid atmosphere also creates hazy star images. Hunting reflection nebulas is best left for dry, transparent nights.

## DARK NEBULAS: SILHOUETTES ON THE SKY

Although composed of the same mixture of gas and dust as bright emission and reflection nebulas, dark nebulas lack any embedded or nearby stars to illuminate or warm them. They appear as nearly starless voids—cold, black patches obscuring whatever lies behind them.

Some dark nebulas can be spotted with the unaided eye. The dark rifts and lanes that split the Milky Way in Cygnus are dust clouds lining the arms of our galaxy about 4,000 to 5,000 light-years away. The five-

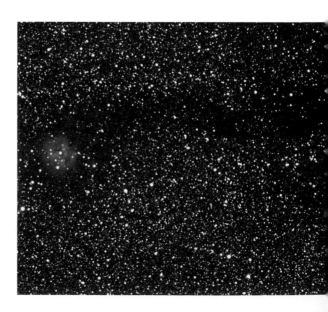

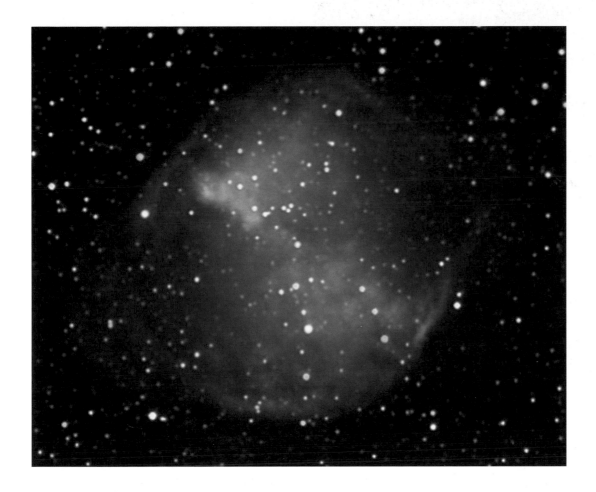

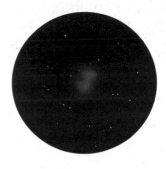

◀ **What Do You See?**
Nebulas often carry fanciful names inspired by their appearance. The Dumbbell, a planetary nebula, also known as M27, perhaps looks like a barbell but might be better called the "Apple Core Nebula." Eyepiece sketch (above) by John Bianchi; CCD image (left) by Chris Schur.

A dark sky is imperative for observing any dark nebula. Unless the Milky Way is a shining river of light, forget about hunting for this elusive class of objects.

# Deep-Sky Tour Four: Where Stars Die

Not all nebulous regions are areas of star formation. Just the opposite. Some are sites of stellar death. We see shells and arcs of gas cast off by dying stars at the ends of their lives. Much of this material pollutes the Milky Way but is eventually swept up by star-forming nebulas, where it goes on to enrich a new generation of stars as part of a galactic recycling program. All the elements heavier than helium, including carbon, oxygen, iron and every element critical for life, were forged inside stars. Look at a planetary nebula or a supernova remnant, and you are seeing examples of where the stuff that makes you possible came from.

## SMOKE-PUFF PLANETARIES

In spite of their name, planetary nebulas have nothing to do with planet formation. They are actually shells of gas expelled by aging stars during an unstable phase in their cycle. Many planetaries appear as small bluish disks through a telescope, reminding pioneer observers of Uranus and Neptune. Uranus's discoverer, William Herschel, was the nomenclature culprit. Ignorant of the role these objects play in the life cycles of stars, he called them planetary nebulas. The name stuck.

Since planetary nebulas are thought to last for little more than 100,000 years, they must be constantly forming. Indeed, a number of strange, compact objects apparent only through giant observatory telescopes are now classified as protoplanetaries, stars in the early stages of casting off a nebulous shroud. The older, well-formed planetary nebulas visible through amateur telescopes typically range from about one-quarter of a light-year to one light-year across and are all in our sector of the galaxy, no more than a few thousand light-years away.

For the deep-sky observer, planetary nebulas fall into one of three broad categories: large and bright; bright but starlike; and large and faint. The differences are partly intrinsic and partly due to distance.

❖ Large and Bright

As an example of a large, bright planetary, the Ring Nebula (M57) is the classic choice. At ninth magnitude and 70 arc seconds across, it has a high surface brightness for a planetary nebula. Its smoke-ring form is easy to see even in a 60mm telescope in a dark sky. However, one feature of the Ring Nebula is far from easy—its 15th-magnitude central star. We have seen it with a 14-inch Schmidt-Cassegrain under superb desert skies, but it is usually rendered invisible by the surrounding nebula.

Another showpiece glows nearby: the Dumbbell Nebula (M27) in Vulpecula, a planetary large and bright enough to be seen in binoculars. A filter-equipped large telescope reveals far more outlying detail than shown in most photos of the Dumbbell.

M27 exhibits a classic double-lobed structure shared by many smaller planetaries. The Helix Nebula (NGC7293) is fainter but spans half the diameter of the Moon.

❖ Stellar Planetaries

Large, bright planetaries like the Ring and Dumbbell are the exception. The majority fall into the category of "bright but starlike," often difficult to distinguish from stars, especially at low power. Most have diameters of well under 20 arc seconds, making them smaller than the disk of Saturn. Few have annular smoke-ring structures like the Ring. Nevertheless, many of these tiny planetaries are worth the search.

A couple of our favorites are the Blue Snowball (NGC7662) in northern Andromeda and the Eskimo Nebula (NGC2392) in Gemini. They are bright enough—magnitudes 9 and 8, respectively—to take high magnification. With a diameter of 30 arc seconds, the Blinking Planetary (NGC6826) in Cygnus is large by planetary-nebula standards. Its notable feature is a bright 10th-magnitude central star. Stare directly at the star, and the nebula seems to disappear; look to one side with averted vision, and the nebula pops back into view.

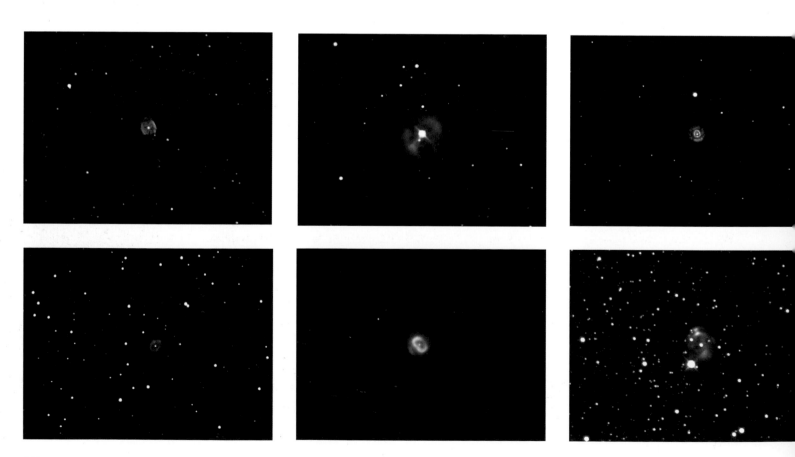

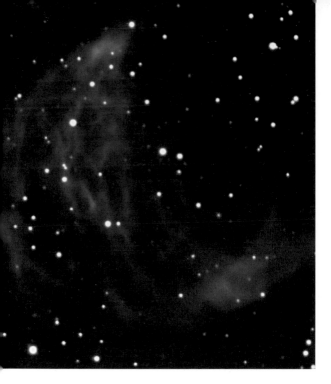

Planetaries with diameters of less than 10 arc seconds are tough to find no matter how bright they are. Even at high power, they can look like blue-green stars. One technique that helps is to hold a nebula filter between your eye and the eyepiece and move it in and out of the light path. While the stars and background sky will dim with the filter in place, the planetary will remain at the same brightness, making it pop out. Try this on two tiny but bright (ninth-magnitude) blue planetaries: NGC6210 in Hercules and NGC6572 in Ophiuchus.

❖ Large and Faint

At the opposite end of the planetary-nebula scale are large (more than 60 arc seconds), dim objects. Two good examples of planetary nebulas are NGC6781 in Aquila and NGC246 in Cetus. Such faint, diffuse planetaries are often difficult to see, if not completely invisible, without a nebula filter. When viewed with a filter under dark skies, they can become showpiece objects in telescopes larger than 10 inches in aperture.

At the extreme end of visibility are planetaries that, as recently as the late 1970s, were assumed to be strictly photographic objects. These large but faint planetaries from George Abell's list or from Perek and Kohoutek's catalog now fall prey to deep-sky hunters armed with light-bucket telescopes. One of the best is Abell 21, also known as PK205 +14.1 and the Medusa Nebula in Gemini. Its disk is more than 11 arc minutes across, huge for a planetary.

When tracking down planetary nebulas, keep in mind that the official magnitude figures may not be reliable indicators of the actual brightness. Large, diffuse planetaries often carry ratings of eighth or ninth magnitude, making them seem like easy targets. But these magnitudes are measures of the total integrated light output of the object; a large, faint planetary could have the same magnitude rating as a compact, bright one.

Also, most magnitudes are measured using a standard set of photometric filters whose passbands do not coincide with the green part of the spectrum where planetaries emit the majority of their light. Small planetaries officially listed as 12th to 14th magnitude can appear brighter than, say, a galaxy of the same magnitude. This is especially true if you are using a nebula filter, an essential accessory for planetary-nebula hunting. This is one area of observing where it is best to ignore all presuppositions of what should and should not be visible through your telescope.

## EXPLODING SUPERNOVA REMNANTS

We are long overdue for a bright naked-eye supernova in our section of the Milky Way. Until one occurs, we must be content to observe the remains of a handful of ancient supernovas that litter the sky. The best example is the Veil Nebula in Cygnus. Viewed with a nebula filter through a large telescope, the Veil's two main arcs appear as intricate, twisted lacework. Without a filter, the Veil is easy to miss, even in a dark sky.

IC443, a similar supernova remnant, appears as a crescent-shaped arc near the star Eta (η) Geminorum. Extremely faint, it is a challenge even for an experienced observer using a filter-equipped 12-inch telescope. In the southern sky, wispy fragments of the Vela supernova remnant can be sighted with large apertures.

An object often wrongly labeled as a supernova remnant is the Crescent Nebula (NGC6888) in central Cygnus. While it resembles the debris of a violent explosion, this nebula is actually a shell of material blown away from a rare type of superhot star, called a Wolf-Rayet star, through intense stellar winds rather than from a sin-

**◀ Elusive Medusa**
In a modest-sized telescope, the Medusa Nebula, one of the faint Abell planetaries, is surprisingly easy to see, provided a nebula filter is used in the eyepiece. Without it, most faint planetaries are invisible. CCD image by Chris Schur.

**▼ Famous Crab Nebula**
Photographs and CCD images, such as Chris Schur's fine image, show the Crab's expanding red tendrils, but in the eyepiece, just the amorphous central glow is visible.

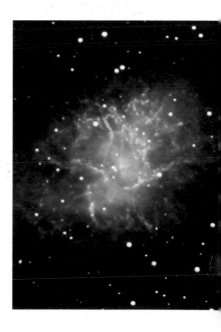

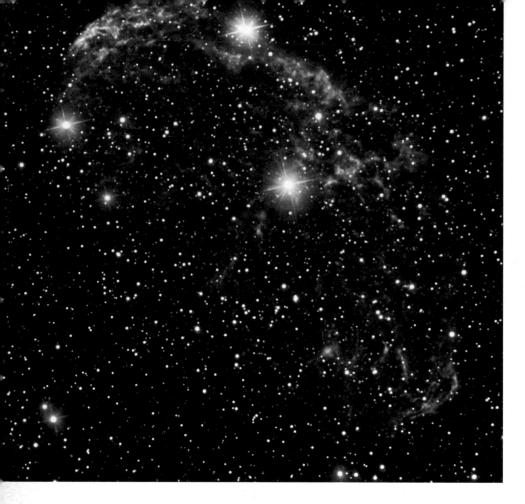

# Deep-Sky Tour Five: Beyond the Milky Way

Everything described so far on our deep-sky tours—stars, clusters and nebulas—resides within our Milky Way Galaxy. Now we leave the Milky Way behind.

Galaxies are by far the most numerous class of deep-sky objects. Several thousand shine brighter than 13th magnitude, the effective dividing line between moderately bright galaxies and those which are just barely perceptible blurs.

However, as with other deep-sky objects, do not put too much stock in published magnitude figures. Most galaxy magnitudes are photographic, which means they were measured in the blue part of the spectrum. These values are generally fainter than yellow-green visual magnitudes. A galaxy with a photographic magnitude of 12.5, for example, might have a visual magnitude of 11.8. Where only a photographic magnitude is given, the galaxy will usually appear brighter than that value.

The best recipe for galaxy hunting is to combine a dark sky with a large-aperture telescope. To see galaxies as more than fuzzy blobs, use at least a 6-inch instrument. Galaxy aficionados will want a 12-inch or larger telescope.

But galaxies are not just for the big-telescope user. Binoculars will show a handful of the brightest galaxies, while many more can be seen even in 3.5-inch telescopes, quite remarkable when you consider that the closest major galaxy, the Andromeda, is so far away, its light takes 2.5 million years to reach us.

## ANDROMEDA GALAXY

Andromeda (M31) is usually the first galaxy anyone observes. Beginners are often disappointed with that initial glimpse. Expecting an eyepiece image that looks like a two-hour time-exposure photograph, they see, instead, a featureless smear. The best way to be introduced to Andromeda is with binoculars under dark skies. More than four

**Impostor Supernova** ▲
The Crescent Nebula (NGC6888) in Cygnus has the appearance of an exploded supernova remnant but is actually a windblown shell of gas puffed off by an active superhot star. CCD image by Robert Gendler.

gle catastrophic explosion. Barnard's Loop (Sh2-276), the large arc of nebulosity on the east side of Orion, is another form of wind-blown bubble, not a supernova remnant.

The best-known supernova remnant is the bright Crab Nebula (M1), the remains of a supernova that was seen to explode nearly 10 centuries ago, in 1054 A.D. We say "seen to explode" because the Crab is 4,000 light-years away, so the actual explosion took place 4,000 years before the light of the explosion reached us in 1054.

Because of its youth, the Crab Nebula has not yet expanded into an open shell or arcs of material, like the much older Veil Nebula. Visually, the Crab is a disappointment to many observers. In small-to-moderate apertures, it resembles an amorphous blur. The wispy filaments that gave it its name can be seen only in large apertures. William Parsons (Lord Rosse) first observed them in 1844 with his 36-inch reflector. Nebula filters make little difference with this object, since the brightest part of the nebula shines with a continuous spectrum generated by the 16th-magnitude pulsar at its center.

degrees wide, Andromeda stretches across most of the field of even 7x35s.

To see the Andromeda Galaxy as more than a diffuse patch, use a 6-inch or larger telescope and look for two dark bands crossing the glow of the central core. These are the dust lanes that separate Andromeda's spiral arms. Now switch to high power, and zoom in on the central core. Look for an intense starlike point of light, coming from masses of stars circling what is thought to be a giant black hole.

As viewing the Andromeda Galaxy demonstrates, small telescopes may not reveal much detail in a galaxy, but they do show its overall shape, a characteristic that depends on the galaxy's morphological type and its orientation to our line of sight.

## THE GALAXY ZOO

Galaxies come in a variety of forms, probably as a result of the initial conditions under which they formed billions of years ago. Older theories that had elliptical galaxies naturally evolving into flattened spiral galaxies have long fallen into disfavor. However, the classification scheme first devised by Edwin Hubble stills remains in use.

❖ Elliptical Galaxies

Andromeda's two close companion galaxies, M32 and M110, are good examples of elliptical galaxies, the most common type in the universe. Ellipticals are also the least interesting to observe. Most elliptical galaxies have no visible internal structure—no dust lanes, mottling or arm structure. They are just amorphous glows that fade from bright cores into the darkness of space.

Depending on the degree of ellipticity, such galaxies can vary from circular comet-like objects to elongated patches. Ellipticals are rated from type E0 to E7: E0 and E1 galaxies are circular; E4s are football-shaped; and E6s and E7s are very flattened. In the Messier catalog, many of the brightest members of the Virgo swarm of galaxies—namely, M59, M60, M84, M85, M86 and the giant M87—are ellipticals.

❖ Spiral Galaxies

The type of object people think of when they hear the word galaxy is the spiral, its graceful curving arms the epitome of deep-sky grandeur. While the majority of bright nearby galaxies are spirals, not all reveal

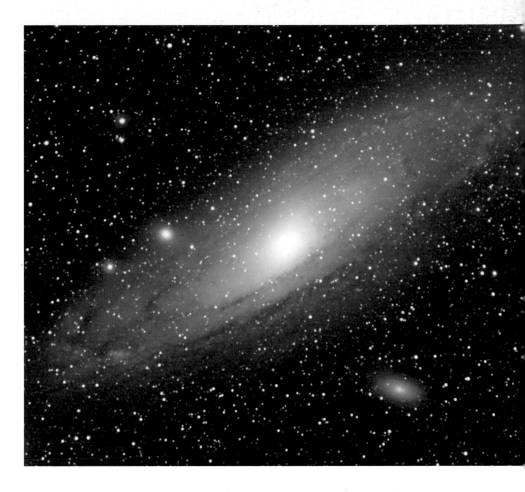

their classic pinwheel structure. It depends in part on whether the galaxy is tilted edge on to us, face-on (the best orientation for seeing spiral arms) or somewhere in between (as is usually the case).

The finest spiral galaxy is M51, the Whirlpool Galaxy. How small a telescope

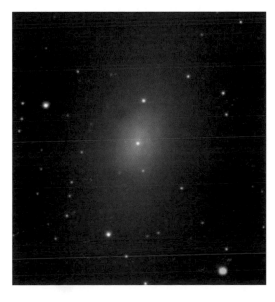

▲ **Milky Way Twin**
Through binoculars from a dark site, the elliptical glow of M31 spans three to four degrees, half the field of most binoculars. The companion galaxies M32 (closest to M31) and M110 (easiest to see) appear as adjacent glows. Through a telescope, the most obvious features are the twin dark lanes separating the spiral arms. CCD image by Steve Barnes.

◀ **Classic Elliptical**
M110, also known as NGC205, is classed as an E5, a moderately elongated, or football-shaped, galaxy. This CCD image by Chris Schur shows some dust clouds, unusual for any elliptical galaxy.

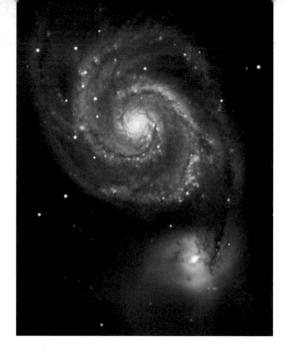

**Whirlpool Galaxy ▶**
The classic spiral is M51, which is accompanied by the small galaxy NGC5195. Because of M51's face-on orientation, the spiral arms are beautifully defined, though it takes at least an 8-inch scope to pick out the spiral structure. CCD image by Robert Gendler.

will reveal its face-on spiral arms is debatable. Most people are so familiar with what this object is supposed to look like, they imagine blatant spiral structure where there is only a hint of a circular glow. But it is safe to say that even novices perceive the suggestion of spiral arms through an 8-inch telescope.

A variation of the spiral is the barred spiral, where the arms start not from a central core but from a long stellar bar protruding from the core. The effect is obvious in photographs but often subtle at the eyepiece. M95 in Leo is a barred spiral, but it looks more like an elliptical in most telescopes. Even NGC1365 in Fornax, one of the best barred spirals, requires at least a 12-

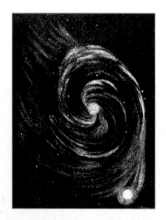

## The Leviathan of Parsonstown

There was a time when the largest telescope in the world sat in misty Ireland, owned not by a government or university but by a wealthy individual. William Parsons, the third Earl of Rosse, constructed the Leviathan, a 72-inch reflector slung on cables between two brickwork walls at his family estate, Birr Castle, in Parsonstown.

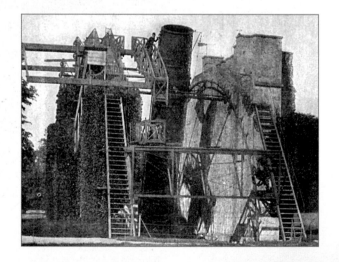

Lord Rosse began observations in 1845 and immediately discovered that Messier 51 had a spiral shape. He and his assistant observers, among them John Dreyer of NGC fame, saw spiral structure in dozens of other nebulas, recording their observations in drawings that today's visual observers praise for their accuracy and beauty. The late 1800s was the Golden Age of visual observing, and for several decades, the Leviathan dominated the discoveries. Many of the names of deep-sky objects so familiar today, such as the Crab and the Whirlpool, were coined by Lord Rosse and his assistants.

The big telescope resolved many nebulous objects into stars. The Leviathan's select users boldly asserted that all nebulas were simply collections of stars. One observer, J.P. Nichol, exclaimed, "The great mirror itself continues baffled and hopeless in the presence of those unfathomed nebulosities, which doubtless also are streams and masses of correlated stars!" The Parsonstown astronomers were partly right. Some of the "nebulas" they found—the galaxies—are made of stars. But it was to take spectroscopy and photography, new techniques that made the Leviathan and the art of eyepiece sketching obsolete, to prove it.

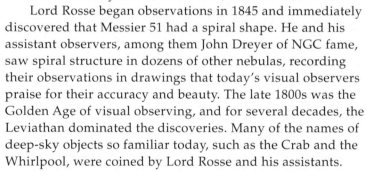

*Shown here in this rare photo from the late 1800s, above, the unwieldy arrangement of the Leviathan allowed an observer, perched on a side ladder, only one hour of viewing an object as it paraded past the meridian due south. Despite the limitation, observers recorded numerous deep-sky objects with the only method of the time: making sketches at the eyepiece, as at center.*

inch telescope to reveal its barred structure.

In contrast, edge-on galaxies are more obvious sights. With disks tilted at an extreme angle, edge-on galaxies appear as thin streaks. Because their light is concentrated into a compact form, edge-ons are good, distinct targets for owners of small telescopes. Most are spirals, but some are elongated ellipticals. Members of a transition type called S0 spirals, such as the Spindle Galaxy (NGC3115) in Sextans, also produce fine edge-ons.

For the best edge-on galaxy, search out the 10th-magnitude NGC4565 in Coma Berenices. Even a 3.5-inch refractor clearly reveals a remarkable sliver of light. A larger telescope shows it bisected by a galaxy-wide dust lane. This galaxy is 16 arc minutes long (half the apparent diameter of the full Moon), large by galactic standards.

A great hunting ground for edge-ons is Canes Venatici. NGC4111, NGC4244 and the neat pair of NGC4631 and NGC4656, the Hockey Stick Galaxy, are all rewarding targets. To the south, NGC4762, on the outer limits of the Virgo galaxy cluster, has the distinction of being the flattest galaxy known.

When selecting candidates for observing, look for galaxies whose cataloged dimensions are asymmetrical—for example, 10 arc minutes long by 1 arc minute wide. This is an indication of an edge-on galaxy that is sure to be an interesting sight.

### ❖ Oddball Galaxies

A few galaxies—irregulars—do not fall into any neat category. This class of galaxies is a minority group whose members are oddly shaped or contain chaotic details, such as patches of nebulosity, mottled dark lanes or straggling appendages. The best example is M82 in Ursa Major, erupting with fountains of material ejected by a chain-reaction burst of star formation. NGC4449 in Canes Venatici looks oddly rectangular.

Another irregular object that doubles as a radio source is the southern-sky galaxy NGC5128, or Centaurus A. It looks like a bright elliptical with a dark band crossing its disk and is probably the product of two galaxies colliding.

The Antennae (NGC4038 and 4039) in Corvus and The Mice (NGC4676) in Coma Berenices are both colliding galaxies whose two participants have yet to merge.

Observers looking for twisted galaxies in abundance can turn to Halton Arp's 1963 publication, *Atlas of Peculiar Galaxies*, a listing of 338 of the northern sky's oddest galactic denizens.

Other less deviant galaxies have distinguishing characteristics. For example, M77 in Cetus is a spiral with a starlike nucleus. It is the brightest example of a Seyfert galaxy, a class with energetic nuclei. Seyferts are one step away from being quasars.

### ❖ Quasars

Most people have heard of quasars, but not all amateur astronomers realize they can see

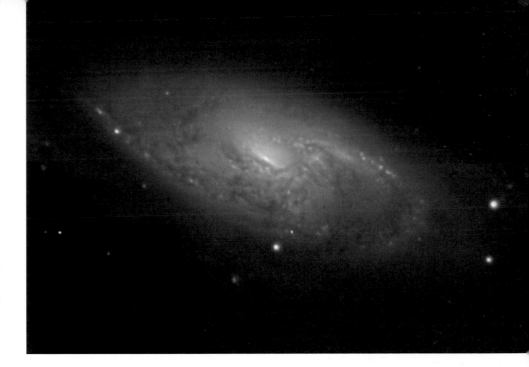

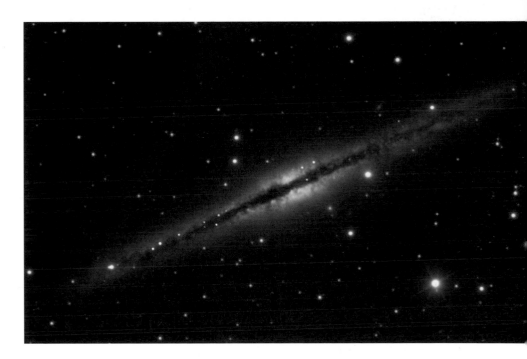

▲ **Spiral Perspectives**
Two of the finest galaxies are the dusty, tilted spiral M106 in Canes Venatici (top) and NGC891 (above), a classic edge-on spiral in Andromeda. To simulate their eyepiece appearance, step back a few feet and squint your eyes to dim the view. CCD images by Chris Schur (top) and Robert Gendler (above).

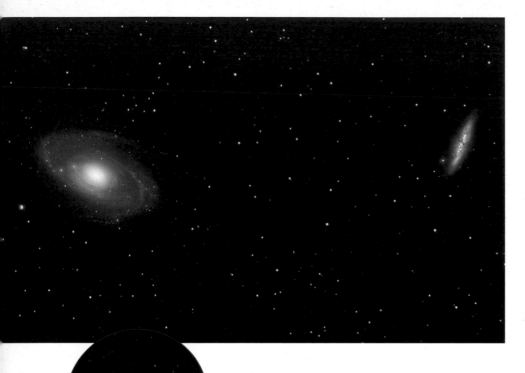

**Ursa Major Galaxy Pair** ▲
M81, the classic spiral, and M82, the cigar-shaped irregular (remember that M82 is an even number but an odd galaxy!), are large and bright enough to be seen in binoculars. A small scope at low power will frame them both in one field, as the eyepiece sketch (above) by John Bianchi shows. CCD image (top) by Robert Gendler.

**Local Group Smudge** ▶
Most Local Group galaxies, like IC10, are irregulars or dwarf ellipticals with low surface brightness. Just detecting their ghostly presence in the eyepiece is an accomplishment. CCD image by Chris Schur.

one. At about 13th magnitude (the brightness varies), the quasar 3C 273 in Virgo is the brightest member of this unusual class of objects. (The next brightest quasars are roughly 14th to 16th magnitude.) All that can be seen, however, is a faint "star." But at an estimated two to three billion light-years away, 3C 273 is one of the most distant objects visible in an amateur telescope. Find it, and you are gazing at the ultralumi-

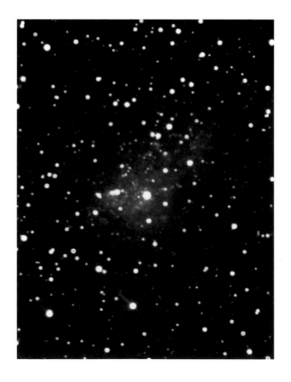

nous core of a distant galaxy energized by matter pouring into a massive black hole.

## THE LOCAL GROUP

Galaxies are gregarious creatures, preferring to live their cosmic lives in groups. The Milky Way is no exception. Our galaxy belongs to a small gathering called the Local Group, whose two largest members are the Milky Way and M31, the Andromeda Galaxy. The only other prominent member in the northern sky is M33, the large spiral galaxy in Triangulum, just south of M31. In the southern-hemisphere sky, the two Magellanic Clouds appear as naked-eye companion galaxies to the Milky Way.

Collecting views of Local Group members is a challenging observing project. Beyond the galaxies mentioned so far, the targets become faint and hard to see. About 30 galaxies make up the Local Group (new members are discovered almost yearly). A few are irregular galaxies, but most are dwarf ellipticals—galaxies that contain few stars and shine with a dim light.

The irregular galaxies NGC6822 (Barnard's Galaxy in Sagittarius), IC1613 in Cetus and IC10 in Cassiopeia carry relatively high magnitudes (about 10th) but are so diffuse, they are notoriously difficult to see even in the darkest skies.

The dwarf ellipticals are just plain faint. Using large-aperture scopes, amateurs have tracked down Andromeda II, Leo I, the Draco Dwarf and the Sculptor Dwarf, but none appear as more than barely perceptible glows, more imagined than seen.

## GALAXY GROUPS

The sky contains other similar families of related, sometimes interacting, galaxies which are not populous enough to be called clusters but which provide interesting fields containing three or more members.

One of the best is the Leo trio. M65 and M66 in Leo are two bright spirals that form a triangle with a large edge-on galaxy called NGC3628. A much fainter target is the NGC5353 group located seven degrees southeast of the Whirlpool Galaxy. Owners of 10-to-12-inch telescopes will find a high-power field containing five 12th-to-14th-magnitude galaxies. Perhaps the most famous

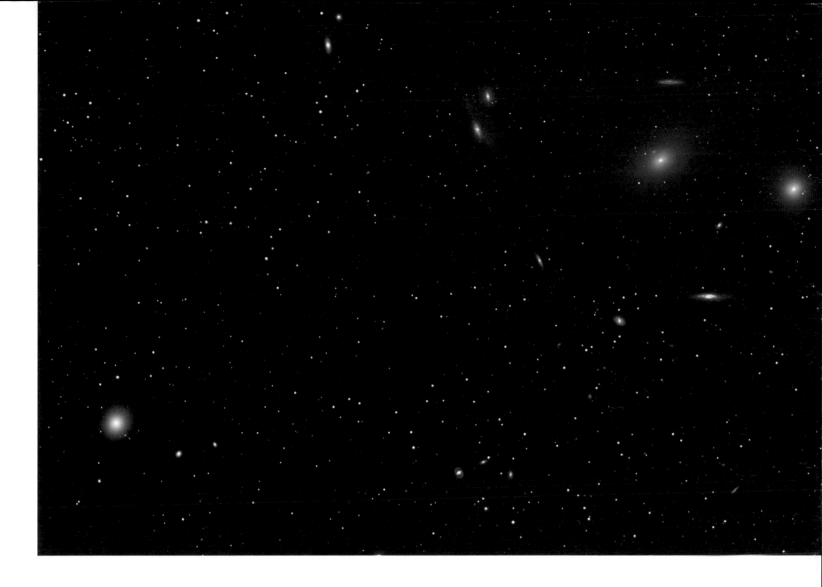

galaxy group is Stephan's Quintet, a gathering of faint 13th-to-15th-magnitude galaxies in Pegasus, just an eyepiece field southwest of the fine spiral galaxy NGC7331.

For southern observers, the field surrounding NGC1399 in Fornax contains no fewer than nine galaxies within a one-degree circle.

The primary listing of galaxy groups is Paul Hickson's 1994 *Atlas of Compact Groups of Galaxies*, a catalog of 100 tightly knit families of four or more galaxies. The brightest members of many Hickson groups also carry NGC designations, but most groups consist of faint 13th-to-16th-magnitude members, making hunting for Hickson galaxy groups big-aperture work.

## VIRGO GALAXY CLUSTER

When we face the northern spring constellations of Ursa Major, Canes Venatici, Coma Berenices, Leo and Virgo, we are looking straight up out of the disk of our galaxy toward its north galactic pole, which lies near the border between Coma Berenices and Virgo. That sight line passes through the least amount of galactic dust, allowing us to see into the throngs of distant galaxies.

The crowd of galaxies in the Coma-Virgo area is a galaxy cluster, the nearest such grand-scale gathering. The center of this galaxy cluster lies about 70 million light-years distant, only a stone's throw away on the galactic scale. In fact, because of this cluster's proximity, member galaxies are scattered over a huge swath of sky, from Ursa Major south to Virgo. Our Local Group lies on the outskirts of this cluster.

The problem with exploring Coma-Virgo galaxies is that there are so many galaxies and so few bright guide stars, it is easy to become lost in a field of anonymous fuzzy spots. A good star chart is essential.

The heart of the "realm of the galaxies"

▲ **Galaxies Galore** Markarian's Chain, the string of galaxies across the top of this image, is one of the sky's greatest wonders. One night, we examined it with a 4-inch f/6.5 refractor using a 16mm Nagler eyepiece that gave 41x and a two-degree field. Ten galaxies floated like tiny, pale snowflakes in the star field. The view included the entire chain and the giant elliptical galaxy M87 to the southeast (bottom left in this image), the true gravitational center of the Virgo galaxy cluster. CCD image by Robert Gendler (ST-10 camera).

263

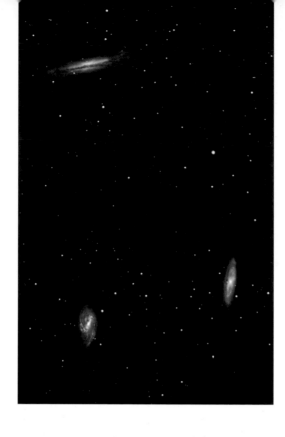

**Leo Trio** ▶
M65, lower right, M66, lower left, and NGC3628, top, can all be framed within a one-degree field of view. CCD image by Robert Gendler.

**Nothing but Galaxies** ▼
At its heart, Abell 1656 contains two bright NGC galaxies, 4874 and 4889, surrounded by dozens of faint galaxies that form the rich Coma Cluster. This map was produced with Guide 8.0 software and includes images from Palomar Observatory Sky Survey (POSS) plates.

lies around M84 and M86. These twin ellipticals are the brightest members of a string of galaxies called Markarian's Chain.

## DISTANT CLUSTERS

If you enjoy observing the Coma-Virgo galaxy cluster, you may wish to attempt other rich but much fainter galaxy clusters. These objects are at the top of the cosmic hierarchy but are among the most challenging of deep-sky targets. Because of their great distance, each is contained within an area only one to two degrees wide at most. Often, the entire cluster can be seen in one field as a collection of faint, ill-defined smudges.

Galaxy clusters usually require a lot of aperture, preferably 14 to 20 inches. Printed finder charts generated by computer programs such as Guide, MegaStar or TheSky are great aids for pinpointing and identifying each member, as even the massive *Millennium Star Atlas* (MSA) charts are barely large enough to identify individual members of the biggest clusters.

A starter cluster is Abell 1656, the Coma Berenices galaxy cluster (MSA chart #653). The brightest members of this cluster are a pair of 12th-magnitude galaxies, NGC4874 and NGC4889, 350 million light-years away. We have seen them in a 5-inch telescope. Surrounding these two giant ellipticals are about 50 very faint 13th-to-16th-magnitude galaxies that require at least a 12-inch instrument.

Nearby is Abell 1367 in the constellation Leo, centered around the 13th-magnitude elliptical NGC3842 (MSA chart #703). Five dozen galaxies brighter than 16th magnitude make up this cluster. An autumn-sky favorite is Abell 246, just two degrees east of Algol, the eclipsing binary star in Perseus (MSA chart #98). This cluster is composed of a chain of 14th-to-15th-magnitude galaxies, with the exploding galaxy NGC1275 at its heart.

The distance recordholder for accessible galaxy clusters is Abell 2065, the Corona Borealis cluster (MSA chart #646). With a 14-inch telescope under pristine skies, amateur astronomers have observed this cluster as a grayish mottling of the sky. Abell 2065 is at least one billion light-years away, nearly 400 times more distant than the

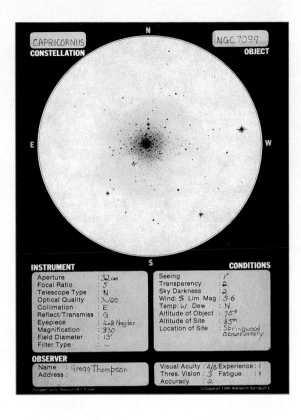

# Sketching at the Eyepiece

*By Gregg Thompson, an expert deep-sky observer who lives in Brisbane, Australia, and the coauthor of* The Supernova Search Charts and Handbook *(Cambridge; 1989).*

A drawing of a celestial object records far more detail and subtlety than can be expressed in words. "But I can't draw," some people exclaim. Drawing astronomical objects does not require the talents of a Michelangelo. Typical drawings are records of simple shapes, with various degrees of shading.

The equipment is ordinary untextured white bond paper and a soft 2B or 4B lead pencil. Use the tip of the pencil for stars and other well-defined objects and the side of the pencil for nebulous objects. Lead pencil on paper provides the easiest medium for the soft smudging needed to give a natural look to deep-sky objects. Smudging is best done with an inexpensive artist's blending stump.

Of course, using pencil on paper means that you are making a drawing with black stars on a white background, much like a photographic negative, but this is of little consequence. It is far more practical than trying to use pieces of chalk or white crayon on black paper.

After more than 30 years of experimenting with astronomical drawings, I strongly recommend two things:

1. Make the circle representing the eyepiece's field of view six to eight inches across. Most observers draw a circle half this size, and it is too small.

2. Apply the highest magnification that permits you to see the object at its best. Contrary to advice in older books, most deep-sky objects reveal much more at high power than at low power, because of their increased size and enhanced contrast against a darker background.

Use most of the area of a page, keeping the bottom for notes about the factors that affect your drawing. Such notes become a valuable reference and encourage consistency. Record the name of the object, the image orientation (mark north and east by watching the drift of the image across the undriven field), the telescope's aperture and magnification, the type of eyepiece, the type of filter (if one was used), the object's elevation in degrees above the horizon, the steadiness of the air (seeing), the darkness of the sky (transparency), the observing site and whether you have made a detailed drawing or merely a rough sketch.

Start by drawing simple telescope objects such as the Ring Nebula (M57). Other good beginning subjects are naked-eye or binocular views of star clusters such as Coma Berenices, the Beehive, the Pleiades or the Hyades. Gradually progress to fainter and more detailed objects.

Always begin by positioning the main features relative to one another—some bright stars or the general shape of a galaxy, for instance. Once you are happy with the overall proportions, fill in the details. Don't be reluctant to draw brighter stars larger or to give them spikes or diffraction rings to indicate the relative brightness.

A proficient observer must learn how to see. Give your eyes time to adapt to the dark field. Novice observers simply glance at an object in the eyepiece for a few seconds and believe that they have seen it. Always inspect the object carefully.

When you make the effort to draw what you see, a wonderful thing happens: You will see far more than you ever imagined you could. Drawing the view in the eyepiece forces you to look for subtle structure. Scrutinizing a celestial object for 10 to 20 minutes often rewards observers with inspiring detail that is invisible to those who merely take a cursory glance.

........................................................................................

*An eyepiece sketch of the globular cluster M30 shows the detail seen by experienced observer Gregg Thompson, using a 12.5-inch Newtonian. Notice the standard observing form Thompson has developed for ease of consistent use at the telescope. Those who have tried sketching deep-sky objects say that it soon trains the eye to detect more detail.*

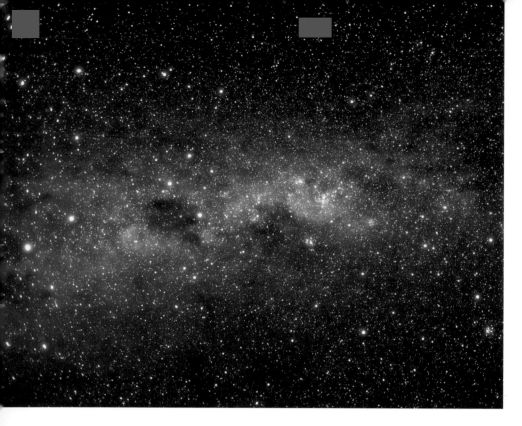

by a landslide. The second best, 47 Tucanae, is strictly a southern-hemisphere target. Then there are NGC6397 and NGC6752, both better globulars than the northern sky's prize contender, M13.

• Best bright nebula. The naked-eye nebula Eta Carinae outclasses the Orion Nebula for size, although Orion is slightly brighter. But you can see Orion from the southern hemisphere as well, so southerners have the best of both celestial worlds.

• Best dark nebula. The Coal Sack stands out like a dark hole in the naked-eye sky. Nothing as prominent exists in the north.

• Best open clusters. In the mass of star clusters around Crux and Eta Carinae are outstanding examples of every type of cluster: bright ones (IC2602), rich ones (NGC3532; John Herschel considered this cluster to be the best in the sky) and colorful ones (the Jewel Box).

**Southern Milky Way ▲**
In April and May, the full extent of the southern Milky Way, from Vela at right of frame to Centaurus, stretches across the southern half of the sky in the early evening.

**Eta Carinae and ▶**
**Clusters**
Zooming in to the center of the above view reveals the huge Eta Carinae Nebula surrounded by a trio of naked-eye star clusters, including NGC3532 at right and IC2602 at bottom. Both photos by Alan Dyer.

Andromeda Galaxy. Except for a few of the brightest quasars, Abell 2065 marks the edge of the backyard astronomer's universe. So for a new frontier, try heading south!

# The Other Side of the Sky

There's a saying in astronomy that God put astronomers in the north but put all the best sky stuff in the south. Until you have been south of the equator, you don't appreciate how true that adage is. From a midlatitude location of around 40 degrees north, any object more than 50 degrees south declination lies forever below the horizon. While the sky from 50 degrees south to the southern pole at 90 degrees south seems like a small chunk of the heavens, those 40 degrees of declination contain the sky's finest sights.

## SOUTHERN-SKY SPLENDORS

Name any type of deep-sky object, and the finest example of it can be found in southern-hemisphere skies.

• Best globular cluster. Without question, Omega Centauri, visible from the latitudes of the southern United States, wins

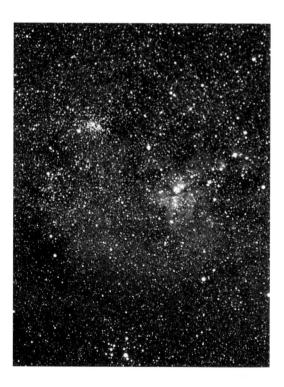

• Best planetary nebula. OK, the north wins here. The south has some fine objects, such as NGC1360 and NGC3132, but none match the Ring and the Dumbbell for brightness and size.

• Second best galaxies. The Magellanic Clouds alone are worth the trip south. These satellite galaxies of the Milky Way are close enough that we can pick out nebulas and clusters in them easily, even with small backyard telescopes. The Tarantula Nebula in the Large Magellanic Cloud displays a stunning wealth of structure and tendrils when seen through moderate apertures, especially with a nebula filter in place.

• Best galaxy. From a latitude of 30 degrees south (southern Australia and central Chile), the center of the Milky Way passes directly overhead on southern autumn and winter nights. With Sagittarius at the zenith, the bulge of the galactic core glows more brightly than you have ever seen it, laced with immense dark lanes you might have thought were detectable only in photographs. On trips to the southern hemisphere, we've spent hours scanning the canyonlike dark lanes of Sagittarius and Scorpius with binoculars or a rich-field refractor. To the naked eye, the spiral arms of the galaxy stretch off symmetrically in either direction, making the Milky Way look like the sky-spanning, edge-on galaxy it really is. It is one of the night sky's most wonderful, jaw-dropping sights and is worth the price of the airfare to Australia, New Zealand, Chile or southern Africa.

If you think you have seen it all in the north and are considering spending $3,000 or more on a new telescope to reexplore the northern sky, think twice. Spend the money, instead, on a trip to southern latitudes, taking along a portable telescope. Step out under those amazing night skies, and when you look up and realize that you don't recognize anything, you'll break out in a big smile! It's just like starting all over again, this time with an even better sky awaiting your telescopic exploration.

# Herschel at the Cape

Imagine being the first person in the world to explore an entire sky with an 18-inch telescope. Imagine having that sky, among the darkest in the world, at your disposal every night from the backyard of your idyllic country estate, located in a warm subtropical climate. A dream for observers today, but from 1834 to 1838, it was how John Herschel spent his time, scanning the southern skies from South Africa. "Whatever the future may be," Herschel wrote, "the days of our sojourn in that sunny land will stand…as the happy part of my earthly pilgrimage."

To extend his father William's catalog of the sky, John, his wife Margaret and their three children packed their belongings and an 18.25-inch reflector telescope and sailed off to the Cape of Good Hope. The Herschels quickly became celebrated citizens of Cape Town's European colony, making the social rounds by day and exploring the sky by night. During this time, John Herschel discovered 2,100 double stars and more than 1,300 nebulas and clusters. In 1837, he recorded the rare explosive flaring of the star Eta Carinae as it briefly shone as one of the brightest stars in the sky.

The years at the Cape marked the height of Herschel's astronomical career. Back in England, he rose to such positions as Master of the Mint, but he rarely looked through a telescope again, and the 18-inch mirror sat tarnished in a cellar. "With the publication of my South African observations," Herschel concluded, "I have made up my mind to consider my astronomical career as terminated."

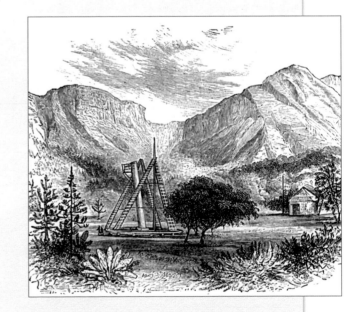

 *John Herschel's main instrument at his site near Cape Town, South Africa, was a "20-foot" telescope of 18.25-inch aperture slung in a crude altazimuth mount in open air. An equatorially mounted 7-inch refractor was housed in a small hut.*

# Astronomical Treasures From Down Under

*By Klaus R. Brasch*

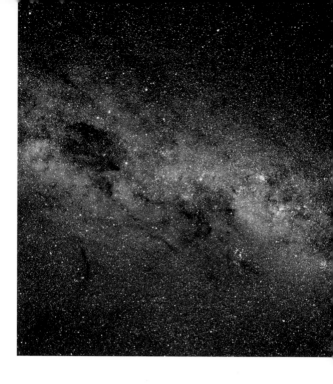

*A lifelong amateur astronomer, Klaus R. Brasch traveled to Australia with the authors and astrophotographer Mike Mayerchak to explore the wonders of the southern skies.*

All sky observers know—or should know—that the great treasures of the night sky are not divided equally between the skies over the Earth's northern and southern hemispheres. A disproportionate majority of the celestial goodies reside in the south, unseen (or not well seen) in the north. The reason for this is that the center of the Milky Way Galaxy, which harbors the densest concentration of clusters and nebulas, is never on full display for northern-hemisphere observers. The best view of the southern glories most of us ever get is a horizon-hazed peek at the Southern Cross and its surroundings from southern Florida, Hawaii or perhaps the Caribbean. But that's just a tease.

The only proper way for us deprived northerners to explore the tantalizingly rich southern skies is to travel deep into the southern hemisphere. It is a pilgrimage every northern amateur astronomer must make.

For most northern-hemisphere stargazers heading south, there is really no question about where in the southern hemisphere you should go. With its favorable climate, familiar culture, first-world infrastructure, stable government and some of the clearest, darkest skies on Earth, Australia is by far the first choice.

When you contemplate your initial trip deep into the southern hemisphere, try to travel in April or May, when it's autumn in Australia and the weather is similar to that in California in October or November. Moreover, this is off-season for vacation travel in most parts of the country, making accommodations and rental vehicles easier to obtain.

But the chief reason for the April/May travel window is the night sky. The richest part of the Milky Way—the core of our galaxy—rides high overhead after midnight. Earlier in the evening, such glorious objects as the Coal Sack, the massive Eta Carinae Nebula, the Jewel Box cluster, Omega Centauri and the two nearby galaxies, the Magellanic Clouds, are ideally placed for observing. It's a dusk-to-dawn feast for astronomy eyes from the north. (Of course, to see all these wonders at their best, be sure

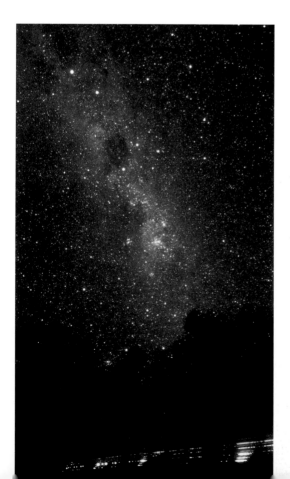

to book your travel around new Moon.)

How to go? Not many choices here. Almost all the flights from North America arrive in Australia's largest city, Sydney, a major metropolis about the size of Boston. For our trip in May 2000, we arrived in Sydney laden with all the "essentials" for an astronomical expedition—more than 500 pounds of cameras, lenses, mounts and telescopes, plenty of film and, of course, warm, comfortable clothing. It is astounding how much disassembled astro-gear, swaddled in clothes, socks and bubble wrap, can be crammed into a suitcase!

Because this trip occurred before September 11, 2001, we had no trouble with customs agents wondering what our funny-looking stuff actually was. Today, the inspections undoubtedly will take much longer. In light of this, plus the general hassle of lugging so much gear through airports, we have decided that for any return trip, we will either cut down considerably or ship our equipment ahead to be picked up when we get there.

We arrived in Sydney on a Sunday morning and were glad we did. The roads were nearly deserted, allowing us to get used to driving our rental van on the "wrong" side of the road. After a few days sightseeing in and around Greater Sydney, we headed northwest toward the beautiful Blue Mountains region and the agricultural and wine districts west of there. With its pastoral charm and numerous small towns, this part of Australia is reminiscent of both southern Europe and northern California.

Australia is a huge country, the same size as the contiguous United States, but with only 7 percent of its population. And the people are not spread out evenly across the vast land. The majority of Australia's 19 million citizens live in urban areas within an easy drive of the ocean. Getting away from light pollution means driving inland—anywhere inland. Once we were a few hundred miles from Sydney, it was like the open spaces of west Texas, sparsely populated and dark at night.

We decided to spend most of our time in and around the town of Coonabarabran, 18 miles from Siding Spring Observatory, the largest collection of research telescopes in Australia. Coona, as it is called locally, is a town of about 4,000 that serves both as the unofficial Astronomy Capital of Australia and as the gateway to the magnificent Warrumbungle National Park. Jutting sharply from the surrounding plains, several peaks of this ancient volcanic range give the area a Jurassic Park quality. In the off-season, there are plenty of motel rooms available in Coona. Some motel owners

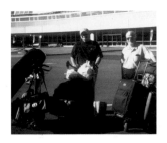

▲ **Anything to Declare?** Packed away in all those cases are several hundred pounds of telescopes, mounts, cameras, lenses, film—and some clothing— all essential gear for an astrophoto junket to the southern hemisphere.

◀ **Overhead Milky Way** The most memorable southern sight of all: the Milky Way overhead. Photo by Klaus Brasch, processed by Tony Hallas.

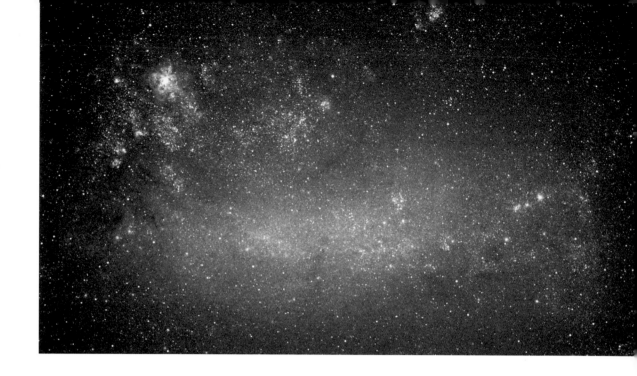

are well aware of the needs of amateur astronomers and can offer helpful advice. One motel on the road from Coona to the big observatories has a sign out front welcoming amateur astronomers.

Perched high in the Warrumbungles, Siding Spring Observatory is Australia's premier optical-astronomy research center. It houses about a dozen instruments, including the 3.9-meter Anglo-Australian Telescope (the largest optical telescope in Australia) and the 1.2-meter UK Schmidt (a twin to the 48-inch Palomar Schmidt).

Our goal was simply to take advantage of the very dark, pollution-free skies around Coonabarabran and capture as many shots as possible of the southern Milky Way and the splendid deep-sky objects that either are not visible from the northern hemisphere or are poorly placed. Our optics included an array of lenses, from a 15mm Super Wide Angle to a 5-inch f/6 Astro-Physics refractor. We shot both 35mm and medium-format films.

Perhaps nothing is more impressive, visually and photographically, than the central bulge of the Milky Way arching overhead. Under the exceptionally clear skies we enjoyed, this portion of our galaxy literally cast shadows. And it is obvious why the Coal Sack got its name: To the naked eye, it is the darkest portion of the sky anywhere, in sharp contrast to the brilliance of the Southern Cross and the nearby Eta Carinae region. Through the 5-inch, even objects familiar to northern-hemisphere observers, particularly the Lagoon and Trifid Nebulas in Sagittarius and the globular cluster M22, appeared far brighter and more contrasty than we see them back home because of their higher elevation in the sky. Likewise, Omega Centauri and 47 Tucanae, the two brightest globulars in the heavens, clearly outshone any other globular cluster.

The Magellanic Clouds are in a league of their own. The subtle structure and contrast features of these mini-galaxies are apparent even to the naked eye, but any optical aid

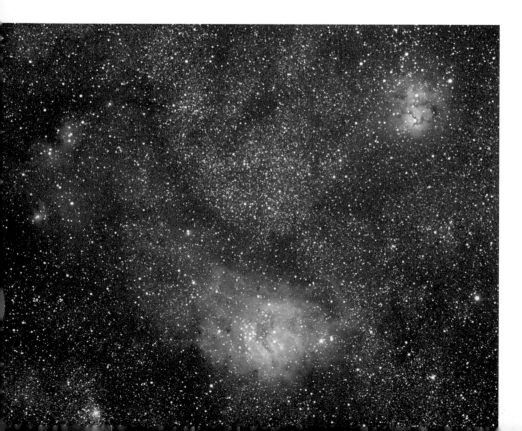

greatly enhances the effect. Photographically, the Large Magellanic Cloud is a true gem. Rich in hot, blue stars and highlighted by numerous red star-forming regions, particularly the Tarantula Nebula, this object appears textured and colorful in both wide-angle and close-up shots.

Probably the ultimate deep-sky object, however, is the Eta Carinae Nebula. Visually, it is stunning with any optical aid, appearing several times larger than the Orion Nebula. Photographically, this complex nebula reveals rich colors and extensive detail even with a medium telephoto lens, and with the 5-inch and medium-format film, it is truly awesome.

Australia is blessed with some of the world's most delightful weather. Except for summer (mid-December to mid-March),

which can be witheringly hot during the daytime, most of the country is pleasant for hundreds of days a year. When people learn of our astrophotography adventures in Australia, we are often asked why we didn't go to what must be the clearest stargazing site: the middle of the outback.

In fact, because Australia has no significant mountain ranges west of the Great Dividing Range, most of the interior of the country has clear skies 40 to 60 percent of the time, depending on the exact location. The Great Dividing Range parallels the east coast, which means that once you are west of those mountains (which we were), the odds of clear skies rise. The problem, of course, is that these are averages. Predicting whether it will be clear when you are there is impossible. We tried! Several times.

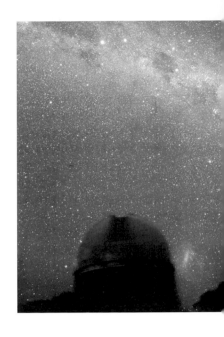

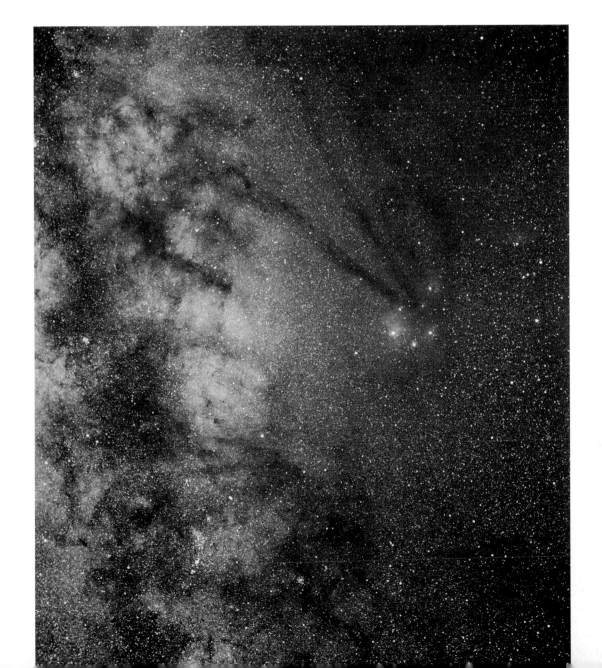

▲ **Awesome Nights**
Well away from any sources of light pollution, the Australian night skies are truly dark and spectacular. In this 24mm tracked shot, the Milky Way rides high while the Large Magellanic Cloud is an easy naked-eye sight close to the horizon (to right of observatory dome). Photo by Terence Dickinson.

◀ **The Great Dark Horse**
Often considered a naked-eye sign of good skies, the Dark Horse is located at upper left in this Mike Mayerchak photo. The horse's foreleg extends down to near Antares at center right. Image processing by Tony Hallas.

271

## PART 3

# Shooting the Sky I: The Stan

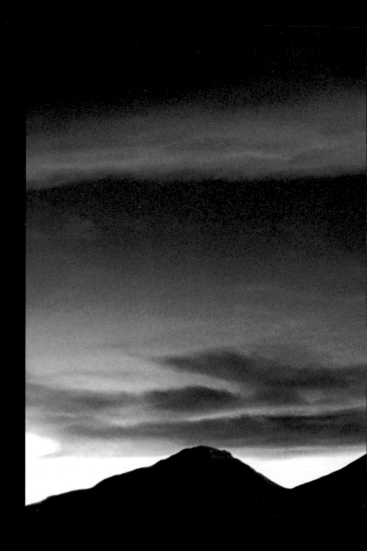

A hobby within a hobby, astro-photography can grow to addictive levels. We have both spent many long nights outside staring into a guiding eyepiece, hoping to bag a memorable shot. Much can go wrong—too much, it seems. But astrophoto addicts keep trying. Their achievements are displayed throughout this book.

Astronomical work pushes imaging technology, both film and digital, to the limit, and that alone provides a sense of accomplishment for those who love to make the most of what cameras and computers can do.

Some astro-images, such as planet, nebula and galaxy shots through a telescope, require practice and proper equipment. However, many stunning astrophotos, such as constellations, star trails and planetary conjunctions, are surprisingly easy to take using the camera equipment you already own. It is with those simple techniques that we begin our survey of astrophotography.

**Everything we see in the sky can be captured on film. Some subjects require a complex array of equipment. But unique photos, such as this scene of a total eclipse of the Sun (from Chile in 1994), can be taken with a simple camera on a tripod. Photo by Alan Dyer.**

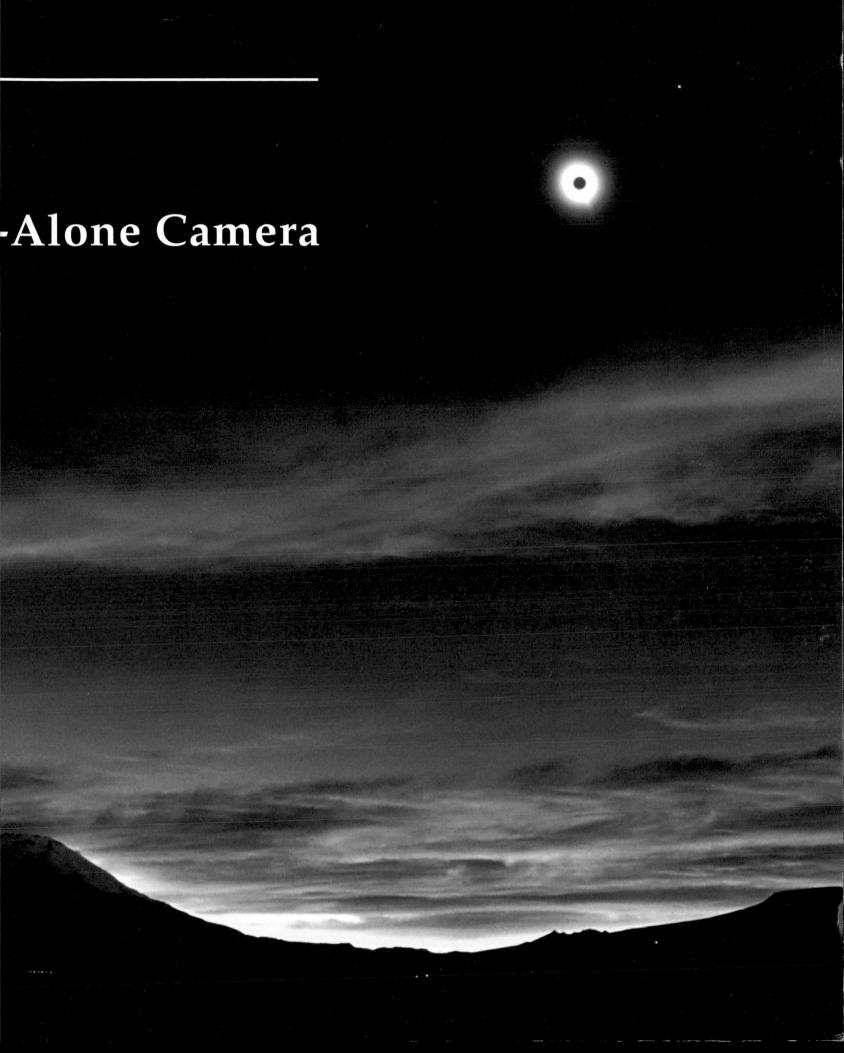

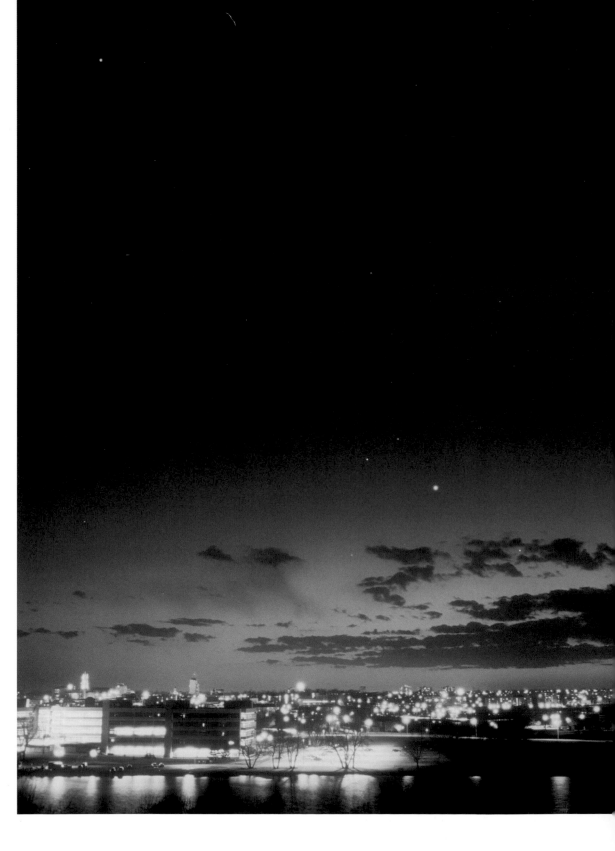

**Five-Plant Lineup** ▶
In late April and early May 2002, the five naked-eye planets were gathered in one sector of the evening sky for the first time in 62 years. Using a 35mm Canon lens at f3.5 and a 10-second exposure on Fuji Provia 400 film, Terence Dickinson took this family portrait on May 3, 2002, overlooking the city of Kingston, Ontario. Venus is the brightest object, just right of center; Saturn and Mars are to Venus's left (Mars above Saturn); and Jupiter is in the upper-left corner. Mercury is peeking out from among the first bank of clouds to the lower right of Venus. Capturing a once-in-a-lifetime shot like this with a simple tripod-mounted 35mm camera would make any amateur astronomer smile.

# Twilight Scenes

Some of the most beautiful astronomical images in our photo collections were among the simplest to capture. They were taken in twilight, either after sunset or before sunrise, using nothing more than a 35mm camera attached to a camera tripod. No telescope was involved. No special lenses, either—just the standard 24mm-to-50mm lenses commonly used for taking daytime snapshots of kids and pets.

# Exposure Guidelines: Fixed-Camera Subjects

| Subject | Lens | ISO | F-stop | Duration |
|---|---|---|---|---|
| Twilight scenes | 20mm to 135mm | 100 to 200 | f2 to f2.8 | 2 to 15 sec. |
| Auroras | 8mm to 50mm | 200 to 400 | f2 to f2.8 | 5 to 40 sec. |
| Constellations | 28mm to 50mm | 400 to 3200 | f2 to f2.8 | 10 to 30 sec. |
| Lunar halos | 15mm to 28mm | 200 to 400 | f2 to f2.8 | 4 to 16 sec. |
| Zodiacal light | 8mm to 28mm | 400 to 1600 | f2.8 | 60 sec.+ |
| Star trails (short) | 15mm to 50mm | 100 to 200 | f2.8 to f4 | 5 to 60 min. |
| Star trails (long) | 15mm to 28mm | 100 | f5.6 to f8 | 1 to 8 hours |
| Meteors | 28mm to 50mm | 400 | f2 to f2.8 | 5 to 20 min. |

The astronomical subject matter in twilight is almost always one or more of the three brightest celestial objects in the night sky: the crescent Moon and the planets Venus and Jupiter. These bodies easily punch through twilight glow that drowns out less brilliant objects. The most consistently striking combination is a conjunction (close approach) of Venus and the crescent Moon; the closer to each other, the better. If this sounds like a bit of a yawn, then you haven't seen these two luminaries just a degree apart. They're gorgeous! And no two conjunctions are the same. The separations and relative angles are always different, plus the sky color and clouds have infinite possibilities too (though just a little cloud is plenty).

A rarer combination is the three brightest celestial objects all within a few degrees of one another. Sensational! A gathering of two or three planets without the Moon is a more subtle scene, but if dazzling Venus is one of them, it's always worth a shot.

In shooting a twilight conjunction scene, whether or not it contains the Moon, there is always the ideal 10-minute window that will yield the best pictures: clear and aesthetically framed astronomical content, with rich color and saturation. Nailing the perfectly exposed shot is largely a matter of experience, but here are a few tips that should keep you in the ballpark.

• Use your camera's light meter on the wide setting rather than spot metering. Most meters at the wide setting consistently read the scene about one stop too bright. This is easily compensated for by resetting the ISO to the next fastest film. Thus if 200-speed film is in the camera, adjust the camera's setting to 400.

• Avoid the temptation to shoot your conjunction photos with telephoto or zoom lenses in tight on your celestial subject. Comfortable framing, with the celestial objects well balanced against the darkening landscape (or lightening at dawn), produces the most memorable shot. Our favorite lenses for such shots are 28mm, 35mm and 50mm.

• Keep in mind that your best shot will likely be taken when the sky is a few minutes darker than your eye would suggest as optimum. Therefore, keep shooting a bit longer at dusk, or start earlier during a dawn shoot.

• Until you gain confidence in your camera's twilight performance, especially using slide film, you should bracket up to one stop over and up to two stops under the exposures suggested above.

For most digital cameras, the light meter reads twilight scenes one stop too bright, sometimes more. But because the images also look too bright on the camera's LCD screen, it's easy to apply intuitive compensation before the next shot. For twilight Moon and planet photos, always set the manual focus at infinity. We usually set the ISO gauge at 100 for the sharpest possible pictures. We have tested this type of shooting with 2-, 3- and 5-megapixel cameras. As you would expect, there are significant improvements in resolution at each

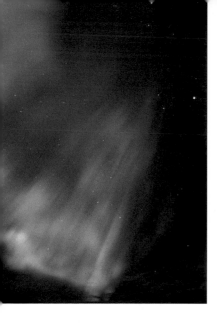

step up. However, even the 5-megapixel camera, as good as it is, is no substitute for the sharpness of modern top-of-the-line film emulsions.

# Auroras and Sky Glows

No other astronomical phenomenon that is visible from just about anywhere is more impressive than a full-blown, sky-filling aurora. The farther north you are, of course,

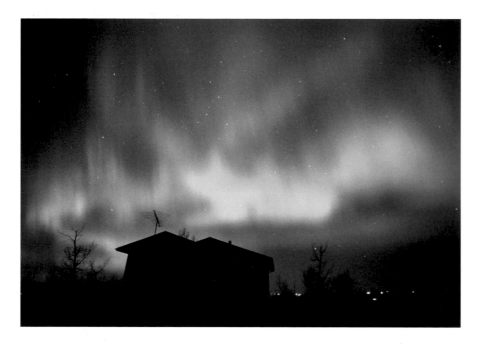

the better. But the limiting factor in grabbing great aurora shots is more often light pollution than geography.

During the three years centered around sunspot maximum, distinct auroral displays are seen a couple of times a decade from as far south as Mexico. Dozens are seen each year from northern-tier states such as Washington, Wisconsin, Michigan and New York. Shooting them is probably the easiest of all types of astrophotography. All you need is a camera and tripod with a cable release for the 5-to-40-second exposure.

We used to recommend high-speed (800 to 3200) color film for aurora portraits, but vast improvements in emulsion over the past few decades have made this unnec-

essary. Today's 200- and 400-speed films (prints or slides) will pick up any aurora glow that is even moderately bright to the eye. A bright aurora will produce slides and prints that will blow your socks off. We especially recommend 28mm and shorter lenses to achieve a wide, sweeping view of the display.

In 2001, we tested the first digital cameras that could begin to record the aurora without looking as grainy as sandpaper. They still have a way to go, but the gap in this type of shooting should close fast.

# Constellations

Camera-on-tripod pictures of the major constellations are a good starting point for deep-sky photography. The technique is simple, making it ideal for a first attempt at celestial portraiture with film or digital cameras. With a digital camera, set the manual focus at infinity. For film cameras, use a fast ISO 400 to 3200 film. Almost any film in that speed range will do a good job here. Use a 50mm normal lens set to f2 or f2.8. With either camera, the shutter-speed dial may allow exposures of up to 30 seconds. Open the shutter for about 15 seconds. In longer exposures, the stars may move enough to become trails, rather than pinpoints (see table on facing page).

**Curtains of Light** ▲
Any ISO 200 to 400 slide or print film can capture northern lights. Use an aperture of f2.8 and exposures of 5 to 40 seconds. Both photos by Alan Dyer.

**Digital Orion** ▶
A Sony DSC-F707 digital camera, described on facing page, took this 25-second exposure of Orion at f2 (zoom lens set to 38mm; 35mm equivalent). Photo by Terence Dickinson.

# Exposure Guidelines: Maximum Exposures to Avoid Trailing

| Lens | Near Celestial Equator | Declination of 45°N or S | Near Celestial Poles |
|------|------------------------|--------------------------|----------------------|
| 28mm | 25 sec. | 40 sec. | 90 sec. |
| 50mm | 12 sec. | 20 sec. | 50 sec. |
| 105mm | 6 sec. | 10 sec. | 25 sec. |

With today's superfast films, it is amazing what can be recorded using this simple technique, provided you shoot from a dark-sky location. An f2 lens will reveal stars of at least eighth magnitude, much fainter than the naked-eye limit. The brightest sections of the Milky Way are within easy reach. Just a few years ago, such images required lengthy guided exposures. Today's emulsions are much improved. Many deep-sky objects can be picked up in wonderful colors. Where your eye sees only white stars, the film will record the reds, yellows and blues that are hovering below the

# Buying a Digital Camera

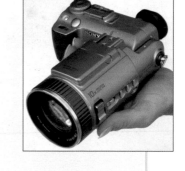

Digital still cameras are fast becoming the camera of choice for all photographers. In 2001, sales of digital cameras exceeded sales of film cameras for the first time. Some outstanding digital still cameras that perform extremely well for long-exposure astronomical shooting are beginning to be available.

In 2002, we tested all the high-end consumer models on the market for astronomy performance. Prices ranged from $650 to $1,100. (We did not test professional photojournalist digital SLR cameras because of their specialized nature and, frankly, exorbitant prices.) Features we looked for were: (a) a fast lens, preferably f2.5 or faster; (b) sharp, aberration-suppressed optics; (c) 3-megapixel image size or higher for detailed resolution; (d) maximum exposure of 30 seconds or longer; and (e) better-than-average sensitivity to low-light subjects.

Among our favorite models were the Nikon 995, Olympus C-4040 and Canon Powershot G2. But standing head and shoulders above any of them in terms of astronomical performance was the Sony DSC-F707, a 5-megapixel camera with a supersharp f2, 5x Zeiss zoom lens. At wide angle (38mm equivalent), the camera records magnitude 7.5 stars in 30 seconds. When at maximum zoom (190mm equivalent), 10th-magnitude stars are recorded in 30 seconds (guided). This is not quite as good as 800-speed film, but it's close. And it is significantly better than any other digital camera on the market as of mid-2002. Of course, digital cameras are a rapidly evolving technology, but the DSC-F707 should set the standard for at least a little while.

*A distinctive feature of the Sony 5-megapixel DSC-F707 digital still camera is the comparatively huge 5x Zeiss zoom lens, which performed superbly in our tests. Even wide open at f2, this lens produces sharp star images at any zoom ratio. Although this camera is bigger and heavier than most digital cameras, its performance day or night is outstanding.*

**Circumpolar Star Trails** ▶

Try aiming at the celestial pole (north or south, depending on your hemisphere). The result: an amazing set of concentric star trails circling Polaris or, as here, Sigma Octantis. This is a 7-hour exposure on ISO 100 film, with the 16mm lens set to f5.6 to prevent sky fogging during the all-night exposure. Photo by Alan Dyer.

color-perception threshold of human vision.

Remarkably, today's best digital cameras can actually begin to match film performance, recording stars almost as faint in 15 to 30 seconds. The images taken by older digital cameras were speckled with electronic noise in exposures longer than a second or two, but each new generation of digital camera produces cleaner, less grainy pictures in long exposures.

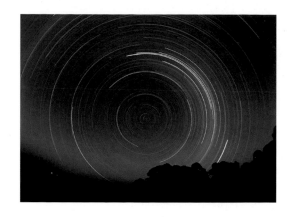

# Star Trails

Here we break the rules for maximum exposure times and purposely let the stars trail across the film. The result is a dramatic photograph of vividly colored celestial streaks. This style of picture is still the exclusive realm of film-based photography. You'll need a 35mm camera that can keep its shutter open reliably for long durations without worry of the batteries dying.

At a dark site, try exposures of 10 to 60 minutes. Even under dark skies, avoid using

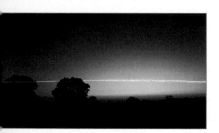

## What Can Go Wrong?

Even with the simplest astrophotography, unexpected gremlins can come out at night to harass astronomers.

### Nothing Shows Up
After an hour-long exposure, you release the shutter. Rather than a satisfying click, you hear nothing. The shutter was set at 1/500 second. It is a mistake everyone makes—once. A lens set to f16 is another blunder. Make a habit of always checking for tension on the rewind knob to ensure the film is loaded properly.

### Kinky Star Trails
If star trails have kinks in them, the fault lies not in the stars. Your tripod or tripod head has moved during the long exposure. Use a heavy-duty tripod, and always let it settle before beginning that multihour star-trail shot.

### Sky-Fogged Pictures
If the sky is white, you know you have exceeded the sky-fog limits for your site. Back off on the exposure times, or in long star-trail shots, use a slower film and stop down the lens to f5.6 or f8. Record exposure times and details so that you can learn from experience.

### Images Sliced in Half
Dark frames covered with dots look like mistakes to the person cutting your roll of 35mm film. Expose frames at the beginning and the end of a roll with nonastronomical shots; even aiming the camera into a flashlight for a second will do.

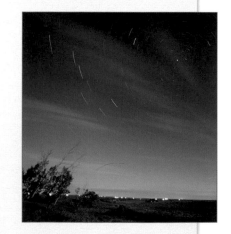

*The star trails in the photo above right gradually widen because the film warped during the exposure, throwing the stars out of focus and proving the astrophotography axiom: The longer the exposure, the more that can go wrong. Above left, an otherwise fine shot of a volcanic sunset was marred by a ghastly scratch, perhaps from the processor or from dirt in the film cassette. Sigh!*

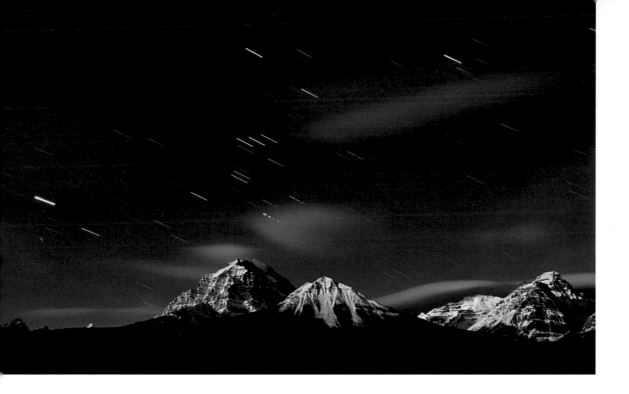

◀ **Orion Over the Rockies**
Light from a full Moon illuminated the landscape and created the blue sky. A 20-minute exposure on ISO 100 film at f8 provided just the right exposure to record sizable star trails without overexposing the scene. The other trick was being in the right place at the right time—such a scene is possible only on a full-Moon night in March when Orion is in the western evening sky. Photo by Alan Dyer.

superfast films and lens apertures, as they will produce only washed-out photos. Instead, use standard ISO 100 to 400 film with the lens set at f2.8 to f4. For ultralong exposures of all-night star trails, use ISO 100 film and stop the lens down to f5.6 or f8. Anything from a fish-eye lens to a normal 50mm works well.

For a unique shot, first take a short exposure of 30 seconds at f2.8. Then cover the lens for a minute or two. Stop the lens down to f4 or f5.6; now uncover the lens for a star-trail exposure of an hour or more. The result is a photo of starlike points, making constellations recognizable, that then streak off the frame. Add some interesting foreground or a landscape scene to create a "nightscape," and you have an award-winning shot.

# Eclipses

Few events inspire more astrophotographs than do eclipses. Everyone wants a souvenir shot of the big event. Some techniques can produce photos you'll be proud to send to friends or post on your personal website.

## LUNAR ECLIPSES

Glowing deep red against a background of faint stars, an eclipsed Moon is an irresistible target. The simplest technique for recording a lunar eclipse is the same method used to photograph untrailed constellation patterns. During totality, use a 50mm-to-105mm lens and fast film (ISO 400 to 800) and expose for up to 15 seconds at f2 to f2.8. A digital camera that produces good star shots will work well here too.

It is tempting to use a longer telephoto, but on a stationary tripod, the sky's motion will blur the lunar image if exposures are more than 2 seconds for a 300mm lens or 1 second for a 500mm lens. During totality, those exposures may not be long enough

*(continued on page 282)*

▼ **A Streaking Moon**
A single frame on 100-speed film records the entire November 29, 1993, lunar eclipse. A 28mm lens was set to f16 for the partial phases, then covered for 1 minute on either side of the mideclipse shot. Opening the lens at mid-totality for 1 minute at f2.8 produced the unstreaked images of the Moon and stars. Photo by Alan Dyer.

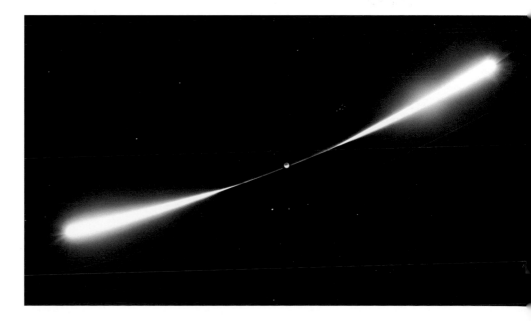

# Buying a Film Camera

For most night-sky shots, digital cameras must have the ability to take exposures of up to several seconds long. While the capability of these cameras is improving yearly, the 35mm camera remains supreme for taking pictures that require exposures of several minutes.

Sadly, most 35mm cameras sold today are poor choices for astro-photography. Some are so automatic that they have no manual over-ride of aperture and shutter speeds. Others allow manual settings but rely on battery power to keep a shutter open for long exposures on a "B" setting. A cold night coupled with a 30-minute exposure is guaranteed to kill the batteries, reducing the camera to a dead high-tech counterweight. For shooting through a telescope, a single-lens-reflex (SLR) camera with removable lenses is essential. Point-and-shoot cameras won't work.

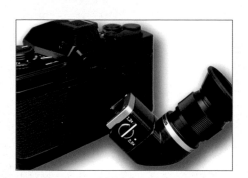

Barely a handful of nonbattery-dependent SLR cameras remain available today. The low-cost Nikon FM10 and most costly Nikon FM2 and Nikon FM3A all have mechanical shutters. The venerable Nikon F3 (in production since 1980) has an electronic shutter that works without batteries at 1/60 second and at the "B" setting. Its interchangeable prisms and focusing screens make it the best of the currently manufactured 35mm cameras for all types of astrophotography.

Most astrophotographers opt for used cameras from the days of mechanical shutters. All that is needed is a camera body in good condition. Lenses can be purchased either used or new. All Nikon lenses will fit older Nikon cameras, but today's Canon lenses may not fit older Canon cameras. With any brand, accessories such as screens and viewfinders may have to be bought used.

The Canon FTb, Contax S2, Minolta SRT101, Olympus 2000 and OM-3Ti, Pentax K-1000 and Yashica FX-3 were mechanical manual cameras that are still adequate for all but the most demanding astrophotography. The high-end Pentax LX had an excellent array of viewfinders and screens. Olympus cameras have long been a favorite of astrophotographers. In 2002, however, Olympus announced that it was discontinuing all 35mm SLR cameras, a sign of the digital age. A good choice for a used film camera is the original 1980s-vintage Olympus OM-1. It featured a lightweight body with a mechanical "B" setting, mirror lockup, interchangeable screens and a selection of good, although not great, Zuiko lenses.

Even so, our first choice in a 35mm system would be a used Canon F-1 (they were last manufactured in 1996), fitted with the unsurpassed L-series aspherical lenses (only Canon's older FD-mount lenses will fit). The F-1 had an excellent array of focusing screens and a superior 6x viewscreen magnifier. However, the viewscreen is both rare and pricey on the used market. A second choice would be a new or used Nikon F3, easier to accessorize than either a Canon F-1 or Nikon's 1960s model F and 1970s F2 cameras. The Nikon F and F2 share the same screens and finders, but those accessories are different from (and harder to find than) the accessories needed by the newer Nikon F3.

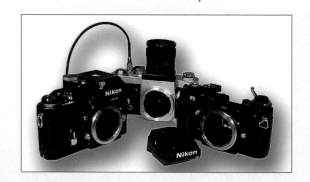

*The high-end Canon F-1 (top) and Nikon F, F2 and F3 cameras (bottom) all have ideal features: nonbattery mechanical shutters on the "B" setting; user-changeable focusing screens; and prisms that can be replaced with a 6x magnifying finder for ease of focusing through a telescope. Other fixed-prism cameras, such as the Olympus OM-1 (middle), require an add-on magnifier that slips onto the pentaprism finder. Either type of finder aid is essential to framing and focusing through a telescope. Only the Nikon F3 is still available as a new camera; all others must be purchased on the used market.*

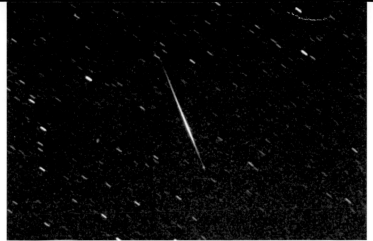

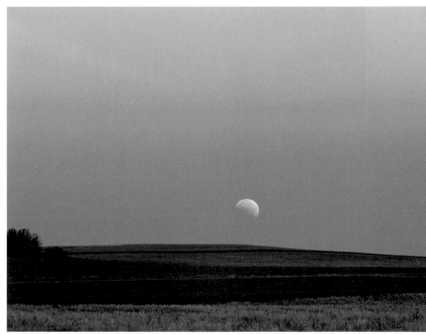

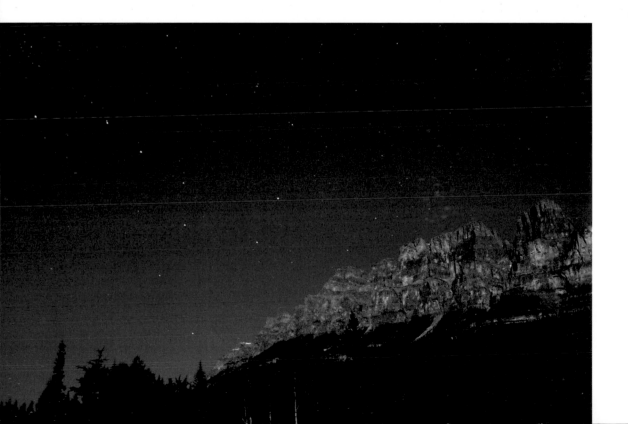

**Astrophoto Gallery**
During an exposure of several minutes, top, the stars trailed and, for just seconds, a brilliant Leonid meteor shot across the frame. Photo by Terence Dickinson. Above: A 200mm telephoto lens captured a partially eclipsed Moon setting into the Earth's blue shadow. Left and above left: A fixed camera with a 28mm lens at f2.8, ISO 400 film and a 40-second exposure records these moonlit scenes like daylight. Three photos by Alan Dyer.

281

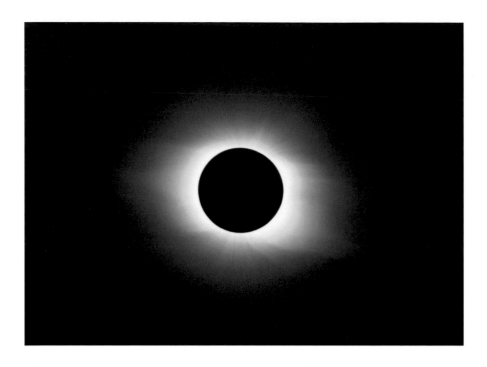

**Caribbean Eclipse ▲**
A telescope was used for this 2-second exposure of the February 26, 1998, eclipse, but a 600mm telephoto lens would produce similar results.

**Eclipse Over Africa ▼**
A tripod-mounted 200mm lens at f4 and Fuji Provia 100F film captured the eclipsed Sun near Jupiter on June 21, 2001, in Zimbabwe. Photos by Alan Dyer.

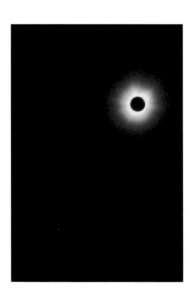

*(continued from page 279)*

to record much, unless you switch to grainy ISO 1000 to 3200 film.

The eclipse-streak photograph is an interesting and easy alternative. Lock the shutter open, and let the Moon move across the frame for the duration of the eclipse. The result is an unusual light streak the width of the Moon that gradually fades from white to red then back to white again. A normal 50mm lens for a 35mm camera has a field of view wide enough to record the continuous motion of the Moon across the sky for 2½ hours, more than long enough to last from the start to the finish of the umbral phases.

To avoid excessive light buildup from sky glow during the eclipse-streak exposure, stop the lens down to f16. Use a slow ISO 50 film. At the beginning or end of the exposure, you might also try a double exposure of an extra 2 to 4 seconds at f2.8 to add foreground landscape details.

A variation of the streak photograph uses the same equipment, but instead of a single long exposure, many short exposures are taken on the same frame at regular intervals. The result is a photograph with multiple Moons in the sky, each recording a different stage of the eclipse. You will need a camera capable of making multiple exposures on the frame. There is no opportunity for bracketing such a photograph, so the

exposure estimates for the partial and total phases must be correct. The motion of the sky carries the Moon its own diameter in about two minutes. Make exposures every 5 to 10 minutes, decreasing from the exposure times needed for a normal full Moon at the start of the eclipse to the long exposures needed during totality.

For an especially creative challenge, combine the streak method with multiple exposures: Have the Moon streak into position, appear singly or in a series of images, then streak off the frame. Hang that photograph on your wall, and it is a guaranteed conversation piece for years.

## SOLAR ECLIPSES

There is no more dramatic astrophoto than one depicting a total eclipse of the Sun. It seems that everyone who makes the effort to see an eclipse wants a photograph of it. Fortunately, total eclipses are among the easiest astronomical events to capture on film. The problem is to contain your excitement enough to shoot the eclipse properly.

The first thought is to use a long telephoto lens. But at several recent eclipses, we have preferred to shoot with no more than a wide-angle lens on a fixed camera. Eclipses with the Sun low in the sky provide an opportunity to include a dramatic landscape to set the scene. The hole-in-the-sky impression created by a total eclipse turns even the most mundane terrestrial setting into a surrealistic image. The eerie twilight glow on the horizon can also be recorded with a wide-angle lens.

Use a slow, fine-grained film, 50mm or shorter lenses and exposures ranging from 1/4 to 4 seconds at f2.8 to f4. This may overexpose the solar corona but will record dimly lit landscape details and bright planets near the Sun.

As with lunar eclipses, the entire sequence of a solar eclipse—whether partial or total—can be captured on one frame. Keep the required solar filter on for every shot during the partial phases. Remove it only for a single image of the totally eclipsed Sun. Make precisely spaced exposures once every 5 to 10 minutes. A 50mm lens should frame the entire sequence, since the Sun and the Moon will move about 30 degrees across the sky during a typical two-hour eclipse. If it is a partial or an annular eclipse, the sequence could be ended with an unfiltered exposure of the whole scene, placing the multiple Suns in a daylit sky. Wait until the Sun moves out of the frame before adding this final touch.

For images of just the eclipsed Sun, use as long a telephoto lens as possible. During totality, use ISO 50 to 100 film. Any film will work, although many photographers prefer print film because of its wide exposure latitude for recording both the bright and the dim parts of the pearly corona.

There is no single correct exposure. Shorter exposures reveal only the brilliant red prominences; successively longer exposures show more of the corona. The best plan: At the onset of totality, start at the fast end of the shutter range to capture Baily's beads and the diamond-ring effect (1/1,000 to 1/500 second at f4 to f5.6), and work down to the slow exposures (1 to 2 seconds), then step back up to the fast speeds for the end of totality.

All the techniques used to photograph total eclipses apply to partial and annular eclipses as well, except that the filter is never removed from the lens or telescope. For telephoto-lens photography up to 400mm, an inexpensive filter is a No. 10 to No. 14 welders' filter. Never use anything lighter than a No. 10 for photography, and *never* look at the Sun for extended periods with a filter lighter than a No. 14. Welders' filters are not optically flat but will provide acceptable images when securely taped over the front of the lens. They produce a green Sun.

Never use photographic neutral-density filters, crossed polarizers or homemade devices. Only approved astronomical solar filters or welding filters block all the harmful infrared and ultraviolet light that can damage the retina, even through a camera viewfinder. Exposures for the partially eclipsed Sun or for an annular eclipse depend on the filter but will probably fall into the range of 1/125 second at f8 with ISO 100 film. Since the Sun's surface is the same brightness whether it is being eclipsed or not, take test shots well in advance of eclipse day to determine the best type of film to use and the proper exposure times.

The secret of a good eclipse photograph is to select one technique, then plan and rehearse the steps over and over again. The limited time available during the big event is *not* when you want to be learning how to do astrophotography, although it inevitably happens. Minutes before totality, a cry goes out: "Hey, what exposures do I use?" or "Can someone help me? I can't get my camera to focus." For want of a little preparation and practice, the hapless photographer misses the chance of a lifetime.

As an alternative or complement to still images, a video camera on a sturdy tripod does a wonderful job on eclipses. And it records all the ambient sounds of excited eclipse chasers, capturing the feeling of the eclipse as no still image can. Camcorders with 10:1 or, better yet, 20:1 zoom lenses are best. The auto-exposure setting usually provides a good image, though the blinding diamond ring just before totality might confuse the camera's auto-focus. Set the focus to manual, and adjust it for infinity.

If there is time during totality, try varying the shutter speed to show more of the outer corona. Or zoom out to reveal any nearby planets. The diamond ring will likely cause the camera's CCD chip to bloom, creating a vertical spike of light. That does not damage the camera, but be sure to filter, cap or turn the camera away after totality, or the returning sunlight could fry the inside of your camcorder.

One final bit of advice: Don't forget to *look* at the eclipse. A binocular view of a totally eclipsed Sun will imprint in your memory with more lasting value than any photograph. In fact, for your first total solar eclipse, consider just looking!

## Solar Eclipse Exposures

........................

ISO 100 film at f5.6

Partial Phases
1/500 second
(filter)

Diamond Ring
1/250 second

Prominences
1/500 second

Outer Corona
1/2 second

Wide-Angle Scene
2 to 4 seconds at f4

Inner Corona
1/30 second

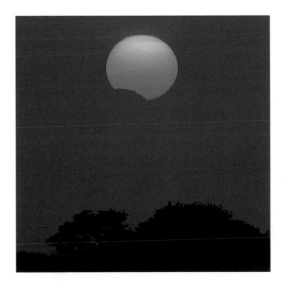

**◀ A Bite From the Sun**
A partial eclipse of the Sun, like this one seen from the Great Lakes area in June 2002, normally requires a solar filter to view or photograph safely. But in this case, summertime haze provided the filter, allowing the event to be viewed and photographed au naturel. Digital-camera image through a 92mm refractor by Steve Barnes.

# Shooting the Sky II: Using a

Many new to the hobby of astronomy want to take pictures through their telescopes. Consider this chapter a guide to doing just that.

Some types of telescopic shots are easy. Most, however, are demanding. While we want to encourage backyard astronomers to take up astrophotography as a rewarding part of the hobby, we offer this counsel: Take it slowly. Learn to swim in the shallow end before leaping into the deep end.

It may be tempting to jump directly into imaging with CCD cameras attached to long-focal-length telescopes, but getting good results is not the turnkey process some manufacturers would have you believe. Even buying all the right gear, at a cost of several thousand dollars, is no guarantee of great results. In our experience, the secret to astrophotography success is to start simple and develop your skills.

We concentrate first on tech-

niques for getting good images through a telescope simply and cheaply. We then move on to piggyback photography, a technique that is often dismissed by aspiring astrophotographers yet is capable of yielding results second to none for beauty and spectacle. Mastering piggyback photography teaches the basics of polar alignment, focusing and guiding, essential lessons for shooting nebulas and galaxies through a telescope, even with the ultimate in imaging technology: advanced CCD cameras.

This outstanding 90-minute guided exposure of the brilliant Milky Way over Chile was taken in April 2002 by Matt BenDaniel. He used a Pentax 6x7 medium-format camera with a 35mm fisheye lens and Kodak E200 slide film.

elescope

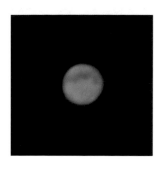

**Mars in Color** ▲
A modern digital camera teamed up with an 8-inch or larger telescope should produce results like this on an average night when Mars is at opposition. The dark feature at right is Syrtis Major. This single Nikon Coolpix 995 image is only slightly contrast-enhanced.

**Moon Shots Are Easy** ▶
With the proper adapter for your digital camera, lunar portraits are a snap. Coolpix 880 image by Alan Dyer.

**First-Night Digital** ▼
Armed with a Nikon Coolpix 995 digital camera, Terence Dickinson took this shot on the first night the camera was attached to a 6-inch Astro-Physics refractor.

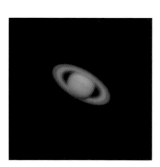

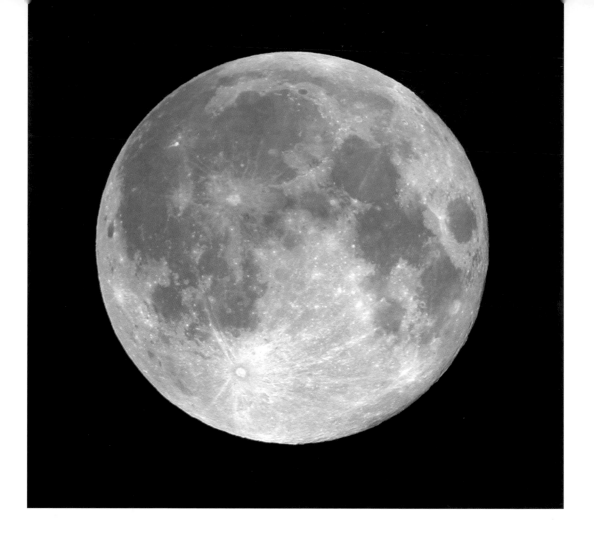

# Shooting the Solar System: Digital

By far the easiest celestial objects to shoot through a telescope are the Sun and the Moon. And by far the easiest way to image them is with a digital camera. Most exposures require only a fraction of a second, and the digital camera displays the results immediately, allowing you to check focus, composition and exposure, make any necessary adjustments and try again. With the Moon, you can expect decent results from your very first session. This is the closest thing in astrophotography to instant gratification.

## DIGITAL ADAPTERS

Acceptable images of the Moon can be taken by simply hand-holding a digital camera up to the telescope eyepiece. Since exposures are short—often around 1/125 second—hand shakes are reduced or eliminated, producing a reasonably sharp picture. Not that we recommend this nontechnique, but it does work.

When hand-holding is replaced with a camera tripod to cozy the camera up to the telescope eyepiece, the chances of bagging a good shot increase. In both cases (handheld and tripod-mounted), the camera lens is held as close as possible to and directly behind the eyepiece. This is called afocal imaging, although without a proper adapter to hold the camera in place, it is also called just fooling around.

To get serious and to make afocal digital imaging much easier, a specially designed adapter is essential. The adapter serves two purposes: to couple your digital camera rigidly to the eyepiece and to align the camera's optics precisely along the same axis as the telescope axis. Because consumer-level digital cameras were never intended to be mated with telescopes in the first place, a number of them require more complicated (some would say clunky) adapters than others.

Since most digital cameras have zoom lenses, you may need only one or two eyepieces to achieve a full range of effective imaging magnifications (e.g., a 30mm Plössl for wide views of the lunar disk and a 10mm Plössl for close-up views of the Moon and planets). Digital-camera designs vary, so experimentation with afocal setups is the only guide. The most common problem is vignetting, the looking-down-a-pipe effect that can appear at either end of the zoom lens's range. Some eyepiece/camera combinations are better than others.

In our tests, digital cameras with fully internal zoom systems (i.e., the camera's lens does not move in and out of the camera body when zooming) are most easily adapted to afocal imaging. The idea is to get the front of the camera's zoom lens fixed as close to the eye lens of the eyepiece as possible to suppress vignetting. Favorite cameras in this regard are the Nikon 990 and 995, both 3-megapixel models, and the upgraded 4-megapixel version, the Nikon 4500. The Sony DSC-F707, recommended earlier for non-afocal imaging, is a beautiful 5-megapixel camera with an internal zoom, but the weight and bulk of its big f2.0 lens work against it when it is affixed to an eyepiece in afocal mode. Expensive SLR-type digital cameras with removable lenses, such as the Nikon D100, can be attached to telescopes as well, using the same prime-focus adapters as for film cameras. However, the results aren't necessarily better than with lower-cost, fixed-lens cameras shooting into an eyepiece.

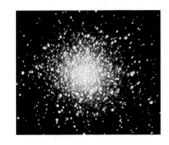

▲ **Digital-Camera M13** This composite of a sequence of stacked 16-second images of the globular cluster M13 was taken with an Olympus 3040Z through a 14.5-inch Starmaster Newtonian. Image by Ken Schmidt.

# Astrophoto Accessories I: Adapters

To attach a 35mm SLR camera to a telescope's focus requires one of two types of camera adapters. One slides into the focuser just as an eyepiece does and is best for refractors and Newtonian reflectors; the other, for Schmidt-Cassegrains, screws onto the threaded plate at the back of the instrument in place of the usual visual back. The camera end of these adapters contains so-called T-mount threads. To attach a camera body to the threads, a T-ring is needed to fit the brand of camera. T-rings are sold in most camera stores and by many telescope dealers. In either case, there is no eyepiece on the telescope and no lens on the camera. The telescope, in effect, becomes the lens, in what is called prime-focus photography.

To couple a fixed-lens digital camera directly to an eyepiece requires a specialized adapter made for your specific camera. Fortunately, several manufacturers offer adapters to fit most popular cameras. However, the other end of the adapter will only fit eyepieces with removable rubber eyecups (see photo, bottom of page 288). Not to worry—rubber eyecups are common on eyepieces manufactured since about 1990. Tele Vue also has digital adapters specifically for its Panoptic and Radian lines of eyepieces, which are particularly well suited for lunar and solar imaging.

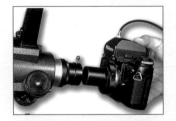
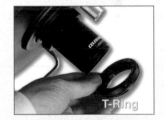
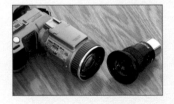
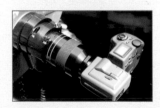

*Single-lens-reflex (SLR) 35mm cameras attach to the focuser of a telescope with a nosepiece-style adapter (top left) coupled to a T-ring made for the user's particular camera brand. Schmidt-Cassegrain telescopes require a special T-adapter tube and T-ring (top right). The ScopeTronix adapters (bottom left) mate digital cameras to various eyepieces (Maxview 40 shown here). Tele Vue's adapters (bottom right) work with its larger Panoptics and Radians (Nikon Coolpix 995 and a Panoptic 35mm shown).*

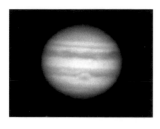

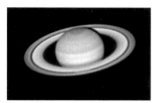

**Imaging Giant Planets** ▲
Two excellent planetary portraits (Jupiter by Gordon Bulger; Saturn by Ken Schmidt) are stacks of one to two dozen images from 3-megapixel digital cameras. Bulger used a 12-inch Schmidt-Cassegrain, and Schmidt used a 14.5-inch Newtonian.

**Attaching a Digital Camera to a Telescope**
Unlike film cameras, digital cameras work best when used afocally (i.e., attached directly to the telescope's eyepiece). The most popular adapter system utilizes the groove that holds the eyepiece's rubber eyecup. In the photo at right, the eyecup has been removed. The thin adapter ring is placed over the groove and secured by the small wrench. A ring with threads to fit the camera's filter thread is then screwed on. Far right, top: A Nikon Coolpix 995 with this adapter in place. Far right bottom: Maxview 40 adapter and Sony F707 combination shown on page 287. ▶

## DIGITAL TECHNIQUES

Digital cameras provide settings for almost every conceivable imaging opportunity, except afocal shooting through a telescope. We have found that experimentation should be the rule to establish what works and what doesn't. For the Moon and properly filtered white-light images of the Sun, most digital cameras will focus themselves with the auto-focus switched on. (Important: *Never* attempt solar photography without reading the section on proper filters on page 178.) Try the macro-focus setting too, as many cameras seem to produce better results on this setting. Other cameras perform better on the Moon with the auto-focus switched off and the manual focus at infinity and adjusted with the telescope's focuser. Approximate exposure will be set by the camera, but for these unorthodox shooting subjects, some camera exposure meters read too "hot." You may want to try to override the meter by adjusting to underexpose from a half-stop to two stops. In all cases, framing is readily achieved in the usual way with the camera's LCD monitor.

## SHOOTING THE PLANETS

Traditional planetary photography—using 35mm SLR film cameras, eyepiece projection and (typically) 1-to-4-second exposures—is one of the toughest assignments in astronomy. Take our word for it. From time to time over the past quarter-century, both of us have attempted to photograph detail on the surface of our neighbor planets. Rolls and rolls of film have been exposed in this elusive and frustrating pursuit. What do we have to show for it? Not much. Because of blurring from atmospheric turbulence during the exposure, images with eyepiece projection onto film are inevitably fuzzy. We'll break with tradition and declare this film technique dead.

Compare that with the first night out with a digital camera. Immediate success! Saturn's rings are well defined, Jupiter displays its belts, and the deserts, polar caps and dark areas of Mars are clearly evident. Why the difference? A big part of it is the ability to see results almost instantly and to react by adjusting focus, exposure or whatever to improve subsequent images. Second, digital cameras have no mechanical shutters or other moving parts like film cameras, thus eliminating opportunities for vibration and blurring during the exposure. (Use the self-portrait timer for hands-off, vibration-free imaging.) Third, digital images have higher intrinsic contrast than does film, and therefore, shorter exposures capture satisfying detail on subjects like planets. Fourth, digital images can be enhanced through the use of software like Adobe Photoshop and/or by being electronically stacked to bring out subtle detail. And the list goes on.

But even though imaging experts were aware of the digital advantage with regard to the planets, no one grasped the full magnitude of what off-the-shelf consumer digital cameras could do until 1999, with the introduction of higher-resolution digital cameras. It was then that amateur astronomers began taking stupendous planetary

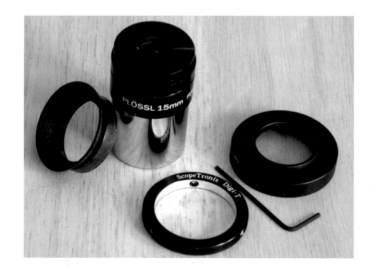

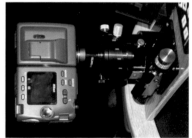

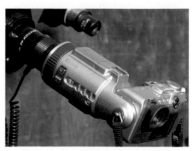

## VIDEO ASTRONOMY

The same afocal technique used for digital still cameras can be used with a video camcorder. However, because of the weight of most camcorders, eyepiece-to-camera coupling rings don't work well here. The best method for shooting astro-movies is to employ a bracket that clamps around the eyepiece and holds the camera by its tripod socket, allowing it to point through the eyepiece and ride along with the telescope. For small palmcorders, the same brackets sold for still cameras work fine. Large video cameras might require a custom-made bracket. A better alternative is to purchase a dedicated video camera sold for astronomy, such as one of the Adirondack units. These low-light surveillance cameras have high sensitivities below 0.1 lux and high resolutions of 400 to 600 video lines.

Now feed your live or taped video into a computer (directly via Firewire or through a video capture card). The video can be stepped through frame by frame with video-editing software such as Adobe Premiere, Sony's MovieShaker or Apple's iMovie to select only the frames when the atmosphere steadied for a fraction of a second. Turn several of those sharp frames into still images, then superimpose them digitally to smooth out noise, and you have created images of the Moon and planets that record far more detail than was ever possible with conventional film. The secret is a night of steady seeing conditions. If the air is perpetually turbulent, no amount of technological magic can produce sharp images.

images compared with their previous results on film. Almost any small telescope fitted with a digital camera using the adapters mentioned above, for example, will easily capture the main belts of Jupiter.

If you have access to it, more aperture is a real advantage here, because the extra image brightness permits faster shutter speed, which in turn increases the chances for you to catch moments of steady seeing and sharp imagery. As off-the-shelf digital cameras continue to improve in the years ahead, backyard astronomers will increasingly enjoy casual and satisfying planetary imaging as a standard part of the hobby, rather than the difficult sideline activity it was for decades.

◀ **Digital-Camera Comet**
This is a single (i.e., non-stacked) image of Comet Ikeya-Zhang taken with a Sony F707 digital camera. The zoom lens was set at 190mm at f2.5, and the camera was tracked for the 30-second exposure. Images by Terence Dickinson.

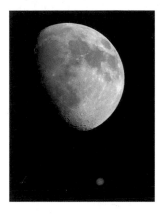

▲ **Conjunction Magic**
When Jupiter was within 0.3 degree of the Moon on February 22, 2002, both objects were imaged with a Nikon 995 digital camera. Later, in Photoshop, the Jupiter image (a mere dot) was replaced with a larger, more detailed image taken that same night.

## Astrovid Camera

The best option for astro-videos is a dedicated low-light camera such as the Adirondack Astrovid 2000. While a separate camcorder is still needed to record the signal, the advantages of these cameras are their tiny size, 12-volt operation, control box (top right) separate from the camera (for remote vibration-free adjustments) and lack of any lens for direct mating to a telescope. Such cameras can record stars down to ninth magnitude.

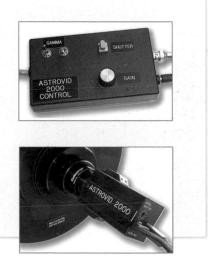

# Shooting the Moon and Sun: Film

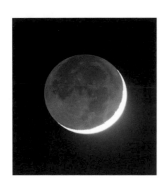

While we do not recommend film for shooting planets, 35mm film cameras offer analog traditionalists the possibility of wonderful shots of the Moon and the Sun with any telescope. Exposures are short enough that a clock drive, even an equatorial mount, isn't essential, though they do help to keep the image centered during a shooting session. But if you have a Dobsonian telescope, this is one type of photo you can take. How long an exposure you can use before the motion of the sky blurs the image depends on the focal length of the telescope. As with fixed-camera constellation shots, the longer the focal length, the shorter the exposure must be to avoid blurring. See the table below for suggested limits.

## THE BEST FILMS AND EXPOSURES

With digital cameras, you can see immediately whether you have selected the best exposure. When using film, the key is to realize that the Moon is simply a sunlit gray rock. If you are photographing a sunlit rock or scene on Earth, follow this rule of thumb: At f16, set the shutter speed at the reciprocal of the ISO speed of the film. For example, ISO 100 film at f16 under full sunlight would be correctly exposed at 1/125 second. The same rule applies to the full Moon. In practice, some light absorption by our atmosphere usually means increasing the exposure by about one f-stop, or one shutter-speed increment. Therefore, the correct exposure for the full Moon at f16 with ISO 100 film is 1/60 second, not 1/125.

## Solar and Lunar Image Size

| Lens Focal Length | Image Size on 35mm Film | Maximum Exposure to Avoid Trailing |
|---|---|---|
| 50mm | 0.45mm | 12 seconds |
| 100mm | 0.9mm | 6 seconds |
| 200mm | 1.8mm | 3 seconds |
| 500mm | 4.5mm | 1 second |
| 1,000mm | 9.1mm | 1/2 second |
| 2,000mm | 18mm | 1/8 second |

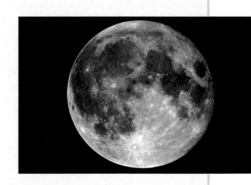

The Sun and the Moon are tiny targets. To capture an image large enough to show features such as sunspots or craters and mountains, use a 500mm or longer telephoto lens or a telescope. To determine the solar or lunar image size with a particular lens or telescope, divide the focal length (in millimeters) by 110 to arrive at the disk diameter (in millimeters) on the film. With an undriven telescope, use exposures no longer than those listed above to prevent the image from trailing.

............................................................................................

*It takes a focal length of 2,000mm, as here with a Celestron Ultima Schmidt-Cassegrain, to make the Moon large enough to fill the short dimension of a 35mm film frame. This image of the full Moon has been processed for high contrast to enhance subtle tonal variations across the lunar disk.*

# Exposure Guide for Lunar Photography (f16*)

| ISO Film Speed | Full Moon | Gibbous | First Quarter | Thick Crescent | Thin Crescent | Earthshine |
|---|---|---|---|---|---|---|
| 3200 | 1/2,000 | 1/1,000 | 1/500 | 1/250 | 1/125 | 2 to 5 sec. |
| 1600 | 1/1,000 | 1/500 | 1/250 | 1/125 | 1/60 | 5 to 10 sec. |
| 800 | 1/500 | 1/250 | 1/125 | 1/60 | 1/30 | 10 to 20 sec. |
| 400 | 1/250 | 1/125 | 1/60 | 1/30 | 1/15 | 20 to 40 sec. |
| 200 | 1/125 | 1/60 | 1/30 | 1/15 | 1/8 | 40 to 80 sec. |
| 100 | 1/60 | 1/30 | 1/15 | 1/8 | 1/4 | N/A |
| 50 | 1/30 | 1/15 | 1/8 | 1/4 | 1/2 | N/A |
| 25 | 1/15 | 1/8 | 1/4 | 1/2 | 1 | N/A |

*Use one shutter speed faster around f11, two shutter speeds faster around f8, and so on. As a general rule, though, always bracket exposures one or two shutter speeds either side of values given to allow for atmospheric absorption and other vagaries. Use a lunar-rate drive for longer Earthshine photos.

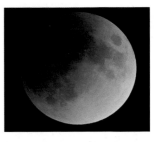

▲ **Total Lunar Eclipse**
This image of the eclipse of January 20, 2000, required a 60-second exposure with the selected Ektachrome SW film, using a 5-inch refractor and a 2x Barlow lens for an effective focal length of 1,600mm and a focal ratio of f/12. Without the Barlow, exposures would have been just 15 seconds, but the lunar disk would have been half the size.

What if another focal ratio is used? Suppose you have an f/10 telescope, which is one full f-stop faster than f/16. (Remember that the photographic f-stops run f22, f16, f11, f8, f5.6, and so on.) A gain of one f-stop means the shutter speed will be twice as fast. For the full Moon, the exposure becomes 1/125 second instead of 1/60. The relationship between f-stop and shutter speed is reciprocal: As the aperture is increased, the exposure is decreased, and vice versa.

What about other lunar phases? A general rule is that a decrease of one lunar phase requires a doubling of exposure. If a full Moon requires 1/125 second at f/10 on ISO 100 film, then a gibbous Moon needs 1/60 second, a quarter Moon 1/30 second, a thick crescent 1/15 second and a thin crescent 1/8 second. Even so, no astrophotography exposure is guaranteed. It is always a good practice to bracket your shots—take one to two shutter-speed increments either side of the recommended exposure.

Shooting the Sun requires similar techniques, but *never* attempt solar photography without a filter. Exposure depends on the density of the filter used. Both Mylar and metal-on-glass filters vary several f-stops in density from filter to filter, even while staying within the range of safety. In general, a good starting assumption is that the filtered Sun is the same brightness as the full Moon.

One or two test rolls should narrow down the correct exposure for your setup.

By using a Barlow lens, you can double or triple the image size of the Moon or the Sun. Keep in mind that images with a 2x Barlow require four times the exposure (an increase of two f-stops) over straight prime-focus images.

Because the Moon and filtered Sun are so bright and offer so much fine detail to record, the best images are captured on fine-grained film. Any ISO 50 to 100 film, either print or slide, will work fine. A favorite of ours is the amazingly fine-grained Fuji Provia 100F slide film.

## SHOOTING ECLIPSES WITH A TELESCOPE

Shooting a total eclipse of the Moon is not much different from taking Earthshine shots. During the partial phases, exposures for the sunlit part are in the same range as exposures for the normal full Moon, although the exposure times increase as the eclipse progresses. Just before the beginning of totality, the exposure for what is left of the bright area of the Moon will be three or four f-stops greater than at the beginning of the partial umbral phase. Exposures for the shadowed half of the Moon are from 5 to 20 seconds with ISO 400 film—the same range

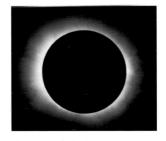

▲ **Total Solar Eclipse**
A clock-driven Questar 3.5-inch Maksutov at f/16 and Kodachrome 25 slide film were the winning combination for the February 26, 1979, total eclipse in southern Manitoba. The motor drive ensured the sharpest images, even during 1/2-second exposures with the 1,400mm-focal-length telescope. Both photos by Alan Dyer.

required to record Earthshine. Exposing for the darkened sector will greatly overexpose the sunlit side. This is fine just before and after totality, but in the early and later stages of the partial phases, it is best to expose for the bright half and let the dark part disappear. Always bracket over and under by at least two shutter-speed increments.

A slower film (ISO 50 to 100) could be used for the partial phases. During totality, switch to a faster ISO 200 to 400 film. For most lunar eclipses, exposures during totality range from 10 to 60 seconds at f/10. With Schmidt-Cassegrains, using a telecompressor to yield f/6.3 is a good idea. That should shorten exposures to well under 20 seconds. For dark lunar eclipses, however, exposures can extend up to 120 seconds, even at f/6.3. In all cases, for the sharpest image of the lunar disk, the mount should be polar-aligned and the drive running at the slower lunar rate. If you are shooting with a telescope on a computerized altazimuth mount, use a telecompressor and fast film to keep exposures as short as possible and to minimize blurring from field rotation.

During total eclipses of the Sun, spectacular shots of prominences and corona details require focal lengths of 800mm to 2,000mm. This means shooting at the prime focus of a telescope. While an equatorial mount and clock drive are not essential, they keep the Sun's image centered during the intense period of activity around totality.

Exposures for prime-focus telescope shots are about 1/500 second for the diamond ring, 1/125 for the prominences, 1/15 for the inner corona and 1/4 to 1 second for the outer corona, assuming f/8 with ISO 100. Fast film is not necessary—the delicate details of the corona and prominences call for fine grain.

# What Can Go Wrong: Short Exposures Through a Telescope

Snapshots of the Sun, Moon and planets through a telescope can be improved by implementing a few simple tips and techniques.

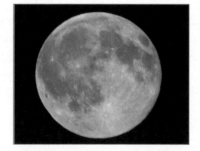

### Out-of-Focus Images
For 35mm cameras that allow this, replace the focusing screen with a fine-grain laser matte screen or an ultrabright Beattie Intenscreen. Even the Moon can look dim and grainy and be hard to bring to a focus using a normal focusing screen. A bright screen is also essential for seeing and framing deep-sky objects.

### Blurry Images
With film cameras, shutter and mirror slap can introduce vibration. Locking the mirror up first (possible only on some cameras) helps reduce vibration. With digital cameras, be sure they are held securely in place with a good adapter; hand-held exposures longer then 1/8 second are bound to blur.

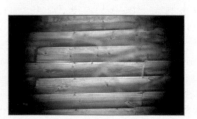

### Vignetted Shots
Lunar shots with digital cameras shooting afocally through an eyepiece often appear cropped or vignetted, as is this digital image of the side of a barn taken through a telescope. The solution is to zoom in or try a different eyepiece.

*To prevent film images being blurred from mirror slap, like this one (top), use the hat trick, which is good only for exposures longer than 1/2 second. Hold a black card over the front of the lens or telescope (nobody really uses a hat), and open the shutter to "B." Wait a few seconds for vibrations to die down. Quickly flip the card away for the duration of the exposure. With the card back in front of the lens, close the shutter. To prevent vignetted images with digital cameras (above), try zooming in.*

# Deep-Sky Piggyback Photography

Lunar and planetary astrophotographers strive to record all the detail visible through the eyepiece. In deep-sky astrophotography, long exposures capture details and objects our eyes cannot see, revealing an otherwise invisible universe. The easiest method for taking deep-sky images is to attach a camera to the side of a driven telescope; the camera then tracks across the sky as it goes along for a piggyback ride. You shoot through the camera's lens, not the telescope.

Perhaps this is an old-fashioned edict, but we feel that all aspiring deep-sky astrophotographers should first graduate from the school of piggyback astrophotography. Only when you have mastered sharp, untrailed piggyback exposures should you venture into the arduous world of prime-focus, deep-sky work—the through-the-telescope shots. But piggyback astrophotography is not just a steppingstone to higher accomplishments. Far from it. It is a powerful technique that can produce stunning results. In fact, piggyback work may give you all the photographs you need to satisfy the urge to capture glowing deep-sky objects in star-filled frames.

## FILM'S LAST DOMAIN

Digital snapshot cameras are getting better every year, able to produce cleaner images in long exposures. Someday, they may replace film completely, but for now, the exposures of several minutes to an hour needed for piggyback subjects demand tried-and-true film. Digital still cameras are too noisy. Even the cooled-chip CCD cameras designed for long-exposure imaging at the telescope have limited use for wide-field photography. For example, Santa Barbara Instruments' big-chip ST-10 camera still produces only a six-by-four-degree field of view when coupled to a 50mm lens. With this lens, the ST-10's 6.8-micron pixels (as small as any camera's) yield an image scale of 10 arc seconds per pixel. That's coarse enough to produce square-pixelated stars under moderate enlargement.

Piggyback shooting is at its best when recording nebula- and cluster-rich regions of the Milky Way, where nebulas are glowing islands of red set in a sea of stars. The Milky Way star clouds can be framed properly only in shots with fields of 10 degrees or more. Some expansive nebulas, such as Barnard's Loop, IC1396 and the Gum Nebula, are also best recorded with 50mm-to-300mm focal lengths. Recording entire constellations usually requires a field of at least 30 degrees, possible only with 50mm lenses or shorter. Some of our favorite shots are ultrawide panoramas of the entire 90-degree sweep of sky from horizon to zenith. These depict the sky as you remember seeing it but require lenses shorter than 20mm.

▲ **Only With Film**
Sharp, high-res images of expanses of the Milky Way and entire constellations are still the domain of film, either 35mm or larger formats, such as 6cm x 7cm. Film photographers, take solace: For this subject, a $20 roll of film outperforms a $10,000 CCD camera and is likely to for many years. This 15-minute exposure of the winter Milky Way was taken on Fuji SuperG 800 film with a 16mm fisheye lens at f3.5. Photo by Alan Dyer.

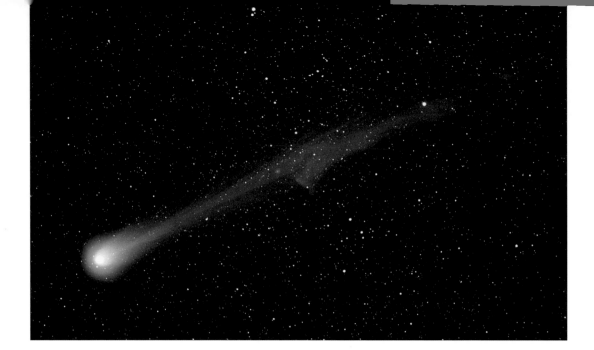

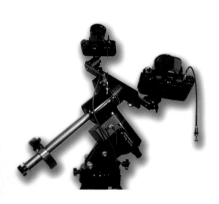

During a total lunar eclipse, a camera riding piggyback on a telescope offers dramatic possibilities. Use a 50mm-to-500mm lens and ISO 200 to 400 film to take exposures of up to 10 minutes during totality. The result records myriad background stars framing a blood-red Moon.

## PIGGYBACK GEAR

The principal requirement for piggyback photography is a clock-driven equatorial mount for your telescope. The mount must be polar-aligned. We prefer German equatorial mounts equipped with polar sighting scopes for quick and accurate alignment.

Mounting a camera on a computerized telescope that is set up in the altazimuth mode simply does not work. Yes, the telescope is tracking the sky, but the entire field of the telescope and of any camera attached to it is slowly rotating around the target object. Instead, fork-mounted telescopes must be placed on a wedge and polar-aligned to the celestial pole.

The mount must have an accurate motor on the polar axis, turning at the sidereal rate. If it turns consistently too fast or too slow, the result will be trailed stars. Ideally, the mount should not suffer from major periodic or random speed errors originating in the drive gears. A periodic-error-correction (PEC) circuit is a good feature to have, assuming that it is properly trained to eliminate the gradual back-and-forth drift induced by even the best motor-and-gear mechanisms. The best mounts, provided they are accurately polar-aligned and PEC-trained, can be left to track unattended and unguided for 30 minutes or more.

The tracking accuracy becomes more demanding in proportion to the focal length of the lens being used. When wide-angle and normal lenses (14mm to 50mm) are used, most mounts can be left to run on their own, at least for short exposures with fast film. Switch to a telephoto lens, and you may notice some trailing creeping in. This could be a sign that more accurate polar alignment is needed, or it may mean a mount with a more accurate drive is required. Polar alignment is unavoidable, but you can coax better performance out of your existing mount by helping it along. This is called guiding.

The required accessories for guiding are an illuminated-reticle eyepiece and a drive equipped with an electric push-button, slow-motion control in at least the right ascension axis (a dual-axis drive is better). Inserted into the telescope, the illuminated eyepiece allows you to position a set of lit crosshairs on a star. Ideally, the telescope's drive should automatically keep the star centered on the crosshairs, ensuring pinpoint images on the film. However, because of mechanical-drive errors and misaligned mounts, the star will wander off. The trick is to jog the star back to the center using the right ascension or declination buttons and keep it there no matter what. The electric slow-motion controls

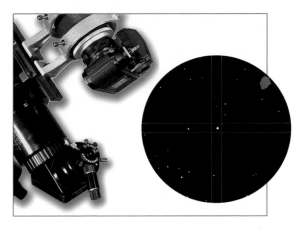

allow the scope to be fine-tuned without having to touch it and thereby introduce more jitters. The longer the focal length of the shooting lens, the more precise your guiding must be. If, when using telephoto lenses and a recommended guiding magnification of 100x or more, the guide star spends any time outside the box, start again.

◀ **Set to Guide**
Guiding eyepieces are essential for long exposures with telephoto lenses. Guiding reticles with two sets of parallel crosshairs that form a central box are best (stars can get lost behind single intersecting crosshairs). Units with movable reticles, like this Celestron model, are also recommended. Avoid models with complex grids and scales, as they will flood the field with light, masking guide stars.

## Astrophoto Accessories II: Mounts and Brackets

Many entry-level telescopes made in China come with tube rings equipped with a 1/4-20 bolt to which a camera can be attached directly. Schmidt-Cassegrain owners can equip their scopes with a standard accessory bracket that bolts to the back of the tube, which also has a 1/4-20 bolt. While these can work as is, the camera's aim point is fixed, with no ability to frame the field or turn the camera vertically.

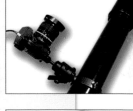

To achieve freedom of movement, buy a sturdy ball-and-socket tripod head (one that has a 1/4-20 bolt hole in the bottom, not the ⅜-inch threaded hole often used to attach some heads to tripod top plates). The Manfrotto #352 head (aka Bogen #3262) does a good job. Don't scrimp. A lightweight ball-head will only sag under the camera's weight, creating another source of trailed stars.

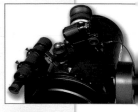

Now you can freely frame the image the way you wish. With wide-angle lenses, be sure to aim the camera so that the telescope tube itself is not in the frame. Indeed, it is not even necessary to have a telescope present. If the mount tracks well enough that guiding is not essential, dispense with the telescope and just use the mount. If you are shooting with a long telephoto, however, the telescope serves as a guidescope. In that case, the lens and telescope should be coaligned fairly closely. If they aren't coaligned, another gremlin called field rotation will be introduced into the final photo.

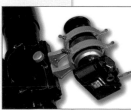

Long telephoto lenses pose a special challenge. Their weight can twist the camera around its attachment point as the camera arcs over the sky. You think it is solidly mounted, but in fact, it is slowly slipping out of position. Telephotos longer than 135mm should be clamped front and back. No one makes brackets specifically for this purpose. The best solution is a ring system such as the one required to mount a separate guidescope.

*Piggybacking a 35mm SLR camera to a telescope can be done directly to an adapter bracket on the tube, but much better is a ball-and-socket tripod head (top and center), which provides freedom to turn the camera to frame the subject. Larger telephoto lenses should be held front and back (bottom) to prevent them from twisting under their own weight.*

# Barn-Door Tracker

Attaching your camera to a commercial equatorial mount is not the only way to take guided astrophotos with normal 35mm camera lenses. An inexpensive alternative, known as a barn-door tracker, costs just a few dollars in materials and can be put together in an afternoon. You need no special equipment to build it and no motors or batteries to run it. It's driven by you, turning a hand crank. This is as low-tech as you can get, but it works.

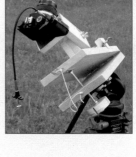

The principle behind the barn-door tracker is that once its hinge parallels the Earth's axis, it mimics an equatorial mount. Turning a hand crank that pushes on one of the two hinged boards compensates for the Earth's rotation. The camera, attached to the moving board, tracks the stars.

The most expensive part of the barn-door tracker is the heavy-duty camera tripod to hold the device. (You will want to have one of these for camera-on-tripod shots in any case.) The other materials are very inexpensive; some might even be found around the house. Specific details of the design are up to you, but here are the basics.

In the model shown here, which features the minimum requirements, a sawed-off section of a piano hinge was used, but two closet hinges will also work. (Large door hinges tend to be too heavy and wobbly.) The drive bolt is a 4-inch-long 1/4-20 carriage bolt. It threads through a 1/4-20 T-nut. The other nuts, bolts and wing nuts are visible in the photos. The only crucial measurement is the position of the hole that contains the drive-bolt T-nut. It should be exactly 11$\frac{7}{16}$ inches from the center point of the hinge axis. Other than that, feel free to be creative.

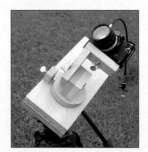

To secure the unit to a standard camera tripod, the bottom leaf needs a 1/4-20 threaded tripod hole, like the one found on the bottom of cameras. Adequate accuracy for a typical 5-minute astrophoto with a 50mm-focal-length lens or shorter is achieved by gently cranking the drive bolt clockwise a quarter-turn every 15 seconds. Quarter-turn marks on the drive-bolt knob and a dim red flashlight make this easy. To work properly, the axis of the hinges must be aimed at Polaris, the North Star, with the drive bolt on the right as you face north. Start each photo with the boards almost closed. The bolt cranks the boards open to compensate for the Earth's rotation.

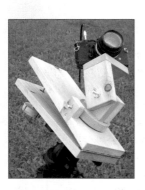

The barn-door tracker pictured here is an ultraeconomy model developed by Kevin Kell and Tom Dean for members of the Kingston Centre of The Royal Astronomical Society of Canada. A desirable upgrade to this unit is a ball-head mount (available at all camera shops) to replace the wood bracket holding the camera.

Barn-door trackers similar to this one are perfectly adequate for lenses in the 24mm-to-50mm range. Today's 400- and 800-speed films will be well exposed in 3 to 5 minutes with the lens at f2.8. For well-tracked photos with longer-focal-length lenses, we recommend a commercial equatorial mount with motor drive and polar-alignment scope to ensure accurate polar-axis positioning on the north celestial pole.

*Anyone who is moderately handy with a saw and a screwdriver can be an astrophotographer using this simple tracking unit. Once the axis of the hinge (bottom photos) is aimed at Polaris, the setup is complete. Turning the hand crank pushes the upper board carrying the camera at the exact rate to compensate for the Earth's rotation.*

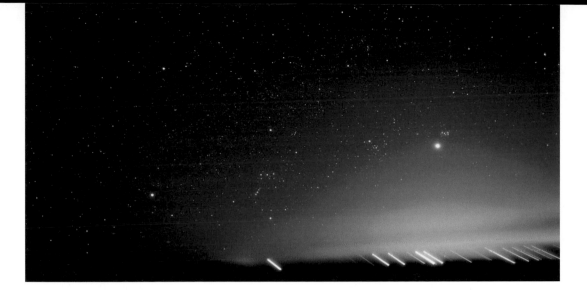

## SKY REQUIREMENTS AND EXPOSURES

For the deepest, richest piggyback shots, a pristine sky is a prerequisite—the darker, the better. In some respects, piggyback shooting is the most demanding of all astrophotography for sky quality. Modest horizon glows can spoil wide-angle images or add an annoying gradient tint across the image that is difficult to remove even with computer processing. Sky fog is always worse when photographing toward a city.

Generally, the longer you can expose a piggyback shot, the more stars and nebulosity you'll record—to a point. That point is usually dictated by the sky darkness. Typically, at a dark site, an exposure of 12 minutes at f2.8 with Ektachrome E200 film is more than enough to pick up reams of nebulosity. An ISO 400 film such as Fuji Provia 400F requires half that exposure. To test your site, use those recommendations as starting points. Very dark skies allow longer, deeper exposures. In light-polluted skies, curtail exposures to avoid sky fogging.

## Field of View
### (for 35mm film cameras)

| Lens Focal Length | Field of View |
| --- | --- |
| 18mm | 90° x 59° |
| 24mm | 75° x 50° |
| 28mm | 66° x 43° |
| 35mm | 57° x 38° |
| 50mm | 40° x 26° |
| 85mm | 24° x 16° |
| 105mm | 19° x 12° |
| 135mm | 15° x 10° |
| 200mm | 10° x 6.6° |
| 300mm | 6.7° x 4.4° |
| 400mm | 5.0° x 3.3° |
| 500mm | 4.0° x 2.6° |
| 600mm | 3.3° x 2.2° |
| 800mm | 2.5° x 1.6° |
| 1,000mm | 2.0° x 1.3° |
| 1,500mm | 1.3° x 0.9° |
| 2,000mm | 1.0° x 0.7° |

## THE BEST FILMS

Almost any film works well for the various techniques described so far, except for piggyback shooting. Many films behave badly when asked to record long exposures where the prime targets are nebulas that emit mostly red light at a wavelength of 656 nanometers. The poor films simply fail to record much nebulosity. They pick up only stars and, even at that, often turn putrid green. Or they are simply too grainy.

Recommending suitable films is a risky business. Kodak and Fuji, the two main film makers, usually have at least one slide or print film that performs beautifully for piggyback and deep-sky photography. Then they proceed to "improve" the film, often without any indication on the packaging, and the film is ruined for astrophotography. Or the product is simply discontinued.

Film spec sheets provide little indication of what films will and will not work. And beware of proclamations of what films are

# What Can Go Wrong: Piggyback

Although easy to do, piggyback photography can be prone to a unique but avoidable set of problems that can mar photos.

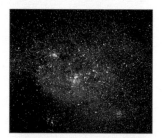

### Aircraft
Any location near airport glide paths is a bad place for piggyback shooting. If you see an aircraft coming, cap the lens until the airplane moves clear of the frame.

### Hazy Photos From Dew
A lens pointed up is a dew magnet. Occasional blasts of warm air from a hair dryer help. The best solution is a heated antidew coil wrapped around the lens. Make sure the coil cannot twist the focusing ring.

### Fuzzy Stars on Part of the Frame
This occurs when the film buckles in the film plane, often the result of the film's absorbing moisture on humid nights. If the film has been in the camera for a few weeks, advance it two frames before a shooting session. Another trick is to turn the rewind knob backwards to pull the film taut in the gate.

### Cold-Weather Precautions
Unless film is advanced and rewound slowly in cold, dry weather, static electricity can be generated that will show up as lightning strokes across the imaged sky. Cold film also becomes brittle and can rip apart at the sprocket holes.

### Film Flaws
Scratches can come from the camera back, the film guides or dirt in the mouth of the film cassette. And processors can introduce scratches, dirt or chemical blobs. To beat the odds, always shoot at least two of everything.

*A camera aimed at the sky for a long period of time can pick up images of UFOs—unwanted film objects—such as aircraft (top), satellites and even fireflies. Dew on a lens (second from top) creates a bluish haze around bright stars, a loss of faint detail and a sky that is not very black. Humidity absorbed by the film causes the emulsion to buckle (third from top), creating blurry stars in parts of the frame. If a piggybacked camera slips out of position (bottom), strange star trails result.*

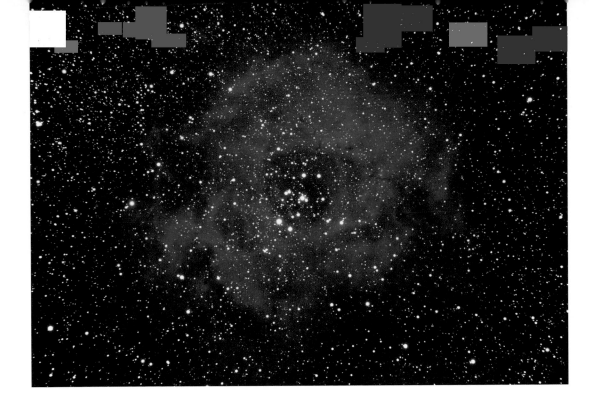

good or bad based solely on indoor tests of faint light sources. The only trustworthy tests are shots of the actual sky.

In slide films, we've had great success with Kodak Ektachrome E200 (EPD), Fuji Provia 100F (RDP III) and Fuji Provia 400F (RHP III). In print films, favorites were Kodak's Royal Gold 200 and Supra 400 until Kodak "improved" them in 2002. What films are the current champs? Check the buzz on astrophoto mailing lists. Or visit www.backyardastronomy.com for our latest recommendations and for details on black-and-white Technical Pan film and hypersensitizing, a film and technique destined to become obsolete in the digital age.

Which should you use, prints or slides? With color negatives, unless you do your own printing, you must rely on someone at an automated machine to prepare your photographs. The results can range from perfect to terrible. With slides, what you see is what you get. Beginners are well advised to shoot slides. We stick with slide film for most astrophotography, including nearly all piggyback shooting. However, for deep-sky shots through a telescope, we prefer print films. They seem to retain their sensitivity better during long exposures, recording faint nebulas well even at f/6 focal ratios and hour-long exposures. Extracting detail from a thin negative is easier than from a dark, underexposed slide.

# Prime-Focus, Deep-Sky Photography

Now we move to shooting deep-sky objects through the telescope itself. Focal lengths are long, and the demands of polar alignment, guiding and focusing are unremitting. Prime-focus, deep-sky work is the point at which many aspiring astrophotographers start their hobby. Unfortunately, it is where many find their enthusiasm coming to an abrupt end. All they have to show for their

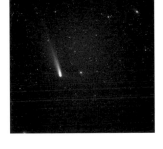

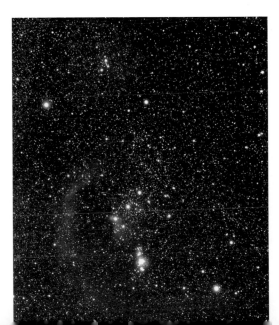

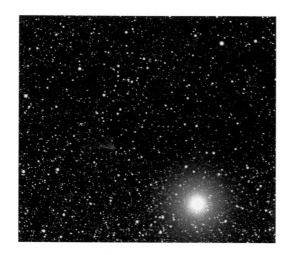

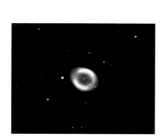

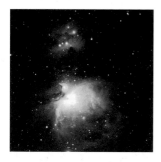

efforts are some blurry photographs, a depleted bank account and an advertisement on the Internet classifieds for a "complete astrophotography outfit, seldom used, mint condition, best offer."

## CCDS OR FILM?

For detailed shots of glowing nebulas and spiral galaxies, however, you enter the minefield of prime-focus photography. It is at this point on the journey that many leap right to digital CCD astrocameras, thinking they are stepping over the minefield —and not without good reasons, as outlined in the next chapter. Our website at www.backyardastronomy.com contains a breakdown of the cost of film versus CCD for deep-sky imaging—surprisingly, it is about the same!

While CCD cameras are clearly superior to film for many telescopic subjects, we're not about to abandon film yet. When shooting deep-sky objects through a telescope, film's prime selling point is its generous field of view. Most affordable CCD cameras (under $4,000) provide fields just a fraction of what a 35mm frame can cover. This makes 35mm film and, more so, 6cm x 7cm film the medium of choice for shooting large deep-sky objects and rich star fields that require fields of two degrees or more. For those subjects, our choice is to combine film with 4-to-6-inch apo refractors.

Then there is ease of use. Yes, film can require exposures of an hour or more, but with auto-guiders, a 60-minute exposure is just as easy as a 10-minute one. Plus, CCD cameras are not the snapshot devices you

might think they are. For the finest CCD images, exposures may have to be 10 to 15 minutes in each color channel, with multiple exposures per channel for noise reduction. Exposures of that length require guiding through one of several methods outlined on the next few pages. Add up all the guided exposures, and you can still spend one to three hours, perhaps all night, taking what will become a single image of one object.

CCDs and their required computers and cables are an awful fuss to set up and power in the field off 12 volts in a temporary setting. A CCD setup is best employed at a home observatory, where most of the gear can be left hooked up and ready to go in moments, all powered off 110 volts AC.

For all these reasons, we aren't about to proclaim film dead yet for deep-sky shooting through a telescope. So we will still cover the techniques and gear needed for deep-sky shots on film, especially since much of this information applies to CCD imaging as well. The first task, no matter what the camera, is to select a good telescope.

## SELECTING AN ASTROPHOTO TELESCOPE

The crucial feature in an astrophotography telescope, more important than the optics, is a solid equatorial mount. As with piggyback photography, a computer-driven alt-azimuth mount will not work. For film shooting, portability is also a prerequisite. CCD cameras can grab images out of light-polluted skies, but film shooting must be done from dark locations, likely far from home. So while the instrument must be sturdy, it should also be compact and easy to set up and accurately polar-align.

The mount must have dual-axis drives, with push-button speed controls on each axis. The R.A. drive motor and gears must be free of significant periodic and random error. When stated at all, drive-error ratings are usually given in arc seconds. An error of less than 20 arc seconds, generally stated as plus or minus 10 arc seconds, is excellent. Periodic-error-correction (PEC) circuits can help reduce those errors even further, to below five arc seconds.

Though once a luxury, a "Go To" mount has fast become a necessity. It makes finding faint targets a snap. Without "Go To," locating targets with a narrow-field CCD camera, let alone a 35mm camera, can be a chore. For a suitable portable mount, we suggest the Vixen Great Polaris DX, a Losmandy GM-8 or G-11 or an Astro-Physics 400GTO. Despite appearances, the Chinese-made EQ mounts are simply not good enough.

Now for the telescope. For deep-sky photography, aperture is not as important as focal ratio. Fast f/4 to f/6 telescopes of any aperture are far easier to use for faint-object portraits than are f/7 to f/10 instruments. Anything slower than f/10 makes exposures torturously long. This may be surprising, but for astrophotography, a larger-aperture instrument will not produce shorter exposures. Rather, deep-sky exposures are determined by focal ratio: the faster the telescope, the shorter the exposure.

What telescope has most of the right features? We can do no better than to recommend what we ourselves have selected and long used with consistent success: fast apochromatic refractors. We suggest one of the units from Astro-Physics, Borg, Takahashi, Tele Vue, TMB or Vixen. A 4-to-6-inch f/5 to f/7 apo refractor has excellent speed and a two-to-four-degree field with a 35mm camera. With small-pixel CCD cameras (under nine microns), image scale is sufficient to provide resolution limited primarily by the sky, not by the camera. Images are sharp across the wide field, vignetting is minimal, the optics don't shift in the tube, and light transmission is high, keeping exposure times to a minimum. The compact tubes can be solidly mounted on portable German equatorial mounts equipped with polar scopes that make polar alignment easy and accurate. Systems of mounting rings and guidescopes are readily available. Some apo refractors also accept medium-format film cameras, such as the Pentax 67.

What about conventional f/4 to f/6 Newtonian reflectors? The mounts and optical tube assemblies of most commercial units simply aren't solid enough. Any photographers who employ Newtonians have done so largely with custom-made or heavily refitted instruments.

This leaves the Schmidt-Cassegrain as the other popular choice for deep-sky photography. The long focal lengths (1,600mm to 2,000mm), while perfect for those small

objects a small apo refractor can't shoot, require dead-accurate polar alignment and precise guiding for successful shots. Lacking an integrated polar sighting scope, fork mounts can be awkward to polar-align. Using the finderscope on the main tube can get you close, but plan to use the drift method (see www.backyardastronomy.com for details) for the necessary precision. In addition, when tilted over on wedges, the fork mounts of lightweight Schmidt-Cassegrains tend to bounce at the slightest touch or breeze. Premium 8-inch models, such as Celestron's NexStar GPS and Meade's LX200 GPS, are much improved in this regard. In fact, if astrophotography is your goal, consider nothing less than a top-of-the-line model in a Schmidt-Cass. Don't forget a heavy-duty wedge, an essential option.

We'll be blunt. We have both used fork-mounted 8-inch Schmidt-Cassegrains for astrophotography and largely abandoned them in favor of apo refractors. We found that our success rate with Schmidt-Casse-

**◀ Apo Astrophoto**
The relatively short 500mm-to-1,000mm focal lengths of most apo refractors are somewhat forgiving of tracking and guiding errors. The wide field of a Vixen 114-ED refractor (with a focal length of 600mm) was a perfect match for framing IC1396, the large, faint wreath of nebulosity in Cepheus. This 50-minute exposure on Kodak Supra 400 print film was taken by Alan Dyer.

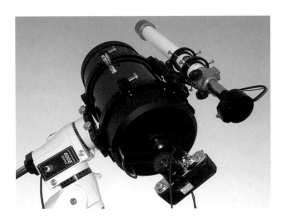

**◀ Schmidt-Cassegrain Astrophoto System**
For targets that require long focal lengths, we continue to use Schmidt-Cassegrains but have chosen to mount Celestron 9.25- and 14-inch tube assemblies on top-class German equatorial mounts to aid polar alignment and stability.

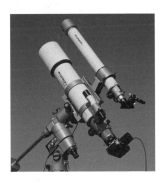

**▲ Good Starter System**
Vixen's Great Polaris mount (shown here with a Vixen 102-ED scope) provides tracking errors of less than 20 arc seconds, remarkable for a mount under $1,000. It can serve as the core of a good budget astrophoto rig. Its main disadvantage is that it does not readily accept an auto-guider.

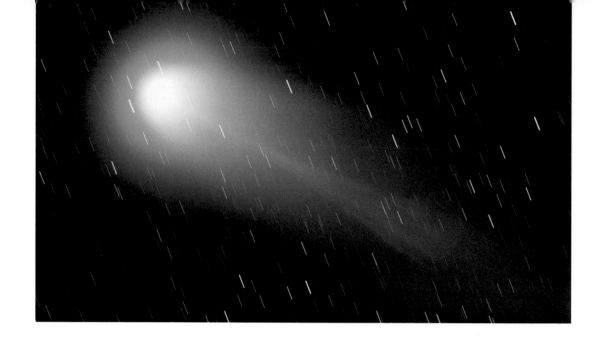

**To Catch a Comet ▶**
Because comets move against the background stars, a telescope should be guided on the comet itself. A separate guidescope that can be aimed independently at the comet head is a must. Or use a CCD camera, such as the Starlight Xpress models, which can guide through the same chip that takes the image. This film portrait of Comet Hyakutake, taken in March 1996, was guided on the comet head with an ST4 guider for 10 minutes on Fuji SuperG 800. All photos by Alan Dyer.

grains wasn't good enough, largely because of the drawbacks outlined above.

Instead, many top photographers needing big scopes for long focal lengths have opted for f/8 to f/10 high-end Ritchey- or Maksutov-Cassegrains from companies such as Astro-Physics, Optical Guidance Systems, Parallax Instruments, RC Optical Systems or Telescope Engineering Company, or they use custom-made Cassegrains. Tube assemblies are paired with massive top-class German equatorial mounts from Astro-Physics, Losmandy, Mountain Instruments, Parallax or other small companies. Such big systems can run at least $15,000 to $20,000. Even for a more modest system based around a small refractor or an 8-inch Schmidt-Cass, plan to spend $5,000 or more for a fully outfitted astrophoto rig with auto-guider. Visit www.backyard-astronomy.com for more details on our favorite telescopes for astrophotography.

## AUTO-GUIDERS: GUIDING SALVATION

It's safe to say that neither of us would touch deep-sky photography through a telescope today were it not for auto-guiders. We've paid our dues hand-guiding exposures all night long. It is a cold, tedious, back-wrenching task. We won't go there again.

Auto-guiders take all the hard work out of guiding. They are CCD cameras designed to control a telescope mount. Once properly calibrated each night, the light-sensitive CCD chip can sense any motion of the guide star away from its original position,

**Flattened Field ▶**
Without a field flattener (right), stars at the frame corners can look comatic or distorted. With a flattener-reducer (far right), stars are pinpoint from corner to corner. The faster focal ratio and shorter effective focal length also reduce exposure time, though at a sacrifice of image scale. These shots were taken through a Takahashi Sky90 refractor.

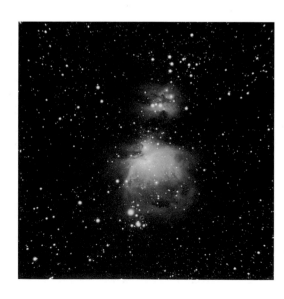

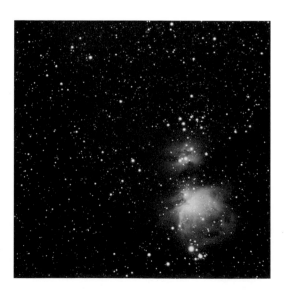

then pulse the drive motors to compensate, automatically ensuring a guiding accuracy far better than fallible humans can achieve. Exposures can run as long as you wish. While your robot does the work, you can enjoy the stars. Or sleep! Visit www.back-yardastronomy.com for more tips on using an auto-guider.

In our experience, the best auto-guiders come from Santa Barbara Instruments Group. Its newest unit, the $2,000 STV, is a marvel. Its 640x480 pixel chip can serve as a decent CCD camera in its own right for exposures up to 10 minutes long. It can store and display images without an external computer, making CCD imaging much easier. The unit can also be used as a low-light video camera, feeding a continuous stream of video frames to an external monitor through a standard composite video

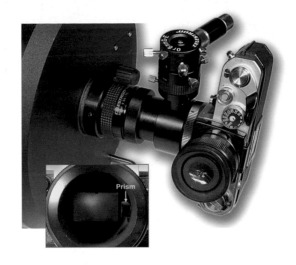

jack. A third role is as an auto-guider. A built-in 5-inch screen displays the field seen by the CCD chip. The STV automatically selects a calibration speed, exposure time

◀ **Off-Axis Guider and Reducer**
Off-axis guiders, like this Celestron Radial Guider, allow shooting and guiding through the same telescope. The guide-star, pick-off prism (inset) usually falls outside the 35mm film frame or CCD-chip area and is invisible to the camera or, at worst, casts a dim shadow at the edge of the image. The f6.3 Reducer-Corrector lens goes between the guider and the telescope back.

# Astrophoto Accessories III: Guiding Options

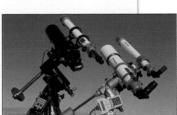

Which guiding option should you choose? An off-axis guider? A separate guidescope? An auto-guider? A self-guiding CCD camera? Or no guiding at all?

All deep-sky film shots through a telescope must be guided. Off-axis guiders are more compact and are a better physical match for stubby-tubed Schmidt-Cassegrains. Even so, we prefer to use a separate guidescope, as it affords the greatest freedom to select a bright guide star. We use Vixen 60mm and 70mm guidescope refractors and recommend them. For Schmidt-Cassegrains, use a Losmandy or Parallax Instruments mounting system to mate the guidescope to the main tube.

Hand-guide your initial shots with a guiding eyepiece. Doing so will teach you how well your mount is tracking. But you'll soon want to add an auto-guider. For a budget guider, we recommend a used ST4. For the best auto-guider, however, an STV cannot be beaten.

For CCD imaging, stacking short 30-second exposures can eliminate guiding completely. But stacked short exposures are noisier than single long exposures. The best CCD shots must be guided. Some CCD cameras can guide themselves, a handy feature. Most of the Santa Barbara Instruments cameras use a separate guiding chip mounted just off the main imaging chip, like an off-axis guider. With the Starlight Xpress system Simultaneous Track and Record, or STAR2000, the same chip serves double duty as an imaging and guiding camera. This allows a greater selection of guide stars but requires doubling the exposure time.

Despite the lure of self-guiding CCD cameras, a separate guiding system has advantages: It is independent of the filters in front of the main camera; it keeps the telescope locked onto a star even when the main camera shutter is closed; and it allows a choice of brighter guide stars or guiding on comets.

*Provided the mounting system is solid (as all commercial setups we've used are), the much-touted problem of separate guidescopes—flexure between the main tube and guidescope—won't be a bother. It shows itself primarily on homemade setups equipped with flimsy brackets or on Newtonian scopes with cardboard tubes. The 60mm guidescope, left, is clamped in two rings, but the single-ring system used by Vixen, right, works quite well.*

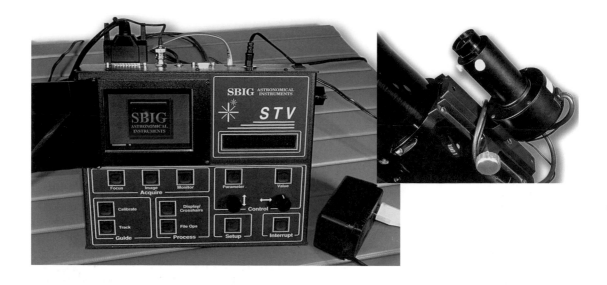

**STV Auto-Guider ▶**
At $2,000, the STV is not cheap, but once you use one, you will never go back to any lesser form of guiding. Its screen displays the guide star and a graph of the recent guiding corrections. An option is the little 100mm-focal-length eFinder lens (inset) that attaches to the CCD head. For guiding telescopes up to 1,200mm focal length, this is all the guidescope the STV needs.

**ST4 Auto-Guider ▶**
For 10 years, the standard auto-guider of choice was Santa Barbara Instruments' ST4. The workhorse ST4 was discontinued in 2001 when key components could no longer be obtained. However, used units are available from amateur astronomers upgrading to the more sophisticated STV.

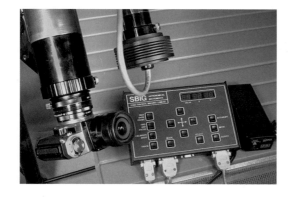

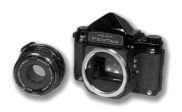

**Pentax 67 Camera ▲**
This is the preferred camera for medium-format astrophotography. While all but the oldest Pentax 67s require battery power to keep the shutter open, its big battery stands up well in the cold. The camera can also be run off external power with an optional power cord.

and guide star. It guides on the brightest star in the field, although you can manually select a star of your choosing. The process is so well designed that guiding requires only a couple of button pushes.

## USING LARGER FILM FORMATS

Some of the most impressive astrophotographs published by amateur astronomers have been taken not on 35mm film but with medium-format 120 (10-exposure) or 220 (20-exposure) film. This is the film size used by cameras such as the Mamiya 645, Hasselblad 6x6 and Pentax 67. The latter, designed like a large 35mm SLR, is the favorite camera for medium-format shooting. The Pentax's film gate holds the big film flat, and its viewfinder can be changed to a right-angle model for ease of use on a telescope.

Most telescopes won't project an image circle large enough to fill a film frame as big as 6cm x 7cm. Medium format is almost the exclusive domain of apochromatic refractors. Astro-Physics, Borg, TMB Optical and Vixen all manufacture telescopes, field flatteners and adapters designed for 6x7 format.

What is the advantage of medium format? Contrary to popular misconception, a 6x7 camera alone does not provide a larger image or finer grain. What it does provide is a larger field of view. Take a 4-inch f/6 apo refractor, for instance: With 35mm film, the field coverage is a nice 3.3 by 2.2 degrees. With 120-format film and a 6x7 camera, the field of view is a whopping 6.4 by 5.5 degrees. Yet any object appears exactly the same size on either film. Use the same emulsion (many films are available in both 35mm and 120 format), and the level of film grain is also the same. But the wider field of the big 6x7 frame takes in much more sky.

Now, put that 6x7 camera on a 6-inch f/7 apo refractor, and you do have a larger image. The field is still a generous 3.5 by 3 degrees, almost the same field provided by a 35mm camera on the small 4-inch telescope, but the focal length of the 6-inch is almost twice as long, providing images twice as large and bringing tiny detail above the film grain. That's where medium format excels: great detail without sacrificing field of view.

A similar gain applies to piggyback photography. A Pentax 67 with a normal 90mm f2.8 lens has a similar field of view to a 50mm lens on a conventional 35mm-format camera. But the longer focal length of the 90mm lens records more fine detail, and the wider aperture of the bigger lens picks up more stars. The downside is the cost of the camera gear, field flatteners and scanners necessary for digitizing the big film frame. For traditional film photographers who want to stay ahead of the CCD crowd, though, medium format is the way to go.

# What Can Go Wrong: Deep-Sky Guiding

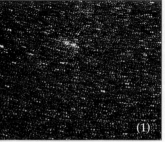

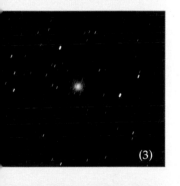

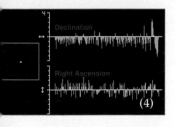

In deep-sky photographs with any focal-length system, stars can be sharply focused and still appear misshapen because of trailing in one or both directions.

### Every Star Looks Double

Commonly called "nose and foot binaries," such images are created by bumping the tripod with your foot or nudging the guiding eyepiece with your nose sometime during the exposure, usually as you nod off to sleep.

### Trailing in Right Ascension

If the guide star takes off suddenly at regular intervals, the motor drive has mechanical periodic errors. If the error is too extreme to compensate for by guiding, the only solution is to replace the entire motor-and-gear-drive mechanism or to buy a new mount. If the star keeps drifting out slowly in the same direction, make sure the drive is set to the sidereal rate. Try running the unit from a different power source.

### Trailing in Declination

Some mounts can oscillate north and south under the constant commands of auto-guiders. Back off the aggressiveness of the auto-guider, reduce the mount's backlash compensation, or reduce the guide speed to half the value with which you calibrated the guider. All will prevent the mount from bouncing back and forth in response to every fluctuation in seeing.

### Rotation Around the Guide Star

Do stars in your guided photographs appear to be trailed in arcs? If so, the telescope has not been accurately polar-aligned, and stars are drifting out in declination during the exposure. Correcting in the usual way for such drift ensures that the guide star is untrailed, while the rest of the frame gradually turns about the guide star.

### Poor Polar Alignment for No Apparent Reason

If you polar-align accurately yet still get trailed stars, chances are that you are aligning with a polar bore-sight scope whose reticle is off-center and is not pointing straight up the polar axis.

### Trailing for No Apparent Reason

If you are using a separate guidescope, it could be shifting with respect to the main scope. Use more solid fittings. In piggyback photography, the camera can slip on its mooring as it changes orientation during the exposure. Use rings or clamps to hold long telephoto lenses both front and back. With Schmidt-Cassegrains, the primary mirror itself can shift as the telescope moves from one side of the sky to the other. Avoid exposures that will cross the meridian.

*Success in deep-sky photography comes by learning from mistakes. From top to bottom: 1. Don't hit the tripod, or you'll get double stars galore. 2. Some forms of trailing can be reduced by purposely throwing the telescope slightly out of balance so that as it turns to the west, the mount has to work to pull the scope up. This ensures the R.A. drive gears remain meshed. 3. Poor polar alignment manifests itself as rotation around the guide star. The problem shows up more when you are shooting near the celestial poles. 4. If you or your auto-guider (as shown here in the top graph of an STV readout) continually have to make declination corrections in one direction, it is a sign that you are not polar-aligned. 5. Use the little setscrews on any polar scope to move its reticle so that anything placed at the center of the reticle crosshairs stays dead center as the scope turns around its R.A. axis.*

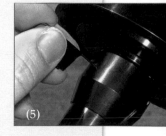

# Shooting the Sky III: The Di

Years ago, professional astronomers gave up on film-based
imaging in favor of electronic
CCD cameras. The same is happening in amateur astronomy.
Even experienced astrophotographers who have been shooting
the sky on film for decades are
amazed at what these high-tech
cameras can do. Results make
film-based shots of just a few
years ago look pale and fuzzy
by comparison.

The switch to CCDs involves
a steep climb up the learning
curve, and it certainly requires a
significant cash investment. But
the end product rivals the best
images from the professional
observatories—all from your
backyard. For example, the
remarkable CCD image of the
Andromeda Galaxy on this page
was taken from a driveway in
suburban Connecticut.

The digital revolution is sweeping every aspect of the field. Even
photographers loyally married
to film find it hard to deny the
incredible tools that computer
image processing brings onto our
desktops. When scanned and
digitized, old negatives and slides
take on new life. Techniques such
as stacking images and stitching
mosaics together, which would
have been torture in a chemical
darkroom, are now just a few
mouse clicks away. It is an exciting
time for astrophotographers.

A special note: Because this
area of the hobby is changing so
rapidly, with equipment, software
and techniques improving by
the month, we urge you to visit
our website at www.backyard-
astronomy.com. There, you will
find more details on camera selection, CCD-imaging techniques
and image-processing tutorials.

..............................

To capture the
full extent of the
Andromeda Galaxy,
Robert Gendler used
a Takahashi FSQ106
refractor and an
ST-10 CCD camera
to take a mosaic of six
frames. Each was an
LRGB four-layer com-
posite with exposure
times of 180/30/30/50
minutes for the re-
spective layers. Total

tal Frontier

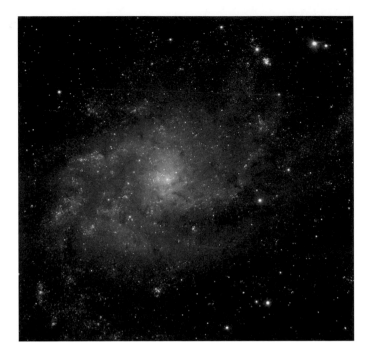 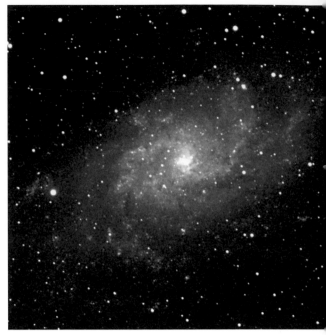

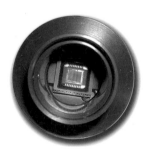

# CCD Advantages

When you see your first CCD image appear on the computer screen just moments after you take it, you will be hooked. In the previous chapter, we made an argument for film. In this chapter, the best pro-CCD arguments are the images themselves, which appear throughout this book. Quite simply, when done right, a CCD image can blow away the best film shots of the same object, often taken by the same photographer.

## HOW CCDs WORK

A CCD (charge-coupled device) is a type of silicon chip that can turn incoming photons (light) into electrons. CCD chips for imaging are made in a matrix of light-sensitive picture elements, or pixels. The electrons generated by each pixel are proportional to the brightness of the light that strikes it. During the exposure, the electrons build up in each pixel. Think of the pixels as little wells filling up with electrons. At the end of the exposure, the electrons are emptied from the wells. The electrons are shuttled off one side of the chip, usually in a bucket-brigade fashion, where they are dutifully counted as they flow off the chip.

When we speak of CCD cameras for astronomy, we mean specialized cameras designed for long exposures. Yes, digital still cameras can expose for several seconds, or even indefinitely if the camera has a "B," or bulb, setting for time exposures. But long exposures are inevitably ruined by electronic noise—a blizzard of colored snow.

This electronic noise is the bane of CCD imagers. Major sources of noise are the residual heat in the CCD chip (thermal noise) and a low level of current always running through the chip (bias noise). Together, these two noise sources generate what is known as dark current—the noise floor that is present even if the camera is just idling and not taking an image. For bright daytime subjects, this noise floor is overwhelmed by the signal. But the incoming photons of faint astronomical objects, like nebulas and galaxies, generate such a low number of electrons that the wanted signal gets swamped by the unwanted noise.

There are two ways to improve the signal-to-noise ratio: Boost the signal, and cut the noise. The more expensive CCD cameras do a better job of converting photons to electrons. A quantum efficiency (QE), as it is called, of 100 percent is ideal. The best cameras sold for amateur use have QEs of 70 to 80 percent. They require the shortest exposures to reach deep to record faint stars and galaxies. Most midrange cameras have QEs of 30 to 60 percent. If that sounds second-rate, consider that film usually has a QE of no more than 4 percent. It's amazing that film records anything!

The other line of attack—reducing most noise to low levels—requires cooling the CCD chip. The colder the chip, the fewer random electrons are stirred up by the thermal activity deep in the semiconductor material. Most hobby cameras use solid-state Peltier-effect coolers, though some use pumps and tubes to circulate water around the chip. Typical coolers lower the temperature of the chip 30 to 40 Celsius degrees below the ambient temperature of the night air, which means that a CCD camera works better in cold weather than on hot, muggy nights.

## WHAT CCD CAMERAS CAN DO

Astrophotographers are attracted to CCD cameras for several reasons. For one, the results are immediate. Just seconds after the exposure ends, the raw result scrolls down the screen.

CCD cameras are also linear in their response. Doubling the exposure doubles the image density. When you are dealing with electrons, rather than silver-halide molecules, there is no reciprocity failure, the effect that causes film to lose sensitivity over long exposures. Their linearity and the greater quantum efficiency of CCD chips make CCD cameras far more sensitive than film. Deep-sky objects that require a 1-hour exposure on film are well recorded in just minutes on CCDs.

Because of the digital nature of CCD images, it is easy to extract an image from the background. With a 30-second exposure, a typical CCD image is still severely underexposed. But with good dark-frame subtraction to get rid of noise and with a boost in contrast, an image emerges that you can put in your "keeper" collection.

This ability to extract images from the murk makes CCD imaging at a light-polluted site possible. Stacking lots of short exposures produces amazing images that would be impossible with film under bright skies. Shooting black-and-white images through a deep red or an H-alpha passband filter helps penetrate sky fog even more, a technique that works well on CCDs, because most chips are very sensitive to the red and infrared portions of the spectrum.

## WHAT CCD CAMERAS CAN'T DO

While CCD chips are getting bigger and cheaper by the year, most in use for hobby imaging are still significantly smaller than a 35mm film frame. This limits their field of view. A camera like a Meade 416XT or the Santa Barbara ST-7 with the common Kodak KAF-0401 chip (just 6.9mm wide by 4.6mm high) provides a field of view of only 19 by 12 arc minutes when used on an 8-inch Schmidt-Cassegrain at f/6.3. That's smaller than the full Moon. A 35mm film frame on the same telescope covers 1.6 by 1.0 degrees of sky, wide enough to take in three full Moons. On the other hand, a field of 19 arc minutes is more than enough to encompass most galaxies and planetary nebulas.

Another drawback is that most CCD cameras cannot record color images. They are black-and-white cameras for a good reason. Many color CCD cameras work by placing tiny color filters over the pixels, grouped in clusters of red-, green- and blue-sensitive pixels (or a matrix of cyan, magenta and yellow pixels). This inevitably produces some loss of resolution, since it takes a group of pixels to contribute all the information needed to create a color image of that little bit of the scene. In addition, the prime subjects of astrophotographers are stars, literal points of light. If a star image falls on just a red pixel and not a green and a blue one as well, it won't be accurately recorded in color. In practice, star images cover more than a single pixel (as they

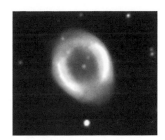

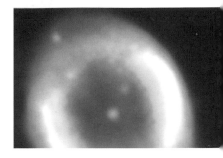

▲ **Pixels on Parade**
If a CCD image is enlarged enough, you can see the square picture elements, or pixels, that make up the image. As a general rule, the smaller the pixels or the more pixels in a picture, the more realistic the image. Too few pixels across an image, and the picture looks blocky, or pixelated, and obviously digital.

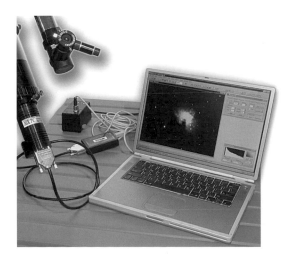

◀ **Computer-Assisted Imaging**
Current CCD cameras all require the assistance of a computer at the telescope to operate the camera and to display and store images. Unlike snapshot cameras, few astro-CCD cameras can operate as stand-alone devices (the STV is an exception, though limited to a small frame size and onboard storage of just 14 images).

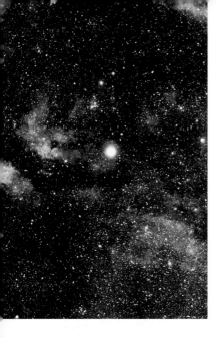

**Wide-Angle CCD** ▲
When coupled to a 300mm or longer telephoto lens or a fast refractor, a large-chip/small-pixel CCD camera becomes a powerful tool for recording large objects such as the Gamma Cygni nebulosity. Robert Gendler took this two-frame mosaic with a 300mm Nikkor lens and a Santa Barbara ST-10 camera. LRGB exposure times were 60/10/10/10 minutes.

**One-Shot Color** ▶
Some CCD cameras are made with a matrix of color filters over the pixels. While still requiring processing steps to reveal and optimize the color, these one-shot cameras don't need exposures through different filters. This is a stack of seven 5-minute exposures of the spiral galaxy M88 (right) taken with a Starlight Xpress MX7 on a 12-inch Schmidt-Cassegrain at f/4. The shot of M16, the Eagle Nebula (far right), is a stack of five 5-minute exposures using an 80mm refractor at f/5. Images by Alan Chen.

should, to avoid square stars), so the problem is not as bad as it sounds. In fact, companies such as Starlight Xpress (the MX Series) and Apogee (LISÄÄ cameras) have specialized in one-shot color cameras. These require a single exposure to create a color image, and the results are more than good enough to satisfy most backyard hobbyists.

Even so, to achieve the highest-resolution images, serious CCD imagers still prefer to take the more difficult route of tricolor photography. The method requires taking three images, one each through red, green and blue filters. This is best done with a motorized filter wheel in front of the camera, which can be controlled by the same software that operates the camera. The three monochrome images are then registered and superimposed later with computer software to create a full-color image. Although the process is time-consuming, the results are superb. Most of the CCD images in this book are tricolors.

Well, not quite. Most are what have become known as LRGB images. The photographer takes a fourth exposure with no filter. This becomes the Luminance (L) channel. In the case of red nebulas, the L channel can be a high-resolution, deep red-filtered image. The L image contributes most of the detail and resolution. The R, G and B images can actually be lower-resolution binned images (more about this later), which require shorter exposures.

# Choosing a Camera

The two most important specifications, and the ones that primarily govern a camera's cost and image quality on your telescope, are the size of its array and the size of the actual pixels. (For an extended explanation of other CCD-camera specs, please visit our website at www.backyardastronomy.com.)

## PIXEL COUNT VS. PIXEL SIZE

A chip with a bigger physical area gives a large field of view, just as a 6x7-format camera gives a larger field than a 35mm camera on the same telescope. But how much detail a camera can record depends primarily on the size of the pixels doing the recording. Some cameras have quite small pixels, less than seven microns ($\mu$) across (one micron = 1/1,000 millimeter). Other cameras have large pixels, 20 to 24 microns across. In a sense, pixel size is equivalent to film grain, with seven-micron cameras analogous to slow, fine-grained film and 24-micron cameras to fast, coarser-grained film. Just like their film counterparts, 24-micron cameras require shorter exposure times to collect enough photons to create an image than do seven-micron cameras. But

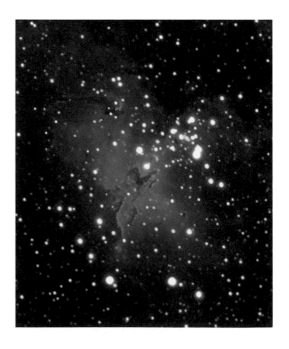

# Picking Pixels

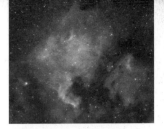

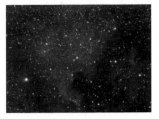

Here we compare the performance of several popular CCD chips on three telescope systems: a 4-inch f/6 refractor, an 8-inch f/6.3 Schmidt-Cassegrain and a 16-inch f/6.3 Schmidt-Cassegrain. For deep-sky imaging, the best choices are marked in **bold blue**.

| Chip Array | Pixel Size | Field of View (arc minutes) | Image Scale (arc sec. per pixel) 1x1 | Binned 2x2 |
|---|---|---|---|---|
| **4-inch f/6 refractor** | | | | |
| 657x495 | 7.4 μ | 25x19 | **2.4** | 4.8 |
| 765x510 | 9.0 μ | 38x25 | 2.9 | 5.9 |
| 1530x1020 | 9.0 μ | 75x50 | 2.9 | 5.9 |
| 1600x1200 | 7.4 μ | 64x48 | **2.4** | 4.8 |
| 2184x1472 | 6.8 μ | 81x56 | **2.2** | 4.4 |
| 1024x1024 | 24 μ | 133x133 | 7.9 | 15.7 |
| | | | | |
| **8-inch f/6.3 SCT** | | | | |
| 657x495 | 7.4 μ | 13x10 | 1.2 | **2.4** |
| 765x510 | 9.0 μ | 19x12 | **1.5** | 2.9 |
| 1530x1020 | 9.0 μ | 38x25 | **1.5** | 2.9 |
| 1600x1200 | 7.4 μ | 32x24 | 1.2 | **2.4** |
| 2184x1472 | 6.8 μ | 40x28 | 1.1 | **2.2** |
| 1024x1024 | 24 μ | 67x67 | 3.9 | 7.9 |
| | | | | |
| **16-inch f/6.3 SCT** | | | | |
| 657x495 | 7.4 μ | 6.4x4.9 | 0.6 | 1.2 |
| 765x510 | 9.0 μ | 9.4x6.3 | 0.7 | **1.5** |
| 1530x1020 | 9.0 μ | 19x13 | 0.7 | **1.5** |
| 1600x1200 | 7.4 μ | 16x12 | 0.6 | 1.2 |
| 2184x1472 | 6.8 μ | 20x14 | 0.6 | 1.1 |
| 1024x1024 | 24 μ | 33x33 | **2.0** | 3.9 |

To calculate the image scale of any camera with your telescope, use the following formula: Image Scale = [pixel size (microns) ÷ focal length of telescope (mm)] x 206. To calculate the total field of view, multiply the Image Scale by the number of pixels along a dimension of the chip, then divide that number by 60 to get the field in arc minutes, as listed above.

seven-micron cameras can record finer detail on a given telescope—to a point. So which is best? It depends on your telescope and on what you are shooting.

For deep-sky imaging, a 24-micron camera is an ideal match for a big 12-to-16-inch observatory-class telescope. With a 1024x1024 pixel chip, the field of view at f/6.3 is a generous 0.5 degree wide. More important, such a combination gives an image scale of two arc seconds per pixel. (Keep in mind that the smallest detail you can see

through most telescopes is just under one arc second.) This is generally considered the optimum image scale for deep-sky imaging. A finer-grained seven-micron camera is overkill on such a telescope, giving 0.6 arc second per pixel. In the jargon, it is said to oversample the image, putting more pixels on a star than needed, with the penalty of longer-than-necessary exposure times.

There are two options in this case. One is binning, in which the pixels are electronically combined to produce the equivalent of

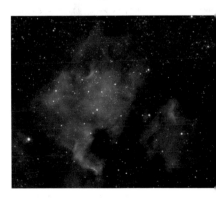

▲ **Three + One = Color** Combining a monochrome hi-res Luminance image with separate red-, green- and blue-filtered images creates an LRGB color image with excellent resolution. This compositing of images is usually done within Adobe Photoshop. Images by Brady Johnson.

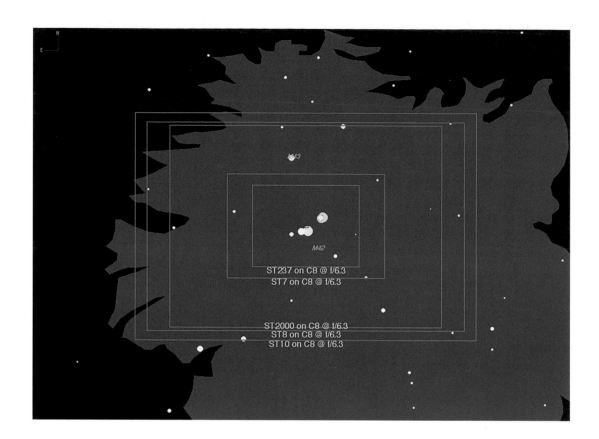

**Celestron FastStar** ▶

In Celestron's FastStar, the CCD camera replaces the secondary mirror, allowing light from the f/2 primary mirror to fall directly on the CCD chip. The image scale is small, making the use of a seven-micron camera a must. While the field is wide, results will never match the resolution of images taken with longer focal lengths and larger image scales.

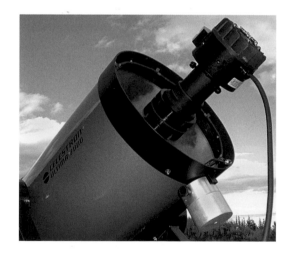

**Digital Planets** ▶

The Moon and planets call for an image scale of 0.5 arc second per pixel (perhaps using a Barlow), which samples the smallest arc-second-level detail. This image by Alan Chen is a stack of eight 0.2-second exposures taken with a Starlight Xpress MX7 one-shot color camera on a 12-inch Schmidt-Cass at f/20.

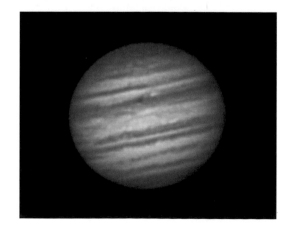

larger pixels. A seven-micron camera binned 2x2 provides the digital equivalent of 14-micron pixels, with the advantage of increased sensitivity and reduced exposure times. The second option is to employ a telecompressor, which will dramatically shorten the effective focal length of the telescope. Units are available that speed up the typical Schmidt-Cassegrain to as fast as f/3. On big scopes, telecompressors provide more ideal image scales while also producing brighter images that require shorter exposure times.

Now what happens if we put that big 1024x1024/24-micron camera on a small 4-inch f/6 refractor? The camera's enormous chip, fully one inch square, is almost the size of a 35mm film frame. Coupled to the refractor, it provides a terrific two-degree-wide field. But it is a poor choice for this telescope. The image scale of eight arc seconds per pixel will produce awful-looking square stars and poor resolution. The image is now undersampled. For the refractor, or any telescope with a focal length under 1,500mm, cameras with smaller seven-to-nine-micron pixels are the best choices. By contrast, to sample planets properly, image scales of 0.5 arc second per pixel are best.

# A Night With a CCD Camera

CCD cameras sound like a snap to use, but the reality is a little less care-free. While grabbing deep-sky images in just seconds sounds enticing, the best results demand carefully guided multiple long exposures.

• Step 1: Setting Up
A CCD setup is best in a home observatory, where equipment can remain hooked up and computers stored snugly and safely. If you are operating in the field away from home, then a laptop computer is the solution. You'll need a hefty source of portable power for the laptop, CCD camera, auto-guider and telescope.

• Step 2:
Finding Things
For centering an object onto a chip with a field of view just arc minutes wide, an accurately aligned "Go To" telescope isn't essential, but it sure is nice to have! For centering an object by hand, a separate sighting tele-scope piggybacked on the main shooting scope can be used to aim the telescope. An alternative is a flip-mirror viewer between the camera and the main scope.

• Step 3: Focusing
First, find a bright star for focusing. In its focus mode, the camera starts snapping a continuous series of short exposures. To speed the refresh rate, most focus modes take an image from just a small subframe, not from the whole chip. With Schmidt-Cassegrains, an add-on focuser may be essential to bypass the usual focus knob that pushes the main mirror, sometimes shifting the image and making focusing a frustration.

• Step 4: Frame and Expose
After swinging to your target, you'll need to take a series of short exposures to center the object by trial and error. Once on target, you have a choice: Track and Stack (also known as Track and Accumulate or Slew and Sum) or long guided exposures. The easiest way to get deep-sky images is to take short 30-to-60-second exposures and stack them together. This boosts the signal (the image) above the camera's noise level, producing a detailed image. The short exposures eliminate the need for any guiding at all.

As well as this works, a stack of ten 30-second exposures is not the equivalent of a single 5-minute exposure. Long guided exposures will beat noise down far more, resulting in a smoother sky background that is less polluted by sparkle and patchiness. The best results come from taking long 5-to-10-minute exposures under dark skies,

then stacking those together for even greater noise reduction. However, in bright, light-polluted skies, a stack of short shots is the only choice.

• Step 5: Take Dark Frames
For every image you take, you will also need an exposure of equal length taken with the lens cap on or the shutter closed.

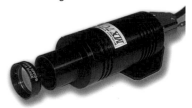

**▼ Cutting Infrared**
When used on refractors, most CCD cameras require an infrared blocking filter to cut out the unfocused infrared wavelengths (to which CCD cameras are sensitive) that can bloat images.

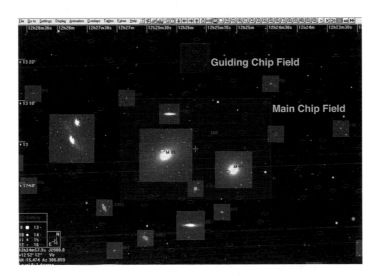

**◀ Previewing the Shot List**
To help you frame a field to include a bright guide star, Software Bisque's TheSky, Megastar and Project Pluto's Guide (at left) can show the outlines of a CCD frame and the guiding chip of self-guiding SBIG cameras.

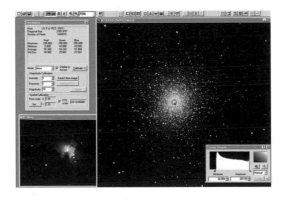

**◀ CCD Software**
A favorite software package of the authors for controlling a camera, filter wheel and telescope is MaxIm DL from Diffraction Limited. The software, which works with most cameras, also provides powerful image calibration and processing routines.

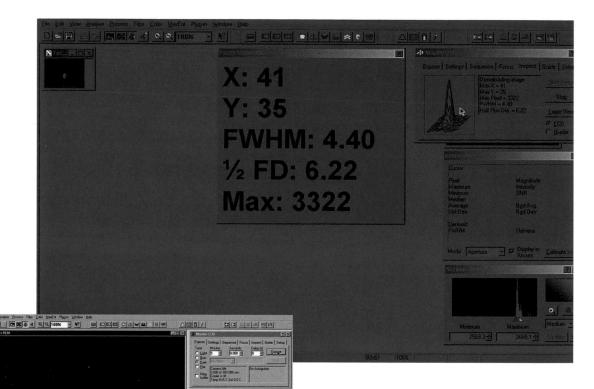

This is called a dark frame. It is effectively a recording of the noise introduced during the long exposure. At an early stage of image processing, this noise image gets subtracted from the light frame to give you an image of just the signal—the sky and your target, unadulterated by flecks of electronic noise. In fact, you should take several dark frames, average them together, then subtract that averaged master dark frame. What you thought was going to be a simple 5-minute exposure has now turned into a 10-to-30-minute process.

The reality isn't quite so bleak. Cameras with well-regulated temperature controls allow a better method. Most CCD users take sets of dark frames at various temperature settings and store them for later use. (Noise varies with the temperature of the chip: the colder the chip, the cleaner the image.)

• Step 6: Take Flat Fields
Though not an essential step, taking a snapshot of a blank field will allow you to clean up the image further. The result is an image of just dust specks, bad pixels and vignetting effects. Unlike dark frames, flat fields can't be archived and pulled out of a library. Flat fields should be taken each night with precisely the same setup you use to take the images—the same focus position, camera orientation, adapters and filters. Each flat field also needs companion dark frames to reduce noise in the flat fields.

There are several methods for acquiring a nightly flat field. One is to take a short exposure of the twilight sky before it is dark enough to show stars. A more consistent method is to aim the telescope at a bright, evenly illuminated surface. Santa Barbara
*(continued on page 317)*

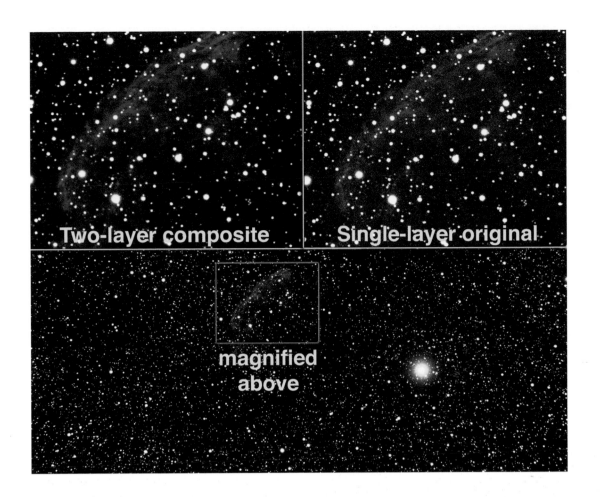

Two-layer composite

Single-layer original

magnified
above

**Benefits of Compositing**
Digital image processing is an essential aid to film-based, deep-sky images. For example, a common technique is to stack two negatives or slides in the computer. This creates a composite image with a smoother-looking background (top left) and less grit and grain than a single slide or negative (top right). This image of the supernova remnant IC433 in Gemini (bottom) is a stack of two 80-minute exposures at f/6 on Kodak Supra 400 film. Photos by Alan Dyer.

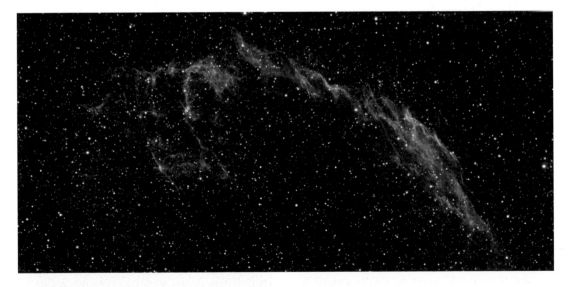

**True-Color Veil Nebula?**
In CCD tricolor images, matching filter passbands to the discrete nebula emission lines makes it possible to achieve a truer color than with film. In this LRGB image by Robert Gendler, the Luminance channel was shot through a deep red H-alpha passband filter.

**Digital Spirals**
With their ability to record more detail and fainter magnitudes than is possible in the best film images, CCD cameras take great galaxy shots. Even so, long focal lengths are essential. Robert Gendler used a 12.5-inch f/9 Ritchey-Chrétien telescope for these portraits of M101 (far left) and NGC6946 (left).

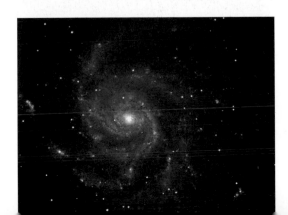
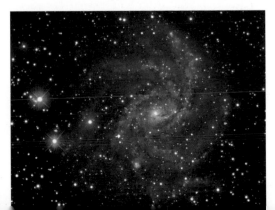

**Whirlpool Wonder** ▶

This is state of the art for amateur astrophotography. Not only does this CCD image record the Whirlpool Galaxy, but it also captures red nebulas lining the galaxy's spiral arms and dozens of faint galaxies far in the background. Robert Gendler registered a white-light, Luminance-channel image (taken with an ST-10 camera) with red-, green- and blue-filtered images (taken with a smaller ST-8 camera). The L channel was a stack of fifteen 20-minute shots, for a total exposure time of 5 hours. Exposure time for the RGB layers was much shorter, just 20/20/30 minutes, respectively. The telescope was a 12.5-inch f/9 Ritchey-Chrétien Cassegrain on a massive Astro-Physics 1200GTO mount.

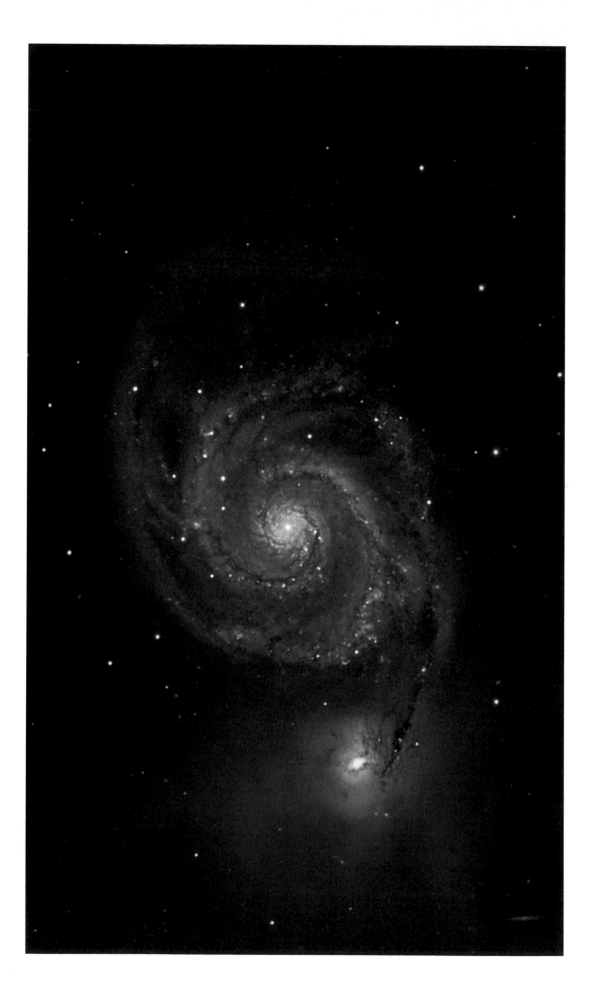

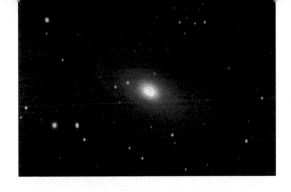

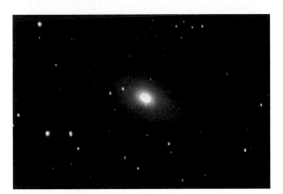

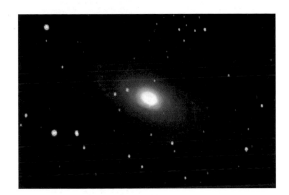

*(continued from page 314)*

Instruments suggests wrapping a white T-shirt over the front of the telescope to diffuse incoming light. And you thought CCDs were high-tech!

**• Step 7: Saving Images**

You'll quickly accumulate dozens of image files a night. When saving, keep in mind that most CCD image-processing programs prefer images in the FITS (Flexible Image Transport System) file format, the standard in scientific imaging that preserves all the image's original 16-bit data. For initial image processing with programs such as AIP4Win, CCDSoft, MaxIm DL and MIRA, images can remain in the FITS format. Moving an image to Adobe Photoshop requires using the camera software to convert the image to a standard TIFF (Tagged Image File Format) that Photoshop can read.

**• Step 8: Calibrating Images**

Before proceeding with making an image look good, the raw camera frames must be reduced, or calibrated, a step unknown to film photographers. This means using either the camera's software or other third-party software to subtract the dark frame and apply the flat field, which itself must be calibrated by subtracting its dark frame. Without calibration, further image processing will only exaggerate the noise and defects embedded in the raw frames. Calibration electronically removes those flaws.

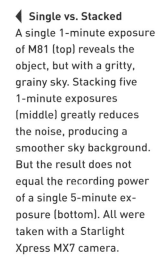

◀ **Single vs. Stacked**
A single 1-minute exposure of M81 (top) reveals the object, but with a gritty, grainy sky. Stacking five 1-minute exposures (middle) greatly reduces the noise, producing a smoother sky background. But the result does not equal the recording power of a single 5-minute exposure (bottom). All were taken with a Starlight Xpress MX7 camera.

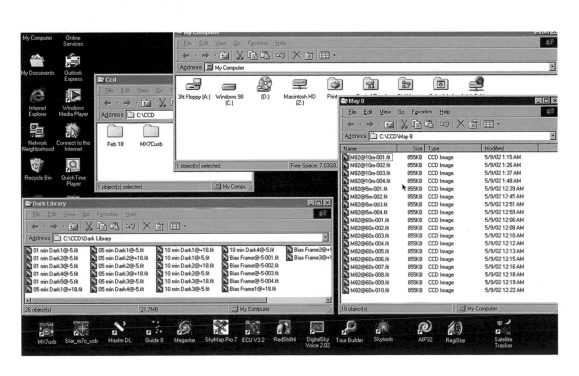

◀ **Organizing a Folder**
A consistent naming system helps sort out what can become, even in a single night, a confusing array of images, dark frames and flat fields. A file name with the object name, followed by the exposure time, the filter used, then the image number will sort by name into an organized and collated list. Make a new folder for each night's shooting.

317

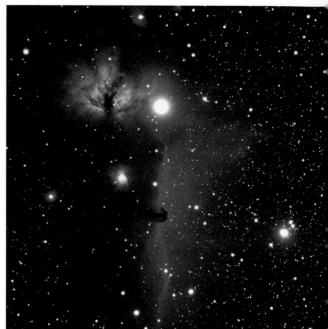

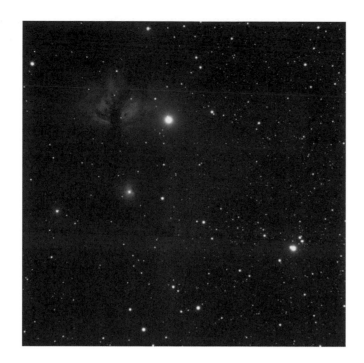

**Before ▶**
**and After**

At right is a typi-
cal original film
image as it comes
out of the camera
and film scanner.
Not too impres-
sive! At far right
is the same im-
age after digital
processing. It is
a stack of two
70-minute expo-
sures, each of
which has had the
contrast boosted
and the sky back-
ground color-
corrected; they
were then regis-
tered and com-
bined with RegiStar soft-
ware. For displaying your
finished works, print on
a photo-quality inkjet
printer. Or take your TIFF
file to a film lab for outputting to a slide or print.

# Processing Images

Getting the sharpest, cleanest images at the telescope requires skill, patience and good skies. But the effort at the telescope is only half the work. Great images are created in the digital darkroom. The processing steps to go from raw to finished product are a lot of fun on cloudy nights and are equally effective on both CCD images and images captured on film. To take advantage of computer processing for film images, the slide or negative must first be scanned. Desktop scanners do a good job. Look for models

with the highest resolution—4,000 dots per inch (dpi) is better than 2,800 dpi—the best dynamic range and the ability to scan at 16 bits per channel for capturing maximum tonal range.

An alternative to scanning your own is to send the images to your local lab for conversion to a Kodak Photo-CD. This costs about $1 to $2 per scan. The standard format provides images up to 2,800 dpi, good enough for publication-quality 8x10 prints. You'll need Adobe Photoshop to read and process the Photo-CD image files.

A complete tutorial on image processing would fill a book—or a CD. Indeed, we recommend *Photoshop for Astrophotographers* by Jerry Lodriguss (for details, see Jerry's website at www.astropix.com). For a mini-tutorial, visit www.backyardastronomy.com. Here, we can provide only a sample of the power of image processing. What used to take tedious hours in the chemical dark-room can now be accomplished in seconds with the click of a mouse. Programs such as MaxIm DL provide powerful sharpening routines for extracting maximum detail from an image. The program RegiStar allows users to stack images on top of each other or to create panoramic mosaics with absolute precision. Adobe Photoshop is the standard program for performing correction of off-color images and for final enhancement of contrast and brightness.

**Resampling the Image ▶**

For most CCD images, use
Photoshop's Image Size
box to increase the number
of pixels. Called resam-
pling, this technique allows
an image to be blown up
larger before showing ugly
pixels. This is the same
image of the Ring Nebula
shown on page 309 but
increased in size by
adding more pixels.

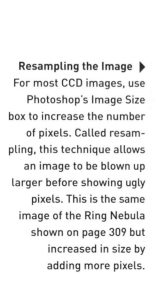

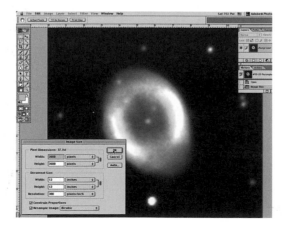

# Is Astrophotography for You?

The digital darkroom has revolutionized astrophotography, making possible amateur color images that sometimes rival in splendor the output of the biggest telescopes in the world. In some instances, stunning photos are just a few mouse clicks away. Telescopes and mounts are better today than ever before, with an array of products enticing astrophotographers every time a new issue of their favorite astronomy magazine arrives.

This is the golden age of astrophotography. If you explored the last few chapters, you may share our excitement about the possibilities. You may also feel overwhelmed and intimidated. Astrophotography is not for everybody who enters the hobby. Many who plunge in drown in a sea of jargon and expensive equipment that is never properly utilized.

We started this section with the advice to go slowly, and we repeat it here. Astrophotos good enough to grace calendars and magazine covers can be taken with just a camera on a tripod. Tracked piggyback techniques offer a lifetime of shooting opportunities. But when it comes to deep-sky imaging through a telescope, think twice. Do you really want to spend a small fortune just to take pictures of the same objects everyone else shoots? We sometimes question

our sanity in pursuing it. But many photographers do, often going from nothing to award-winning photos in just two to three years. And thank goodness they do. Their pictures inspire us all with remarkable portraits of the beautiful universe around us.

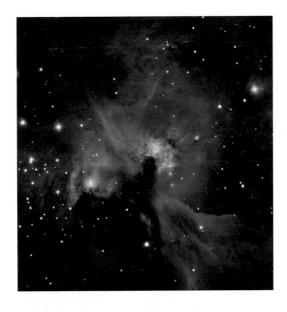

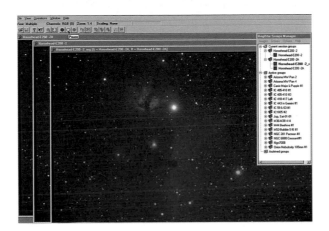

**◀ Preserving Detail**
The Orion Nebula is notoriously difficult to photograph. Long exposures burn out the central Trapezium stars, while short exposures fail to pick up the outer wings. Compositing a short exposure and a long exposure creates an image that more closely resembles the eyepiece view. Robert Gendler used a Finger Lakes IMG1024 camera for this composite of 5- and 20-minute exposures.

**◀ RegiStar Software**
RegiStar (shown here registering an image) allows precise star-for-star alignment of any number of deep-sky images to a master image, allowing you to create multiple-image composites or mosaics. Though inexpensive, this program from Auriga Imaging opens up image possibilities that cannot be achieved in a conventional darkroom.

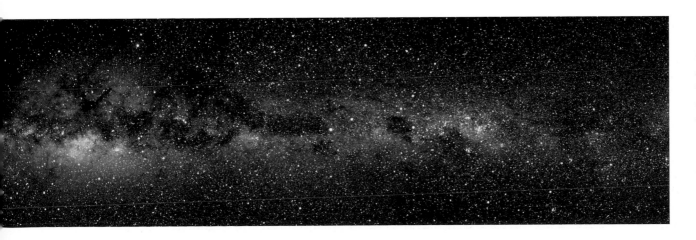

**◀ Milky Way Mosaic**
A painstaking task in the dark(room) ages, the creation of multi-image mosaics is now easy and seamless using digital software. Mosaic by Alan Dyer.

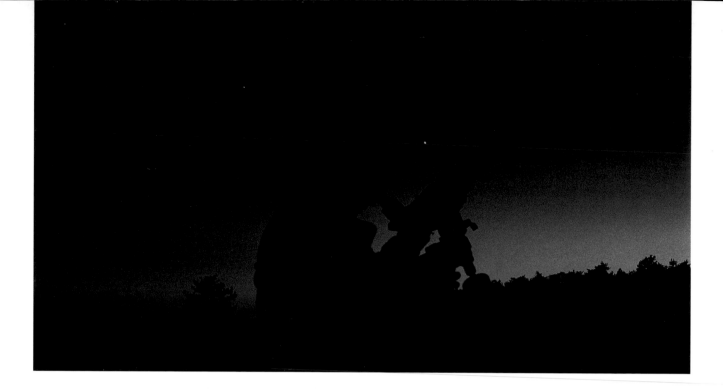

# EPILOGUE

Every August, a small band of dedicated skywatchers navigates the tortuous 12½-mile road up Mount Kobau, in southern British Columbia, in search of perfect skies. Sometimes, they are rewarded: The weather cooperates, and the black canvas of the sky is painted with the delicate brush strokes of the Milky Way. But other years, the normally dry summer weather turns foul. Isolated at the top of a 5,000-foot mountain, the troop of observers is forced to wait out a thunderstorm's torrential rains, hoping that the next night, or perhaps the next hour, will reveal the stars.

And yet even when the weather turns cold and wet, everyone leaves the Mount Kobau Star Party saying, "See you next year." They know they will be back. And so it goes at virtually every star party and amateur-astronomer gathering. The great thing about backyard astronomy is that it can extend much further than your backyard. There is a vast community of thousands of like-minded lovers of the sky. Perhaps as you pursue your interest in the stars, you will find yourself becoming a part of that community. You, too, may discover a place such as Kobau or Stellafane or one of the many other dark-sky observing meccas that have emerged across the continent.

On the other hand, your personal mecca may always be as close as your backyard, and your community of fellow skywatchers no larger than your family and friends. But no matter: Wherever you observe, it is the same limitless sky overhead, a sky we hope this book helps you to explore.

## THE UNIVERSE AWAITS

Throughout *The Backyard Astronomer's Guide*, we have tried to emphasize certain aspects of recreational astronomy that are often glossed over by other guidebooks. For example, we have talked in specific detail about equipment. We have done this because we find that most of the inquiries we receive from beginners are variations on a single question: What should I buy? Our emphasis on hardware may lead to the impression that amateur astronomy is nothing more than collecting equipment. For some, this is, indeed, the case. But those collectors rarely sustain their enthusiasm for the hobby. Which brings us to the concluding topic—one that is seldom discussed in amateur-astronomy literature—why people lose interest in astronomy.

In Chapter 1, we said you cannot buy your way into astronomy. But some newcomers try. They purchase the best and most prestigious equipment on the market but never get around to investing the time to learn how to use it properly. Are they backyard astronomers? Not in our view. Real backyard astronomers learn how to use the instruments and to appreciate what they can reveal. Even in the era of computerized telescopes, a full appreciation of the universe cannot come without developing the skills to find things in the sky and understanding how the sky works. This knowledge comes only by spending time under the stars with star maps in hand and a curious mind.

Since the first edition of this book appeared in 1991, equipment for backyard astronomy has evolved at an incredible rate. The high-tech gear is enticing. Yet our advice remains the same: The primary reason people lose interest in astronomy is that equipment absorbs their attention, and they neglect the stars. The sky never becomes a friendly place. The star patterns remain anonymous, and the locations of the sky's attractions remain hidden. Instruments that are too complex make setting up the gear an onerous chore. What could have been a lifelong interest becomes a passing phase. We have seen this happen many times, which is why we recommend binoculars, rather than a telescope, to most first-time buyers.

There are other reasons people lose interest in astronomy. Some leap into astrophotography, and now CCD imaging, too quickly. Three chapters of this book deal with imaging techniques, because many beginning amateurs express an interest in taking astrophotos. Even so, don't think that astrophotography is something you must do. Both of us began our astronomical careers as avid astrophotographers, and we both gave it up. We reached the limits of our tolerance and found ourselves expending too much effort and too many dollars for the results we were getting. So we returned to visual astronomy for the best part of a decade. But with the improvement in equipment and films during the 1980s and 1990s, we both cautiously inched back into astrophotography and, most recently, CCD imaging. From experience, we advise others to start simply and go slowly.

Developing stargazing skills takes time, more time than many people can find. It seems the more leisure hours people have, the more they are filled with demanding activities that are far from leisurely. Finding time to be under the stars is, in our opinion, more a matter of attitude than scheduling. For us, the time spent pursuing this hobby is a quiet respite from life in the fast lane. And backyard astronomy does not have to be a solitary pursuit. Involving the family will make your moments under the stars all the more meaningful.

After the initial novelty wears off, some amateurs drift away from the hobby for lack of a purpose. To rekindle the interest, take on a project and work toward a goal, such as observing all the Messier objects, sketching the planets or photographing the constellations. Or schedule your next trip to include a few nights at a dark-sky location. There are many possibilities.

Every now and then, we all need a shot of inspiration to recharge our batteries. Sometimes, a casual observation of a planetary conjunction or an exceptionally clear night is enough to remind us of how inspiring the sky is. We have both been in the hobby for decades yet never cease to be amazed at what new sights and wonders the sky presents us each year, if not each night. Keeping abreast of celestial happenings is essential to maintaining your interest and sustaining the feeling that you are a true naturalist of the night.

Other times, inspiration comes from a group therapy session, such as a star party, a motivating lecture at a club meeting or just touching bases with skywatching friends. A word of warning, though: Beware the Internet. The virtual clubs created by special-interest news groups and chat rooms can, indeed, be a source of helpful answers to specific questions. But the ratio of useful signal to annoying noise is often extremely low. We've found many astronomy e-groups rife with so much misinformation and immaturity that we avoid most of them. They are not a true reflection of the great people who populate the community of backyard astronomers and who make the hobby all the more enjoyable.

Like every other leisure pursuit, astronomy can be taken seriously or casually. It is entirely up to you. Our task has been to provide advice on the tools and an introduction to the techniques of sky observing. Armed with that advice, you are ready to explore a hobby—and a universe—that can provide a lifetime of amazement and fun. Welcome to backyard astronomy.

# APPENDIX

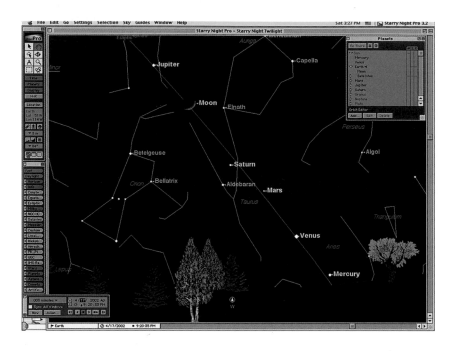

**Starry Night Pro ▲
(Mac version 3.1)**
This program offers
realistic sky simulations
and the ability to fly far
into deep space with ease.

**TheSky ▶
(Mac version 5.06)**
Level II and above, including the Mac version, can
seamlessly control all of
today's "Go To" telescopes.

## Computer Software

The computer software described here is the desktop option that backyard astronomers can access to learn the sky, figure out what's up in tonight's sky and print custom star charts. The capabilities of these programs are so amazing, we often wonder how we ever conducted the hobby without them. For example, all the sky diagrams in Chapter 11 were created using these programs.

Our mini-reviews describe only the most popular programs, dividing them into two main categories: sky simulation and star atlas. Sky simulators will suit most observers' needs, while the star-atlas programs are more for hard-core deep-sky observers. After purchasing software, always visit the program's support website for updates and supplementary plug-ins.

### SKY-SIMULATION PROGRAMS

The common feature of this class of software, often called planetarium programs, is the realistic depiction of the night sky, complete with a horizon landscape, twilight glows, lunar-disk details, even light-pollution simulation.

❖ Starry Night (Space.Com)
Available for both Mac (including OS X) and Windows, Starry Night Pro ($150) sets the standard for its ease of use and well-designed interface. Yet it is extremely powerful. Its large database of objects allows Starry Night Pro to act as an advanced star atlas as well as a naked-eye planetarium. Its principal flaw is that most nebulas are shown only as circles, not as accurate shapes. As of version 3, its main weakness is the lack of integrated telescope control. To operate a "Go To" scope, Windows users must run third-party software, the Astronomer's Control Panel, included with the program's CD. Mac users must get plug-ins (mostly limited to Meade scopes), which are downloadable from third-party websites. A stripped-down but still adept version is offered as Starry Night Backyard ($60), also for Mac and Windows.

❖ TheSky (Software Bisque)
Available for Mac and Windows, TheSky ($130 and up) provides realistic simulations and, at the same time, includes object databases complete enough to make it a powerful star-charting program. Level IV (Windows only) also integrates with Software Bisque's other software packages for CCD-camera control (CCDSoft), precise polar alignment of observatory telescopes (T-Point) and automating a robotic tele-

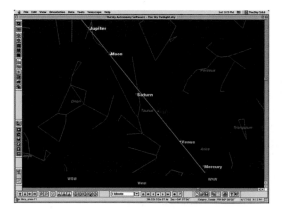

scope (Orchestrate). A surprising drawback of TheSky is its awkward user interface. Simple functions such as changing the date, time or location force you to go through multiple dialog boxes.

❖ Voyager III (Carina Software)
Voyager III ($120), available for both Mac and Windows, falls between Starry Night and TheSky for ease of use. A welcome feature is the ability to search for conjunctions of various combinations of solar system worlds. As with Starry Night and

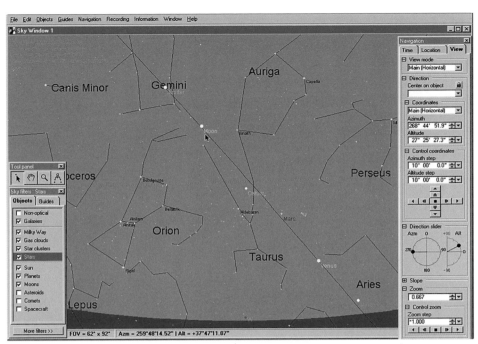

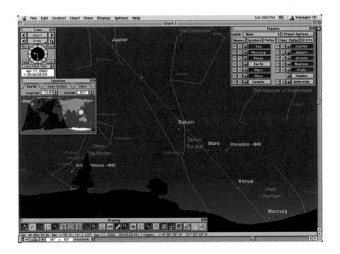

TheSky, Voyager III users can control the fonts, type size and colors of just about every label and class of object, which is great for producing custom charts to your liking. Using an optional plug-in program (Sky Pilot), Voyager can control popular computerized telescopes.

❖ SkyChart III
(Southern Stars Systems)

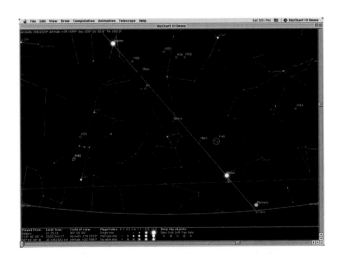

Don't let the $40 price fool you. SkyChart III for either Windows or Mac platforms, including OS X, is a powerful program, pre-

senting gorgeous views of the dark night sky with photographic realism. The principal downside is a crude interface, with functions such as zooming and time-stepping available only through pull-down menus or awkward keyboard shortcuts—no convenient control panels.

❖ Redshift (Cinegram Media)
Although long available for both Mac and Windows, Redshift abandoned the Mac with the release of version 4 and became a Windows-only program ($60). Redshift's strong points are its ability to generate almanac-like tables of celestial events and its excellent control over a vast database of asteroids and comets. It offers on-line dictionaries and educational QuickTime movies that most users will view once then ignore. As of version 4, Redshift is the only program described here that cannot control a "Go To" telescope.

## STAR-ATLAS PROGRAMS

These programs make little effort to simulate the actual appearance of the night sky, perhaps only tinting the sky blue if the Sun is up. The horizon is often drawn simply as a line, and planets might be featureless circles or dots, even at high power. These programs are designed to be electronic star atlases, plotting every conceivable object and able to draw eyepiece and CCD-cam-

▲ **Redshift**
**(Windows version 4.01)**
Redshift uses control panels in abundance, but operating them proves overly complex.

**Voyager III (top left)**
**(Mac version 3.2)**
Though an otherwise excellent program, Voyager has a poorly maintained website and is tardy providing upgrades to fix bugs.

◀ **SkyChart III**
**(Mac version 3.5)**
SkyChart includes well-integrated control of Meade and Celestron "Go To" telescopes, all for a bargain price.

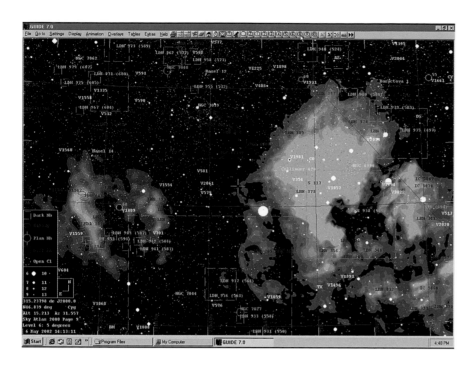

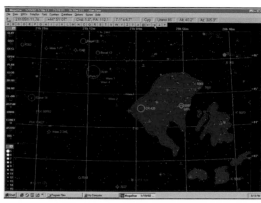

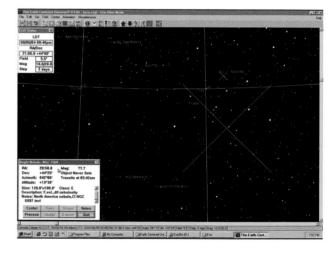

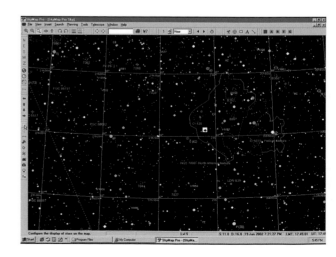

**Guide (version 7.0)** ▲
Upgraded regularly, Guide deserves consideration from enthusiasts looking for the ultimate atlas.

**MegaStar (top right) (version 4.0.36)**
Compared with version 4 (shown here), MegaStar version 5, released in 2002, offers an improved interface and scope-control functions.

**Earth Centered Universe** ▶ **(version 3.2)**
ECU plots objects from all the major catalogs, though with cruder symbols for objects such as nebulas.

**SkyMap Pro** ▶ **(version 7.0)**
SkyMap's charts have a cartoonlike appearance compared with the more professional-looking MegaStar or Guide.

era fields, so you can frame exactly what you will see or photograph at the telescope. All these programs can control a "Go To" telescope through an RS-232 serial-port connection. All are Windows-only programs but will run on a Mac using Windows-emulation software, such as Virtual PC.

❖ Guide (Project Pluto)

At $90, Guide is a bargain. Its deep-sky databases plot everything contained in the printed *Millennium Star Atlas* and more. Not only does it plot more emission and dark nebulas than any other program, but it depicts them beautifully, with realistic isophote contours that closely match the appearance of nebulas on long-exposure images. In addition, Guide presents a semi-realistic horizon and detailed images of planets when zoomed in. Its primary drawback is a sluggish screen redraw.

❖ MegaStar (Willmann-Bell/E.L.B. Software)

Developed by amateur astronomer Emil Bonanno in the early 1990s, this program has evolved to become one of the most popular with deep-sky observers. Now distributed by publisher Willmann-Bell, Mega-Star ($130) is well supported, with frequent maintenance releases to registered owners. One unique selling point is the inclusion of Mitchell's Anonymous Catalogue. Compiled by amateur astronomer Larry Mitchell, this database contains more than

20,000 galaxies not listed in any other advanced galaxy catalog.

❖ Earth Centered Universe (Nova Astronomics)

Developed by Canadian astronomer David Lane, ECU features essential amenities such as scroll bars that other Windows programs leave out. Despite the bargain price of $60, ECU's telescope control is the best of the lot, with more flexible choices for displaying telescope position and centering targets. That and its choice of several night-vision modes make this a favorite program of ours for use at the telescope.

❖ SkyMap Pro
  (SkyMap Software)

British amateur astronomer Chris Marriott continues to hone the abilities of his Sky-Map Pro software ($100). In addition, loyal fans have developed plug-in databases of special-interest objects and lists that can be downloaded from the software's website. Other selling points include functions for planning observing sessions, as well as on-line logging of observation notes. Unique to SkyMap is its ability to provide local circumstances for eclipses and to plot solar eclipse tracks on world maps.

## SPECIALIZED PROGRAMS

A host of other software programs perform specialized tasks, providing amateur astronomers with powerful new tools for studying the sky. Programs such as Astronomical Image Processing for Windows (AIP4Win), AstroArt and several others can measure the position of moving targets, like asteroids and comets, with arc-second precision and make accurate measurements of an object's brightness.

For the less serious-minded, programs such as Deep Space 2000, SkyTools and NGCView serve as observation planners, providing filtered lists of deep-sky objects visible that night and the order in which they should be viewed. Observers can then log in their impressions and notes. Small utility programs, often available as share-

ware or for free, help observers track Moon phases, calculate tables of rise and set times or simply put the current satellite view of the Sun or of Earth from space on your desktop as a backdrop.

## USING A COMPUTER AT THE TELESCOPE

Avid deep-sky observers, even those without computerized telescopes, often have a laptop handy at the scope running an astronomy program as an electronic atlas (most programs have a red-screen feature for preserving night vision). When connected to a "Go To" telescope, a computer can display precisely where a telescope is pointed. The computer screen becomes a star atlas linked to the telescope. This makes it easy to see what other targets lie nearby,

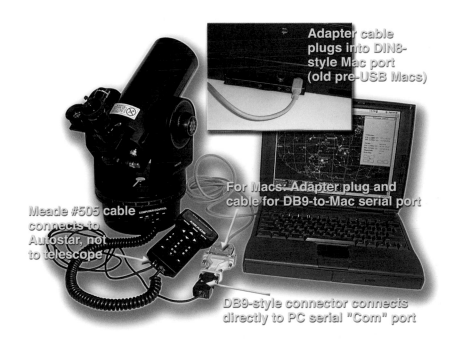

Meade #505 cable connects to Autostar, not to telescope

Adapter cable plugs into DIN8-style Mac port (old pre-USB Macs)

For Macs: Adapter plug and cable for DB9-to-Mac serial port

DB9-style connector connects directly to PC serial "Com" port

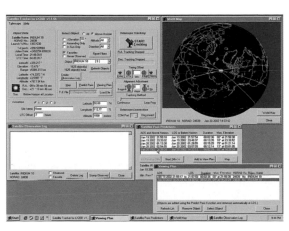

encouraging random exploration of unfamiliar objects; with most "Go To" hand controllers, you have to know what object you want to look at and punch in its catalog numbers. The programmed tours of many "Go To" telescopes also bounce you around the sky in what seems like random order; using a portable computer at the scope allows you to stay in one area and thoroughly explore a constellation. It's all great fun and more entertaining for a family or group—others can look at the computer screen and

▼ **Getting Connected**
Most scopes connect to computers via old-style RS-232 serial ports, using the D-shaped DB-9 connector shown here on the optional Meade cable. These connect directly to older PCs, but users of older Macs will need an adapter cable (also shown) to go from DB-9 to the Mac's round serial port (a DIN8M used for modems and printers). Computers with only newer USB ports (PC or Mac) require yet another adapter to go from USB to the older-style serial plug required by telescopes.

◀ **Tracking Satellites**
With a program such as SatTracker, "Go To" telescopes can be programmed to follow Earth satellites automatically as they travel at high speed across the sky. This requires the download of the latest orbital elements (Two-Line Elements, or TLEs) from NASA Internet sites.

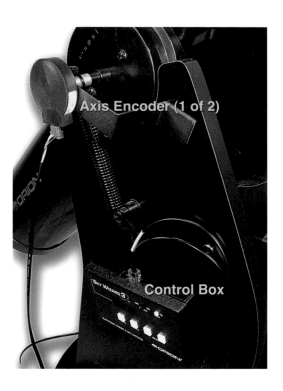

Axis Encoder (1 of 2)

Control Box

see where the scope is aimed, click on an object to learn more, see a picture of the object or take charge of sending the scope to its next destination. The downside is all the technology that has to be able to work in the cold and damp, to which batteries, screens and hard drives are not partial.

## PALMTOP ASTRONOMY

As of 2002, many simple computer programs were available for palmtop computers and Personal Digital Assistants (PDAs) running the PalmOS or Windows CE operating systems. Limited by the hardware, almost all are as crude as desktop planetarium programs were in the late 1980s. They are handy for looking up Moon phases, Jovian moon positions and other almanac information. Software Bisque even released a pocket-computer full-color Windows CE version of TheSky that can control a telescope. Other simple star-chart programs for PalmOS systems, such as Planetarium, also feature telescope control. But the confusion of cables, adapters and gender changers necessary to connect a handheld computer to a telescope (with no guarantee that it will all work) seems to have put most people off handhelds as telescope controllers. No doubt this will change as handhelds become more powerful and the connection issue is solved via wireless technology. Future high-end telescopes will probably have the equivalent of color-screen PDAs built in or communicating through some wireless protocol, replacing the simple two-line displays of today's Autostar and NexStar controllers.

# Using "Go To" Telescopes

The hot ticket for finding things in the sky is a "Go To" telescope, such as Meade's ETX models or Celestron's NexStar Series. Punch M-4-2 into the control paddle, and the scope dutifully whirs off to find that object for you, stopping with the Orion Nebula (M42) neatly centered in the eyepiece. For those who opt for a more traditional telescope, computerized finding

can be incorporated later through a number of add-on accessories.

## DIGITAL SETTING CIRCLES

Several manufacturers offer digital setting circles that can be retrofitted to most of today's popular telescopes. Examples include Celestron's Advanced Astro Master, JMI's NGC-Max, Lumicon's Sky Vector and Orion's Sky Wizard, all essentially the same units made by Tangent Instruments of California. A particular favorite of deep-sky observers is Sky Engineering's Sky Commander, with its excellent deep-sky catalogs (it includes Barnard dark nebulas and obscure Trumpler and Collinder open clusters) and its welcome ability to select objects within a single constellation. This feature is so obvious and useful, yet most computerized scopes ignore it. These devices sell for $300 to $500 and are good choices for Dobsonian owners.

## USING A "GO TO" TELESCOPE

As good as digital setting circles are, they still require moving the telescope manually. The next step up in technology is the "Go To" telescope. The slewing motors can point the telescope at high speed to any one of thousands of objects.

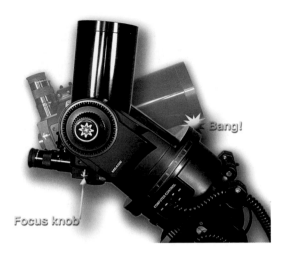

Focus knob

Bang!

To do this, "Go To" scopes contain a virtual map of the sky in their computer chips, along with the ability to calculate the positions of the Sun, Moon and planets. For the telescope to match its virtual sky map with the real sky, it must know several key facts: where it is on Earth, what time it is and where it is pointed. Most scopes require an

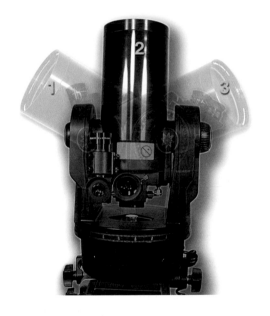

initial input of your location. Usually, this needs to be done only once—the computer remembers that site and defaults to it upon start-up. With Meade Autostar and Celestron NexStar scopes, the time must be loaded in with every power-up; others, such as Vixen's SkySensor and Astro-Physics' GTO mounts, have internal clocks that run even when the scope is turned off. No input of time is necessary with these models.

All "Go To" telescopes then require an initial alignment on two stars. The procedure is simple. First, place the telescope into some required standard position (usually level and aimed due north). Now slew the scope over to the first alignment star (or at least where the computer thinks that star should be). You can select the star yourself from the computer's internal list or let the computer select it for you. The telescope then points close to the selected star. Center the star in the eyepiece using the push-button motion controls, then hit the Enter or Align button. This tells the computer how far off the star was.

A precise alignment now requires sighting a second star. So off the scope goes to another bright star. Center it, hit the button, and the alignment is done. The telescope now knows how the real sky deviates from its virtual map and, therefore, how to point to other objects in the sky.

Once aligned, a "Go To" scope can keep objects centered for many minutes, if not hours, at a time. "Go To" scopes aren't the first telescopes which can track objects—any telescope on a polar-aligned equatorial mount can do that—but they are the first scopes that can track without the need for traditional polar alignment (i.e., lining up one rotation axis with the celestial pole). Yet, as you can see, an alignment of sorts is still required, taking two to five minutes at the start of each night. But be warned: Weak batteries, a loose connection, a blown power supply or a fried computer chip can leave you lost in the stars. Even on a good night, a computer crash or a kicked tripod

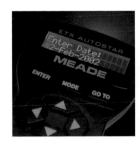 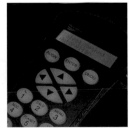

can necessitate a reboot and a realignment from scratch. But when working correctly, a "Go To" telescope allows for effortless exploration of the night sky, taking you to targets you might never have seen.

**Polar-Aligned ETX**
"Go To" telescopes can be polar-aligned and used as conventional fork-mounted scopes. This is necessary only for long-exposure shooting with a piggy-backed camera. However, some visual observers prefer to polar-align regardless. When this ETX-90 is polar-aligned, the tiny focuser knob is easier to reach even with the telescope aimed high in the sky. A drawback is that the tube cannot aim low in the south because the base gets in the way.

**Altazimuth ETX**
Here, the ETX is set up as an altazimuth scope—the way most observers use it. As the sky moves, the computer pulses the motor on each axis at a speed sufficient to track objects. A tricky maneuver comes when tracking an object through the zenith. The scope must rise up from the east (1), quickly do an about-face (2), then continue descending into the west (3). Not all "Go To" scopes can accomplish this trick. Meade Autostars manage the task fine.

**Picking a Star**
If you cannot see the alignment star the scope is asking for, skip to the next star on the pick list. On a Meade, this can be done by hitting any direction button or the Down scroll key anytime during a slew. If a slew has finished, just hit the Down key. On a Celestron, hit any direction button to stop a slew, then hit Undo. Or simply hit Undo once a slew has finished.

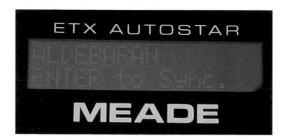

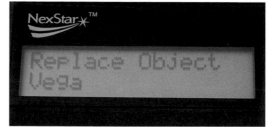

**Resyncing on the Sky ▶**
To improve pointing accuracy in one region of the sky, realign on a bright star in that area. On Meades (left), slew to the new star, then center. Once there, hold the Enter button for two seconds. Then hit Enter again as instructed. On Celestrons (right), first center the new star in the eyepiece, then select that star under List: Named Stars. Press the Align key. The display will ask which of the two previous alignment stars you want to replace. Choose the one closest to your new star, and hit Enter.

**Calibrate Declination Circle**
With an ETX sitting on a flat surface, level the tube by using a bubble level as a gauge. Loosen the large knob on the fork, and turn the numbered Declination circle so that it reads 0 degree (if it doesn't already). Then tighten the knob. In future, with the mount level, leveling the tube requires simply turning it until the circle reads 0. ▶

**Hitting the Stops**
On the ETX-90 to ETX-125 models with mechanical stops, place the scope with the control panel aimed to the west. Unlock the azimuth (horizontal motion) axis, and turn the scope counterclockwise until it hits the hard stop. Then turn the scope clockwise about one-third of a turn, until the fork arm with the numbered dial is over the control panel. Lock the axis, and start the alignment process. ▶

# "Go To" Tips and Techniques

This section is adapted from material that originally appeared in *Sky & Telescope* magazine and is used with permission.

Even low-cost computerized telescopes can be remarkably accurate in pointing to targets—if you take care when performing the initial setup. Be sure to read the instruction manual. Here are some tips the manual might not mention.

## FOR ANY TELESCOPE

• If your city is not on the telescope's list, enter your latitude and longitude, available from atlas maps or from websites such as www.mapblast.com. Enter the numbers correctly: a minus latitude is in the southern hemisphere; a minus longitude is west of England's Greenwich meridian (true for all of North America). Locations in Europe have a positive longitude.

• Enter time zones correctly. Eastern time is Greenwich time minus five hours. Do not make any correction for daylight time here; there is a separate choice for this.

• No need to be ultraprecise, but take care to level the tripod, then level the tube.

• Aim the telescope as close to true north as you can, toward Polaris. Don't aim at magnetic north, where a compass points. The more accurately you can level and aim north, the closer the scope will get to its initial alignment stars and the more accurate the subsequent pointing will be.

• Know the sky. Aligning on Castor when the scope is asking for Pollux, just 4.5 degrees away, will cause the telescope to point several eyepiece fields from your target.

• Try to pick alignment stars that are widely separated in the sky.

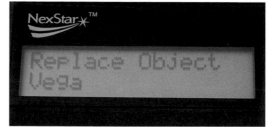

• Ensure that any add-on finderscope or device used to sight stars is aligned with the main optics. Do that during the day by sighting a distant terrestrial target.

• When a scope aims at an alignment star, always wait for the beep before hitting any motion button to center the star.

• Take care to center each alignment star in the main eyepiece.

• During the night, do *not* bump the tripod or telescope tube.

• If pointing at the Moon is consistently off, check whether you have entered the appropriate daylight-time correction.

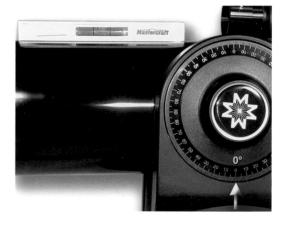

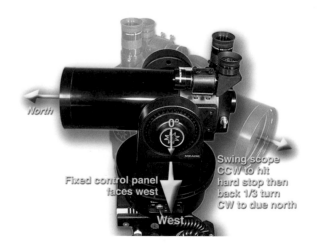

North

Fixed control panel faces west

West

Swing scope CCW to hit hard stop then back 1/3 turn CW to due north

## FOR MEADE AUTOSTAR TELESCOPES

• In order to activate the tracking motors, the telescope must be set for Targets: Astronomical (found in the Setup menu). The scope may not be set for this by default out of the box.

• An observing location may have already been entered into your telescope at the store. To check, go under Setup: Site, then hit Select to see for which location the scope is programmed. Hit Add to choose a new location if necessary.

• Be sure the axes are snugly locked down. A loose axis will slip and lose alignment under motor control.

• When selecting a new target, always hit Enter (to put the object name in the top line of the display) before hitting Go To.

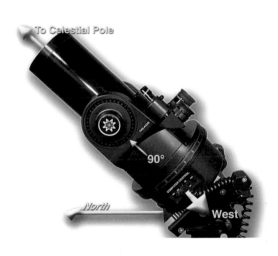

• Hitting Go To after arriving at an object (or where the ETX thinks the object should be) initiates a clever square dance: The telescope starts to scan around in an ever-widening search pattern, a handy way to locate an object just outside the field.

• You may have to Calibrate Motors (an option found under Setup: Telescope) after slewing problems caused by low batteries. If problems persist, follow the instruction manual for the steps required to train the drives. Always train the drives with the scope set to the mode in which it will be used, either polar or altazimuth.

• If you can leave the telescope set up, just hit Utilities: Park at the end of the night, then turn off the scope. This eliminates the need for a star alignment the next day.

## FOR CELESTRON NEXSTAR TELESCOPES

• Under the Menu button, the Tracking should be set to Altazimuth, and under Model Select, be sure your telescope model

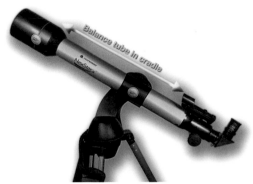

is chosen (if it isn't, scroll to it and hit Enter).

• Be sure to enter the date and time in the right format (mm/dd/yy).

• If the telescope seems to run on or to jerk backwards after jogging a motion button, go into Alt Backlash and Az Backlash (hit the Menu button and scroll down) and set the values to 050. If that does not eliminate the errant motion, try other values.

• While pressing one direction button, hold down the opposite direction button to accelerate a slew to full speed, a handy shortcut to Rate 9 speed when aligning.

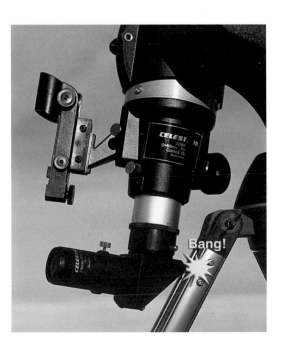

▲ **Fresh Power**
All "Go To" scopes behave badly with weak batteries. Subfreezing temperatures sap battery power. In winter, operate off AC power or a larger battery pack.

◀ **Take Time to Balance**
On NexStar refractors and reflectors, be sure to balance the tube in its cradle, or the scope may not move up and down correctly.

◀ **Polar Home Position**
If you set up an ETX telescope as a polar-aligned equatorial, start with the scope in the home position: Set to 90 degrees declination, with the base turned so that the control panel faces west.

◀ **Collision Zone**
Celestron's small NexStar reflectors and refractors can hit the tripod when aimed up high. If a collision seems imminent, don't panic. Hitting any direction key stops a slew in progress, preserving your alignment.

# FURTHER READING

• Star atlases are described in more detail in Chapter 11.

• Our website (www.backyardastronomy.com) contains a more complete bibliography for each chapter, as well as one of the hobby's most extensive listings of clickable links to manufacturers of hardware and software and to hundreds of websites of interest to amateur astronomers.

## Chapter One
## General-Interest Astronomy Guidebooks

*NightWatch* by Terence Dickinson (Firefly Books; Willowdale, Ontario; 1998). One of the best-selling guides to the hobby for beginners. *The Backyard Astronomer's Guide* is an advanced sequel to *NightWatch*.

*Summer Stargazing* by Terence Dickinson (Firefly Books; Willowdale, Ontario; 1996). A guide to the constellations and deep-sky objects of the summer sky, with plenty of charts.

*Starlight Nights: The Adventures of a Star-Gazer* by Leslie C. Peltier (Sky Publishing; Cambridge, MA; 1999). Written by one of the 20th century's most gifted amateur astronomers, this wonderful book chronicles one man's odyssey in backyard astronomy. To show a skeptic why astronomy is so compelling, give him or her this charming book.

*Skywatching* by David Levy (Nature Company/Time-Life; San Francisco; 1994). A lavishly illustrated introduction to astronomy as a science and a hobby. Good monthly star charts and individual constellation maps.

*Advanced Skywatching* (aka *Backyard Astronomy* in softcover) by Robert Burnham, Alan Dyer, Robert Garfinkle, Martin George and Jeff Kanipe (Nature Company/Time-Life; San Francisco; 1997). A sequel to *Skywatching*, with more detailed hobby information and excellent star-hopping charts.

*Astronomy: The Definitive Guide* by Robert Burnham, Alan Dyer and Jeff Kanipe (Weldon-Owen; Sydney; 2002). A beautifully illustrated introduction to astronomy, with good seasonal star maps and star-hopping charts.

*David Levy's Guide to the Night Sky* by David H. Levy (Cambridge University Press; Cambridge; 2001). A revision of a fine personal guide to all types of amateur observing and to the wonder of the night sky.

*The Beginner's Observing Guide* by Leo Enright (The Royal Astronomical Society of Canada; Toronto; 1999). An introduction to the night sky for novice observers.

*365 Starry Nights* by Chet Raymo (Prentice Hall; New York; 1982). A classic work that is a tour through interesting sky sights for every night of the year.

*Seeing in the Dark* by Timothy Ferris (Simon and Schuster; New York; 2002). A first-rate science writer examines the realm of the scientific amateur astronomer.

*The Universe and Beyond* by Terence Dickinson (Firefly Books; Willowdale, Ontario; 1998). Lavishly illustrated summary of current astronomical knowledge and "armchair" topics such as black holes and extraterrestrial life.

## Chapter Two
## Binocular Guidebooks

*Binocular Astronomy* by Craig Crossen and Wil Tirion (Willmann-Bell; Richmond, VA; 1992). A comprehensive guide to deep-sky objects for binoculars.

*Touring the Universe Through Binoculars* by Philip Harrington (John Wiley and Sons; New York; 1990). No charts but lots of information on suitable targets.

## Chapters Three, Four and Five
## Telescopes and Hobby Hardware

*Star Ware* by Philip Harrington (John Wiley and Sons; New York; 2002). Tons of brand- and model-specific information with recommendations on telescopes and accessories by an experienced telescope user.

*Choosing and Using a Schmidt-Cassegrain Telescope* by Rod Mollise (Springer-Verlag; London; 2001). A user's guide to SCTs and Maks with many model-specific tips.

*Using an ETX Telescope* by Mike Weasner (Springer-Verlag; London; 2002). A user's guide to Meade's popular ETX models, with suggested observing targets.

*Star-Testing Astronomical Telescopes* by Harold Richard Suiter (Willmann-Bell; Richmond, VA; 1994). Technical but authoritative guide to star-testing telescope optics.

## Telescope Making

*Build Your Own Telescope* by Richard Berry (Willmann-Bell; Richmond, VA; 2001). Plans for five simple scopes of various sizes and styles, both Dobsonian and equatorial.

*The Dobsonian Telescope* by Richard Berry and David Kriege (Willmann-Bell; Richmond, VA; 1997). Detailed construction plans for advanced builders, from two telescope-making experts.

## Astronomical Travel

*Astronomical Centers of the World* by Kevin Krisciunas (Cambridge University Press; Cambridge; 1988). Background information on the world's great observatories.

*Star and Sky: Discovery Travel Adventures*, Robert Burnham, ed. (Discovery Communications; London; 2000). Visitor's guide to North America's best observatories, planetariums and space centers.

## Chapter Seven
## Naked-Eye Astronomy Guides

*Seeing the Sky* by Fred Schaaf (John Wiley and Sons; New York; 1990). A fine compilation of neat things to look for in the sky.

### Atmospheric Phenomena

*Color and Light in Nature* by David K. Lynch and William Livingstone (Cambridge University Press; Cambridge; 1995). Beautifully illustrated guide to sky phenomena.

*Exploring the Sky by Day* by Terence Dickinson (Firefly Books; Willowdale, Ontario; 1988). An introduction to day-sky phenomena; aimed at all ages.

*Light and Color in the Outdoors* by Marcel Minnaert (Springer-Verlag; London; 1993). Revision of the classic work on optical sky phenomena.

*Peterson's Field Guide to the Atmosphere* by Vincent J. Schaefer and John A. Day (Houghton Mifflin; New York; 1981). A pocket reference to weather and clouds.

*Rainbows, Haloes, and Glories* by Robert Greenler (Cambridge University Press; Cambridge; 1980). Still the classic work on these phenomena.

*Sunsets, Twilights, and Evening Skies* by Aden and Marjorie Meinel (Cambridge University Press; Cambridge; 1983). A fine companion volume to *Rainbows, Haloes, and Glories*.

## Aurora

*Aurora: The Mysterious Northern Lights* by Candace Savage (Greystone Books; Vancouver; 1994). The best popular-level work on the aurora borealis in recent years.

## Meteors and Meteorites

*Cambridge Encyclopedia of Meteorites* by O. Richard Norton (Cambridge University Press; Cambridge; 2002). A comprehensive volume detailing types of meteorites and many falls and finds.

*Meteors* by Neil Bone (Sky Publishing; Cambridge, MA; 1992). A guide to making scientific observations of meteors.

## Chapter Eight
## Light Pollution and Observing Sites

*Light Pollution: Strategies and Solutions* by Bob Mizon (Springer-Verlag; London; 2001). About its effects, how to minimize them and how to conduct a campaign in your community to combat light pollution.

## Chapter Nine
## Guides to the Sun, Moon and Comets

### Lunar Observing

*Atlas of the Moon* by Antonin Rukl (Hamlyn; London; 1991). This superb atlas has had an unfortunate publishing history of frequently going out of print. Worth searching through used-book outlets to find.

*Atlas of the Lunar Terminator* by John E. Westfall (Cambridge University Press; Cambridge; 2000). A lunar atlas made of mosaics of CCD images taken along the lunar terminator at various phases.

*Epic Moon* by William P. Sheehan and Thomas A. Dobbins (Willmann-Bell; Richmond, VA; 2001). A history of telescopic exploration of our Moon.

*The Hatfield Photographic Lunar Atlas*, Jeremy Cook, ed. (Springer-Verlag; London; 1999). A reprint of the classic atlas with photos taken by Henry Hatfield with a 12.5-inch telescope in the 1960s.

### Eclipses

*Eclipse!* by Philip Harrington (John Wiley and Sons; New York; 1997). A thorough guide to all upcoming eclipses, with paths, times and weather prospects.

*Eclipse* by Duncan Steel (Headline; London; 1999). A good summary of the history and science of solar eclipses.

*Fifty Year Canon of Solar Eclipses* (NASA RP 1178) and *Fifty Year Canon of Lunar Eclipses* (NASA RP 1216), both by Fred Espenak (NASA Reference Publications/Sky Publishing; Cambridge). Definitive technical information on upcoming eclipses.

*Glorious Eclipses* by Serge Brunier and Jean-Pierre Luminet (Cambridge University Press; Cambridge; 2000). A coffee-table book filled with magnificent eclipse images.

*Totality: Eclipses of the Sun* by Mark Littmann, Ken Willcox and Fred Espenak (Oxford University Press; New York; 1999). All about eclipses and why people travel the world to see them.

### Transits

*June 8, 2004: Venus in Transit* by Eli Maor (Princeton University Press; Princeton; 2000). Good summary of the history of transit observations.

*Transit: When Planets Cross the Sun* by Michael Maunder and Patrick Moore (Springer-Verlag; London; 2000). A guide to the history and mechanics of transits.

### Comets

*Comets: Creators and Destroyers* by David H. Levy (Simon and Schuster; New York; 1998). The history and significance of comets by a comet expert and discoverer.

*Great Comets* by Robert Burnham (Cambridge University Press; Cambridge; 2000). The facts and lore surrounding the brightest comets seen in historical times, including Hale-Bopp and Hyakutake.

*Comet of the Century* by Fred Schaaff (Copernicus; New York; 1997). One of the best comet books. Excellent reference.

## Chapter Ten
## Guides to Planet Observing

*Introduction to Observing and Photographing the Solar System* by Thomas A. Dobbins, Donald C. Parker and Charles Capen (Willmann-Bell; Richmond, VA; 1988). The imaging information is now out of date, but not the observing advice.

*The New Solar System*, Beatty, Peterson and Chaikin, eds. (Sky Publishing/Cambridge University Press; Cambridge; 1999). A fact-filled summary of what we know about the solar system's worlds.

*Mars: The Lure of the Red Planet* by William Sheehan and Stephen James O'Meara (Prometheus Books; New York; 2001). An engaging history of what we thought we knew about Mars.

*The Planet Observer's Handbook* by Fred Price and John Westfall (Cambridge University Press; Cambridge; 2000). How to make detailed and useful planetary observations.

## Chapter Eleven
## Astronomy Guidebooks

### Annual Reference Works

*Observer's Handbook*, Rajiv Gupta, ed. (The Royal Astronomical Society of Canada; Toronto). Indispensable annual guide to the year's celestial events. Used worldwide.

*Astronomical Calendar* by Guy Ottewell (Universal Workshop; Middleburg, VA). Superbly illustrated annual guide to sky events; in large format.

### Star-Hopping Guidebooks

*The Observer's Sky Atlas* by Erich Karkoschka (Springer-Verlag; New York; 1998). An overlooked but excellent guide to finding a choice selection of celestial targets.

*Stars and Planets: A Viewer's Guide* by Gunter Roth (Sterling Publishing; New York; 1998). A fine guide to selected deep-sky objects, with charts.

*Star Hopping: Your Visa to Viewing the Universe* by Robert Garfinkle (Cambridge University Press; Cambridge; 1994). The author provides 14 star-hopping tours for the telescope owner.

*Star Hopping for Backyard Astronomers* by Alan M. MacRobert (Sky Publishing; Cambridge, MA; 1993). Good charts and illustrations take you on 14 star-hopping tours of selected regions of the sky.

*Turn Left at Orion* by Guy Consolmagno and Dan M. Davis (Cambridge University Press; Cambridge; 2000). A great star-hopping guide to the sky's best 100 objects, with finder charts and eyepiece sketches of object appearances.

## Chapter Twelve
## Deep-Sky References

### Introductory Works

*Barron's Nature Guide: Stars and Planets* by Joachim Ekrutt (Barron's; Munich; 1992). Contains excellent monthly hemisphere star maps for northern and southern latitudes.

*Collins Gem: Stars* by Ian Ridpath and Wil Tirion (Harper Collins; Glasgow; 1985). A constellation guide for a shirt pocket; a gem of a little book.

*Eyewitness Handbooks: Stars and Planets* by Ian Ridpath (Dorling Kindersley; London; 1998). One of the best of many constellation guidebooks titled *Stars and Planets*.

*Great Atlas of the Stars* by Serge Brunier and Akira Fuji (Firefly Books; Willowdale, Ontario; 2001). Although it is not a complete sky atlas, this book contains a wonderful gallery of wide-angle sky photos with the best targets highlighted with transparent overlays. A fine coffee-table book.

*Peterson Field Guide to the Stars and Planets* by Jay M. Pasachoff (Houghton Mifflin; New York; 2000). A comprehensive pocket field guide to the stars and constellations.

*The Universe and How to See It* by Giles Sparrow (Reader's Digest Books; New York; 2001). A lavishly illustrated guide to finding and learning about celestial wonders.

**More Advanced Works**

*Burnham's Celestial Handbook* (three volumes) by Robert Burnham Jr. (Dover Publications; New York; 1978). A lifetime's compilation of lore, mythology, poetry and hard science information on all the finest deep-sky objects. A must for any observing library.

*The Deep Sky* by Philip Harrington (Sky Publishing; Cambridge, MA; 1997). An experienced observer provides his guide to the best deep-sky objects.

*Deep-Sky Wonders* by Walter Scott Houston (Sky Publishing; Cambridge, MA; 2001). A wonderful collection of writings by the dean of deep-sky observers.

*The Night Sky Observer's Guide* (two volumes) by George Robert Kepple and Glen W. Sanner (Willmann-Bell; Richmond, VA; 1998). Guides to thousands of objects, with charts, descriptions, photos and sketches. A monumental and essential work.

*Messier's Nebulae and Star Clusters* by Kenneth Glyn Jones (Cambridge University Press; Cambridge; 1991). Reprint of the classic 1968 work providing descriptions from Messier and many historic observers for each object.

*The Messier Objects* by Stephen James O'Meara (Cambridge University Press; Cambridge; 1998). Finder charts and detailed information for all the Messiers as seen through a small telescope.

*The Year-Round Messier Marathon* by Harvard C. Pennington (Willmann-Bell; Richmond, VA; 1997). A great guide with charts to all the Messiers, in the order they need to be found for a one-night Messier marathon.

*Observing Variable Stars* by David H. Levy (Cam-

bridge University Press; Cambridge; 1989). An expert guide to variable-star observing.

*Star Names: Their Lore and Meaning* by Richard Hinckley Allen (Dover Publications; New York; 1963). This is an authoritative guide to the origin of star names.

**Chapters Thirteen and Fourteen**
**Astrophotography Guidebooks**

*Astrophotography for the Amateur* by Michael Covington (Cambridge University Press; Cambridge; 1999). A guide to all types of imaging, with the emphasis on film-based imaging.

*Splendors of the Universe* by Terence Dickinson and Jack Newton (Firefly Books; Willowdale, Ontario; 1997). A book-length version of the material covered in Chapters 13 to 15 of *The Backyard Astronomer's Guide*. Includes CCD-imaging information from CCD-master Jack Newton.

*Video Astronomy* by Steve Massey, Thomas A. Dobbins and Eric J. Douglas (Sky Publishing; Cambridge, MA; 2000). An excellent guide to shooting video through a telescope.

*Wide Field Astrophotography* by Robert Reeves (Willmann-Bell; Richmond, VA; 2000). A thorough guide to piggyback photography, with an emphasis on film and traditional darkroom techniques.

**Chapter Fifteen**
**CCD Imaging**

*The New CCD Astronomy* by Ron Wodaski (New Astronomy Press; Duvall, WA; 2002). Published by the author, this is the most thorough and up-to-date guide to CCD imaging available. Visit the book's website at www.wodaski.com, where you can download a sample chapter. The complete book can be ordered on-line or through dealers and booksellers.

*The Handbook of Astronomical Image Processing* by Richard Berry (Willmann-Bell, Richmond, VA; 2001). Teaches image-processing techniques and technicalities. Comes with a CD of images and the very capable AIP4Win software program (for Windows) to run the book's tutorials.

**Special Publication**

*Amateur-Professional Partnerships in Astronomy*, John R. Percy and Joseph B. Wilson, eds. (Astronomical Society of the Pacific). A compilation of papers presented at a conference of the Astronomical Society of the Pacific in 1999. Essential background reading for anyone considering venturing from recreational astronomy into scientific amateur astronomy.

# INDEX

# The Authors

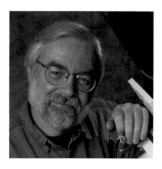

Terence Dickinson

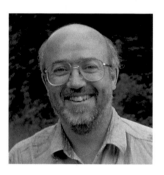

Alan Dyer

## Terence Dickinson

Terence Dickinson is an astronomy editor, writer and educator. In the 1960s and 1970s, he was a staff astronomer at two major planetariums and, since then, has written 14 astronomy books and more than one thousand articles on the subject. His lifelong fascination with astronomy began at age 5, when he saw a brilliant meteor from the sidewalk in front of his home. At age 14, he received his first telescope (a 60mm refractor), a Christmas gift from his parents.

He has received numerous national and international awards for his work, among them the New York Academy of Sciences book of the year award and the Astronomical Society of the Pacific's Klumpke-Roberts Award for contributions in communicating astronomy to the public. Asteroid 5272 Dickinson is named after him. He and his wife Susan, who has been production manager and copy editor for all his books, live in rural eastern Ontario with a menagerie of cats, dogs and horses— and an array of telescopes housed in two observatories.

## Alan Dyer

Alan Dyer is a writer and producer of multimedia shows for the Discovery Dome theater at the Calgary Science Centre in Alberta, Canada. His programs have played in planetarium theaters throughout North America. A former associate editor with *Astronomy* magazine, Dyer currently serves as associate editor of *SkyNews* magazine and is a contributing editor to *Sky & Telescope*. Dyer is widely recognized as an authority on commercial telescopes, and his reviews of astronomical equipment appear regularly in those publications. He is the author or coauthor of several books, including *Advanced Skywatching*; *Astronomy: The Definitive Guide*; and *Pathfinders: Space*, a children's astronomy book.

Dyer recalls, as a child, asking his parents' permission to stay up late to watch the stars. At 15, using money he had earned from delivering newspapers, he purchased his first telescope—a 4.5-inch Newtonian reflector. His telescope collection has since grown to take over his house in the big-sky country of rural Alberta.

## Acknowledgments

We want to thank the readers of the first edition of *The Backyard Astronomer's Guide*, who, through their questions and comments, richly contributed to the development of this new edition. We continue to be convinced that the primary focus of the book should be the topics being actively discussed by today's backyard astronomers.

A work of this magnitude can be brought to fruition only by a team of skilled professionals. Topping the list is graphic designer Robbie Cooke, who masterfully redesigned the book into this new enlarged version. Equally dedicated was our production manager, Susan Dickinson, whose attention to detail amazes all who know her. Thanks also to Michael Webb, Bookmakers Press editor Tracy C. Read and proofreaders Catherine DeLury, Charlotte DuChene and Christine Kulyk.

Alan Dyer appreciates the contributions by Dennis DiCicco of *Sky & Telescope*; Simon and Megan Hum of ScienceWorks in Calgary; Walter MacDonald of Winchester Electronics; Don Hladiuk and the Calgary Centre/RASC (for the great eclipse trips); and Bill Peters, Brad Struble and the Calgary Science Centre for their ongoing support.